ETHICAL CHALLENGES IN ONCOLOGY

ETHICAL CHALLENGES IN ONCOLOGY

PATIENT CARE, RESEARCH, EDUCATION, AND ECONOMICS

Edited by

COLLEEN GALLAGHER PHD, MA, LSW, FACHE

Executive Director and Chief, Section of Integrated Ethics in Cancer Care, and Professor, Department of Critical Care and Respiratory Care, Division of Anesthesiology, Critical Care and Pain Medicine, The University of Texas MD Anderson Cancer Center, Houston, TX, United States

MICHAEL EWER, MD, MPH, MBA, JD, LL.M

Professor of Medicine, The University of Texas MD Anderson Cancer Center; and Guest Professor of Law, University of Houston Health Law and Policy Institute, Houston, TX, United States

Academic Press is an imprint of Elsevier
125 London Wall, London EC2Y 5AS, United Kingdom
525 B Street, Suite 1800, San Diego, CA 92101-4495, United States
50 Hampshire Street, 5th Floor, Cambridge, MA 02139, United States
The Boulevard, Langford Lane, Kidlington, Oxford OX5 1GB, United Kingdom

British Library Cataloguing-in-Publication Data
A catalogue record for this book is available from the British Library

Library of Congress Cataloging-in-Publication Data
A catalog record for this book is available from the Library of Congress

ISBN: 978-0-12-803831-4

For Information on all Academic Press publications
visit our website at https://www.elsevier.com/books-and-journals

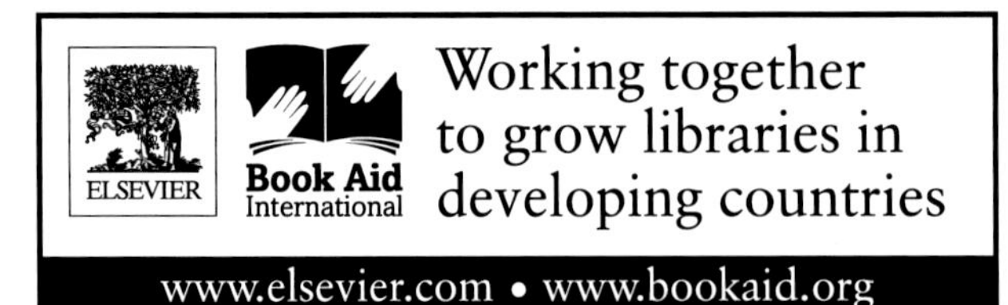
ELSEVIER
Book Aid
International
Working together
to grow libraries in
developing countries
www.elsevier.com • www.bookaid.org

Publisher: Mica Haley
Acquisition Editor: Rafael Teixeira
Editorial Project Manager: Tracy Tufaga
Production Project Manager: Lucía Pérez
Cover Designer: Mark Rogers

Typeset by MPS Limited, Chennai, India

Contents

6. Integrative Oncology and Ethics

RICHARD T. LEE AND JESSICA A. MOORE

7. Clinician Experience: Some Current Ethical Considerations

ROBERT S. BENJAMIN AND NANCY B. BENJAMIN

8. The Ethical Imperative of Providers to Engage Patients and Families in Improvement

JOHN BINGHAM AND DORIS QUINN

9. Ethical Considerations in Human Subjects Research: Emerging Issues in Cancer Research

TYRON C. HOOVER AND AMAN BUZDAR

10. Ethical Concerns When Cancer Patients Become Human Research Participants and Are Treated on Clinical Trials

RICHARD L. THERIAULT

11. Training Current and Future Leaders

JOLYN S. TAYLOR AND DIANE C. BODURKA

12. Ethics in Cancer Patient Education

LORENE PAYNE

13. Commercialism in Cancer Care

MICHAEL S. EWER

14. A Perspective on the Cost of Cancer Care

JEFFREY S. FARRONI

15. Mergers, Acquisitions, and Consolidations in the Health-Care Industry: Does Bigger Have Ethical Consequences?

ROBERT (BOB) BRIGHAM

16. Legal and Compliance

HOLLY O. RUMBAUGH, KRISTA M. BARNES AND AMARJYOT S. PUREWAL

17. Diversity and Health Care

HARRY R. GIBBS AND COLLEEN M. GALLAGHER

18. An Ethical Framework for Disclosing Individual Genetic Findings to Patients, Research Participants, and Relatives

JEFFREY S. FARRONI, JESSICA A. MOORE AND COLLEEN M. GALLAGHER

19. Ethical Issues of Cancer Center Administrators

FRANK R. TORTORELLA

20. Ethics and Information Technology

LESLIE A. KIAN AND SCOTT D. EASTMAN

List of Contributors

Krista M. Barnes The University of Texas MD Anderson Cancer Center, Houston, TX, United States

Nancy B. Benjamin US District Courts, Houston, TX, United States

Robert S. Benjamin The University of Texas MD Anderson Cancer Center, Houston, TX, United States

John Bingham The University of Texas MD Anderson Cancer Center, Houston, TX, United States

Diane C. Bodurka The University of Texas MD Anderson Cancer Center, Houston, TX, United States

Robert (Bob) Brigham The University of Texas MD Anderson Cancer Center, Houston, TX, United States

Aman Buzdar The University of Texas MD Anderson Cancer Center, Houston, TX, United States

Kathy Denton The University of Texas MD Anderson Cancer Center, Houston, TX, United States

Scott D. Eastman The University of Texas MD Anderson Cancer Center, Houston, TX, United States

Daniel E. Epner The University of Texas MD Anderson Cancer Center, Houston, TX, United States

Michael S. Ewer The University of Texas MD Anderson Cancer Center, Houston, TX, United States

Jeffrey S. Farroni The University of Texas MD Anderson Cancer Center, Houston, TX, United States

Colleen M. Gallagher The University of Texas MD Anderson Cancer Center, Houston, TX, United States

Harry R. Gibbs The University of Texas MD Anderson Cancer Center, Houston, TX, United States

Tyron C. Hoover The University of Texas MD Anderson Cancer Center, Houston, TX, United States

Leslie A. Kian The University of Texas MD Anderson Cancer Center, Houston, TX, United States

Richard T. Lee Case Western Reserve University School of Medicine, Cleveland, OH, United States; UH Seidman Cancer Center, Cleveland, OH, United States

Paula Lewis-Patterson The University of Texas MD Anderson Cancer Center, Houston, TX, United States

Kevin Madden The University of Texas MD Anderson Cancer Center, Houston, TX, United States

Jessica A. Moore The University of Texas MD Anderson Cancer Center, Houston, TX, United States

Guadalupe R. Palos The University of Texas MD Anderson Cancer Center, Houston, TX, United States

Lorene Payne The University of Texas MD Anderson Cancer Center, Houston, TX, United States

Amarjyot S. Purewal The University of Texas MD Anderson Cancer Center, Houston, TX, United States

Doris Quinn Consultant in Process Improvement, Emmitsburg, MD, United States

Maria Alma Rodriguez The University of Texas MD Anderson Cancer Center, Houston, TX, United States

Holly O. Rumbaugh The University of Texas MD Anderson Cancer Center, Houston, TX, United States

Jolyn S. Taylor The University of Texas MD Anderson Cancer Center, Houston, TX, United States

Richard L. Theriault The University of Texas MD Anderson Cancer Center, Houston, TX, United States

Frank R. Tortorella The University of Texas MD Anderson Cancer Center, Houston, TX, United States

Eli Weber The University of Texas MD Anderson Cancer Center, Houston, TX, United States

Angelique Wong The University of Texas MD Anderson Cancer Center, Houston, TX, United States

Donna S. Zhukovsky The University of Texas MD Anderson Cancer Center, Houston, TX, United States

Patrick Zweidler-McKay The University of Texas MD Anderson Cancer Center, Houston, TX, United States

Foreword

Medical practitioners have been concerned about ethics since before the ancient Greeks identified and codified such principles. To this day, maintaining the highest level of ethical concern is a goal that identifies and distinguishes those among us who are privileged to be part of the healing arts. Practitioners also have a solemn responsibility not to allow conflicts of interest—social, financial, or political—to interfere with providing information, recommendations, and treatments for those seeking their counsel and advice.

The University of Texas MD Anderson Cancer Center has stood at the forefront of defining ethical care for cancer patients. It was among the first to establish an ethics committee, and among its initial administrative charges was to create the Institutional Code of Ethics. The code is not rigidly etched in stone, but rather one that reflects the intuition's philosophy to evolve and change as the patients' needs adjust based on new interventions and strategies.

This volume incorporates the unique perspectives of many who have helped to achieve ethical care of the cancer patient. The editors—one a former chairperson of the Institutional Ethics Committee and the other the present executive director of Integrated Ethics—have sought to bring the special ethical concerns of cancer patients to a broad readership, with the goal of helping practitioners understand care that is not only timely and scientifically proven, but also compassionate, sensitive and ethical.

I trust you will find this volume instructive and a valuable addition to your library and reasoning mind.

Ronald A. DePinho
The University of Texas MD Anderson Cancer Center

Introduction

A reader taking this volume in hand may well ask why, with all that has been thought about and written about medical ethics for perhaps as long as people tried to influence and intervene in the natural course of illness, do we need yet another book on medical ethics. The basic premises of ethics have not changed; as healthcare professionals we still strive to do good, avoid doing harm, and respect a patient's preferences and desires. Yet the environment in which medicine is practiced has changed dramatically, and perhaps in no subcategory of the healing arts has it changed to the extent we see as we care for the cancer patient. Modern cancer care involves the utilization of highly toxic or invading treatment modalities, often used in combinations that can no longer be administrated by a single practitioner. Cancer patients, in the course of their path to cure, stabilization, or death are likely to see many physicians with such diverse specialties as surgery, oncology, radiation therapy, and palliative or supportive care; even among these groups there are likely to be sub-specialists with expertise in transplantation, surgical sub-specialists who have developed expertise with regard to specific body areas or organs, or nononcologic specialists who provide expertise in fields such as onco-cardiology or onco-nephrology. While each encounter strives to assure that treatment remains ethical, the goals and perspectives of one may be juxtaposed with another, and ethical considerations must be considered, balanced, and in some instances deliberated; even then ethical questions may remain, and differences of opinion reconciled. Care of the cancer patient involves many uncertainties, and in some instances the degree of uncertainty is far greater than in others. Many of our patients have developed considerable understanding regarding the uncertainties of their illness, and make decisions; as ethicists we must consider these choices and provide our patients with the wherewithal to make choices that are rational, meaningful, and yet remain true choices. While too little choice may be thought of as paternalistic, choices that are excessively broad may be irrational, counterproductive, and, in some instances, wasteful. It is these dilemmas that face members of the cancer treatment team, and that have been considered in this volume.

But ethical care of the cancer patient must take factors beyond therapeutics into consideration. Clinical research plays a huge role in bringing new treatment strategies to our patients; while Institutional Review Boards (IRB) are charged with the protection of human subjects, and components of the Health Insurance Portability and Accountability Act (HIPAA) are intended to insure privacy of protected health care information, ethical questions arise and require insight; compassion and understanding remain important, but an understanding of ethical principles as they relate to the cancer are crucial.

Cancer care is provided within a system financed through an extraordinarily complex payment system that involves payment by governmental and private payers. Both access to care and responsibility for payment are aspects of cancer care where ethical considerations and conflicts abound. Along with cancer care's shift from *practitioner* to *health care team*, and along with the shift from *individual* payer to a payment system from *pooled* sources, the concept of *micro ethics*, i.e., the zealous advocacy of patient interest is now increasingly weighed against *macro ethics* concerns, i.e., concerns that also take the broader societal impact of what we do into account. This book looks at these intersections.

More than 30 years ago, the first Chairman of the Institutional Ethics Committee at The MD Anderson Cancer Center, Dr. Jan Van Eys, led the team that created a Code of Ethics specifically related to cancer care. Some asked why we need a code of ethics. Others commented that any code of ethics looks only at how we wish to provide care at the time of creation, and that such documents should be scrutinized and updated so that they keep pace with evolving practice. While such codes of ethics for cancer patients were unusual at the time our Code of Ethics was created in 1984, they are common place; both of the editors of this volume have served as chairpersons of the MD Anderson Institutional Ethics Committee under the overriding ethical umbrella of the Institutional Code of Ethics.

The various authors whose work and thoughts are reflected in this volume primarily have written to educate. But beyond that they share the goal of helping our colleagues who care for patients with cancer by providing a resource that will stimulate further thought, provide perspective, and recognize the enormous difficulties and controversies that are involved in the ethical care of the cancer patient.

The editors take this opportunity to thank Dr. Maria Alma Rodriguez for her encouragement and mentorship, and especially to Timoteo A. Gonzales III, who ensured that we keep on track, and the (hopefully) the i's got dotted and the t's got crossed.

Colleen Gallagher
Michael S. Ewer

CHAPTER

1

Ethics and Organizational Identity in a Cancer Care Setting

Eli Weber and Colleen M. Gallagher

The University of Texas MD Anderson Cancer Center, Houston, TX, United States

INTRODUCTION

Whenever we question our duties or obligations, what sorts of outcomes would be good or bad, valuable or not, or what sorts of persons we want to be, we have entered the domain of ethics [1]. These same sorts of questions permeate the landscape of health care [2]. For example, physicians may find themselves considering whether they are obligated to provide, or perhaps obligated to not provide, a requested treatment. Patients may question whether a particular course of treatment would be good or bad for them, and health care institutions may ask themselves whether a certain policy change would be good for their patients, or whether it is consistent with the sort of organization that they strive to be. Professional health care organizations may ponder whether they ought to take an official position on a controversial issue in health care, and if so, what that position should be. At both individual and organizational levels, ethical questions inevitably arise within the health care setting.

The unavoidably ethical nature of health care takes on unique features within the context of cancer care. For example, cancer patients tend to be especially vulnerable in a number of different ways; the uncertainties of cancer care and the rapidly evolving management strategies make participation in research at almost all stages of treatment highly desirable. Furthermore, cancer care often takes on a singular focus, where the overall health of patient may take a secondary position to that of eliminating the patient's cancer. In this chapter, we initially will identify some of the features of cancer care that are perhaps more prominent than in other areas of health care, and that give rise to ethical considerations which, though not necessarily unique to cancer care, are somewhat distinctive, in that they are both more common and more pressing than in other areas of health care. The relationship between organizational identity and ethics, with a focus on some of the mechanisms health care organizations devoted to cancer care might utilize to ensure a prominent place for

Ethical Challenges in Oncology.
DOI: http://dx.doi.org/10.1016/B978-0-12-803831-4.00001-4

ethical thinking within their organization will then be discussed. We will conclude by identifying a potential challenge associated with putting an organization's ethical vision into practice so that it influences the decisions and actions of individuals who represent that organization, as well as several areas where further research is needed. Our goal is to identify some of the ways that health care organizations devoted to cancer care can incorporate ethics into their organizational identity, in order to adequately address the ethical issues that are distinctive of cancer care.

WHAT MAKES CANCER CARE ETHICALLY DISTINCT?

Ethical issues in health care are typically analyzed in light of the ethical principles thought to be important for thinking about ethics within this domain. Before discussing some of the ethical challenges that are distinct to cancer care, these principles will be briefly identified, and their significance for health care ethics discussed more generally.

We begin with the principle of autonomy, which is typically associated with the notion of self-determination or self-rule. Autonomous agents are those whose actions are truly their own, in that their actions follow from, or are in accordance with, the agent's own reasons and values [3–5]. The principle of autonomy requires that competent individuals have the right to make decisions for themselves, without undue interference from others. The principle of autonomy as it pertains to health care thus implies that patients have a right to make decisions for themselves about their treatment, without undue coercion from others. Additionally, many purport this right implies a duty on the part of health care providers to honor the autonomous wishes of patients, even when they do not agree with the patient's autonomous choice. In general, many ethical issues in the health care of cancer patients have an element of autonomy.

The principle of beneficence is also an important ethical principle in health care. The principle of beneficence states that we should promote the interests, welfare, or well-being of others, and that in general, we should try to "do good" [6,7]. The obligations that follow from this principle depend, of course, on what we take the notion of interests, welfare, or goodness to involve. Despite this, the idea that we should try to pursue good outcomes is a helpful source of ethical guidance in health care.

The principle of nonmaleficence is often considered to be a corollary to the principle of beneficence. The principle of nonmaleficence states that we should avoid doing harm to others [1,6,8,9]. As we saw with the principle of beneficence, the obligations that follow from the principle of nonmaleficence depend on how we understand the notion of harm. What is more, we should not interpret this principle to imply that we should avoid actions where the benefits outweigh the harms that may result, or that we act unethically when our actions cause unintended or unanticipated harms. We should avoid intentional or willful harm when there is no significant corresponding benefit. The edict that we should "first, do no harm" is regarded by some as the most important ethical principle in health care.

The principle of justice is another important ethical principle in health care. The content of this principle, just as we saw with the principles of beneficence and nonmaleficence, will depend on how we construe the concept of justice [6,8,10–13]. There are many different types of justice that may be ethically relevant, both in general and within a health care setting. For example, commutative justice pertains to what is owed between individuals, e.g., in conducting business transactions. Contributive justice refers to what individuals owe to society for the sake of promoting the common good, while legal justice has to do with the rights and responsibilities of citizens to obey laws, respect the rights of others, and protect the social order. Distributive justice refers to the fair allocation of resources. In general, the principle of justice requires that we treat others fairly, such that all individuals are granted equal consideration when we make ethical decisions, and none are asked to bear more than their share of the burdens associated with a particular course of action.

How then does one apply the ethical demands of these principles to particular situations in a health care setting in general and in the care of the cancer patient in particular? This analysis can be complicated, but one useful strategy is to consider the burdens and benefits of the available courses of action and determine which has the most favorable balance between the two. For example, suppose a physician is considering whether to offer their patient a treatment option that is known to have harmful side-effects. The physician might consider exactly what these side-effects are, how harmful they might be for the patient, and how likely they are to occur. These sorts of considerations are weighed against the possible benefits of the treatment being considered. If the treatment offers a high probability of significant benefit, with only a slight risk of substantial burdens for the patient, one is ethically justified in offering the treatment to the patient. However, if the treatment offers a low probability of benefit, with substantial risk of serious burdens, such that the treatment is disproportionately burdensome, it may not be ethically justified to offer the treatment as an option for the patient. In most cases, considering the burdens and benefits of various available courses of action with patients can help us to determine which option is ethically best.

Now that we have a sense of the sorts of considerations that tend to be relevant for health care ethics in general, we can better appreciate the ways in which cancer care gives rise to distinct ethical challenges. Some of these challenges are related to the sorts of treatments utilized in cancer care. Depending on the type and severity of a patient's cancer, as well as other features of their personal and medical history, their treatment may include chemotherapy, radiation, biotherapy, surgery, or some combination of these interventions. For an increasing number of malignancies, stem cell transplantation has become an important modality. While these treatments differ significantly from one another in many respects, they all have two features in common that stand out as ethically significant. First, many of these treatments have the potential to be burdensome and especially harmful for the patient [9,14–16]. In some cases, the patient may actually be made worse off in some way than they were prior to receiving treatment [17–21]. In other cases, they may be harmed, either temporarily or permanently, in exchange for an uncertain but an anticipated benefit in survival or quality of life. Therefore, patients often must decide whether to pursue a potentially harmful course of treatment with only a modest chance that it will be beneficial [18]. In some cases, such as those where patients are considering a stem cell transplant, they must decide whether to pursue an extremely aversive, life-altering treatment with a three-year survival rate of less than 50% [22]. These conditions give rise to ethically difficult decision scenarios, since it becomes challenging to identify a treatment option that is not disproportionately harmful for the patient. In many cases involving patients with cancer, disproportionately harmful treatments are the only curative options available. This is further complicated by the fact that for some cancer treatment options, the potential benefits are largely to future generations rather than the patient currently receiving the treatment. What is more, it is unavoidable that for many patients with cancer, their illness is such that their capacity to make autonomous decisions is seriously impaired [23–25]. This can make it difficult to interpret the obligation to respect patient autonomy. While other medical conditions may also give rise to these sorts of high-risk, low-reward decision scenarios, such scenarios are relatively common within a cancer care setting.

Social and cultural attitudes about cancer also give rise to unique ethical challenges in cancer care. For example, many cultural groups view a cancer diagnosis as a "death sentence" [26]. Patients from these cultural groups, as well as their family members, may insist that information about a patient's cancer diagnosis be withheld from the patient or struggle to adhere to the proposed treatment plan [27–29]. Cultural beliefs and values regarding cancer and cancer care can make it more difficult to provide adequate treatment for patients with these values and beliefs and they may require that we rethink or revise our understanding and application of relevant ethical principles. Cultural beliefs about cancer may make it difficult, for example, to interpret the demand that we respect patient autonomy. We may also struggle to evaluate the benefits and burdens of a particular treatment option for a patient who is not fully informed of their prognosis.

Cancer care also gives rise to distinct ethical challenges for cancer care institutions [30,31]. Most prominent cancer care institutions have a strong commitment to the advancement of cancer research as cancer care often involves the use of experimental therapies and participation in clinical trials. As such, cancer care tertiary facilities must balance the need to provide the best patient care possible with the demands of their research program. This balancing effort may require, for example, that an institution think seriously about whether the clinical trials they approve are in the best interests of their patients, or whether allowing a research protocol might involve a conflict of interest, and if so, whether that conflict can be managed effectively. Cancer care institutions with a strong research focus must also find ways to empower their physicians as both researchers and care providers, in order to ensure that both aspects of their institutional commitments can be equally realized. This requires, among other things, that we think about harms and benefits for individual patients who may be asked to participate in a research study as well as potential benefits and burdens for possible future cancer patients. In some cases, a separation of those involved in research from those involved in direct care of the patient may be ethically justified. These sorts of considerations can be extremely difficult to formulate in meaningful ways but they are essential to evaluate the ethical justifiability of things like institutional involvement with clinical trials.

Social and cultural factors give rise to other distinct ethical challenges for cancer care institutions. Many cancer care institutions, for example, must balance the international scope of care and research with the localized nature of individual care centers. A large-scale, globally focused cancer care institution must find ways to serve an international, culturally diverse patient population while also meeting the cancer care needs of the local population from which many of their patients may come. Both patient populations may receive types of cancer care, but they may also have differing psychosocial needs that require culturally informed solutions. Such institutions must also consider the ethical demands of conducting research within an international context and find ways to ensure that high ethical standards for research are maintained even in settings where such standards are not the norm.

These are just some of the ways that cancer care can give rise to distinct ethical challenges. But what can cancer care institutions do to try and meet some of these challenges? To answer this question, we need to address the relationship between organizational identity and ethics, and to identify some of the mechanisms health care organizations devoted to cancer care might utilize to ensure a prominent place for ethical thinking within their organization. In doing so, we will advance the claim that cancer care institutions can begin to meet the distinct ethical challenges of cancer care by incorporating ethical thinking into their organizational identity.

ORGANIZATIONAL IDENTITY AND ETHICS

We have suggested that cancer care institutions can meet the distinct ethical challenges of cancer care by incorporating ethical thinking into their organizational identity. How exactly does this work? Organizational identity is an important initial consideration that can be defined as a shared understanding of an organization's values and commitments [32]. Organizational identity is what informs a response to the organization's own question of "who we are" as well as "who they are" for those outside the organization [33]. Organizational identity is what defines an organization's character [34].

We propose that an organization's mission statement, code of ethics, institutional decision-making process, and leadership structure all present opportunities to incorporate ethical thinking into an institution's organizational identity. In what follows, we will briefly explain how each of these entities can be structured so as to promote ethical thinking within an organization.

Mission Statements

First, an organization can incorporate ethical thinking into its mission statement. Consider some examples. The Mayo Clinic Cancer Center's mission is as follows:

> Mayo Clinic Cancer Center combines personalized cancer treatment with leading-edge research to provide you with unparalleled cancer care. Our integrated practice brings together a multispecialty cancer treatment team of experts to ensure you get the best care available. Mayo Clinic Cancer Center also offers medical professionals the opportunity to refer and follow patients and to access medical education offered by key specialists [35].

This is an excellent example of how an organization's mission statement can help incorporate ethical thinking into its organizational identity. For instance, from the above statement, one can clearly identify this institution's response to the ethical challenge of balancing the goals of both research and patient care. This institution, via their mission statement, is committed to a personalized approach to patient care that integrates cutting-edge research into the patient's treatment plan. This sort of mission statement thus allows the institution to define itself in direct response to the sorts of ethical challenges that cancer care institutions face.

Cancer Treatment Centers of America offers another example of how an institution can incorporate ethical thinking into their mission statement. Their mission is stated as follows:

> Cancer Treatment Centers of America is the home of integrative and compassionate cancer care. We never stop searching for and providing powerful and innovative therapies to heal the whole person, improve quality of life and restore hope [36].

Here, one can again see how a mission statement can help to shape organizational identity as we can again clearly identify this institution's response to the ethical challenge of balancing the goals of both research and patient care. This institution, via their mission statement, is committed to research aimed at treatments that improve quality of life and support a whole-person approach to cancer care. While this mission statement differs from that of the Mayo Clinic Cancer Center in subtle but important ways, the point here is that ethical thinking can be easily incorporated into the mission statement of an institution. This allows cancer care institutions to formulate their organizational identity in ways that are responsive to the sorts of ethical challenges that are distinct to cancer care.

It is also worth noting some of the differences that exist between the mission statements of health care organizations that focus on cancer care and health care organizations more generally. For example, let's compare the above mission statement from the Mayo Clinic Cancer Care Center with the mission statement of the Mayo Clinic qua provider of general health care services. Their mission is "to inspire hope and contribute to health and well-being by providing the best care to every patient through integrated clinical practice, education and research" [37]. This mission reflects similar values and commitments when compared to the mission statement of the Mayo Clinic Cancer Care Center, but there are also some noteworthy differences. The mission of the Mayo Clinic Cancer Care Center makes the integration of cutting-edge research and multispecialty perspectives into patient care a focal point of its value commitments. The role of research in patient care is mentioned but not emphasized in the mission of the Mayo Clinic. This subtle but important difference reflects the fact that balancing research goals with the need to provide quality patient care is a distinct and more frequent ethical challenge in cancer care.

Mission statements can also help cancer care institutions to address the challenge of balancing the international scope of cancer care and research with the localized nature of individual care centers. For example, the mission statement of MD Anderson Cancer Center is

> to eliminate cancer in Texas, the nation, and the world through outstanding programs that integrate patient care, research and prevention, and through education for undergraduate and graduate students, trainees, professionals, employees and the public [38].

Here, one can clearly identify a commitment to providing cancer care, research and prevention, and educational opportunities at both local and global levels. This commitment, in turn, is reflected in MD Anderson's efforts to expand its reach into the surrounding areas of Houston, where the main campus is located, and into other parts of both the United States and the world via various partnership arrangements with existing health care entities in those areas.

One might ask how incorporating ethical thinking into an institution's mission statement can give rise to ethical thinking at the level of institutional decision making. To see how this might work, consider the following example. Suppose that a cancer care institution is faced with a decision about whether to participate in a multicenter clinical trial. There are many questions that will need to be answered before a decision can be made and one of those questions may be "Is this consistent with the sort of organization that we strive to be?" For someplace like the Mayo Clinic Cancer Center which identifies the integration of cutting-edge research into personalized cancer care as a key component of their mission, answering this question may require determining whether the experimental arm of the proposed clinical trial can be successfully integrated into the patient care that is typically provided. For Cancer Treatment Centers of America whose mission emphasizes the importance of research that will improve patient quality of life, answering this question may require determining whether the experimental regiment in question is likely to do so or whether it is consistent with the whole-person approach to cancer care that is a significant component of their stated approach to research. Consistency with an organization's mission is not the sole determinant in these sorts of institutional decisions nor is it the sole ethical consideration. However, having a well-formulated mission statement that incorporates ethical thinking can provide helpful ethical guidance when tasked with these sorts of institutional decisions.

Finally, we note the importance of a mission statement that is individualized to the entity whose organizational identity it is intended to express. We have here focused on large, internationally focused cancer care centers with commitments to patient care, research, and education. These sorts of entities will, by their very nature, define themselves in terms of mission statements that reflect and balance these commitments. However, smaller, more regionally focused cancer care institutions may be better served by missions that emphasize their commitment to patient care within the area that they serve. Oncologists in private practice, or who are part of a medical group with other physicians, may be best served by mission statements that reflect their own values and commitments, whatever they may be. Oncologists who practice within an institution that is not explicitly devoted to cancer care may need to adopt an interpretation of their organization's mission that fits the distinctive demands of cancer care. The point here is that while we have focused on ways that large-scale, research-focused cancer care institutions can utilize mission statements to integrate ethical thinking into their organizational identity, one should not infer that only these sorts of institutions can utilize mission statements in this way. Mission statements are a useful tool for integrating ethical thinking into an organization's identity. This is rue no matter the size, scope and commitments of the organization.

Codes of Ethics

An organization can also incorporate ethical thinking into its organizational identity by developing and adopting a code of ethics. One of the best examples of this is the code of ethics for The University of Texas MD Anderson Cancer Center. There are several reasons that this is an especially good example of incorporating ethical thinking into a cancer care institution's organizational identity. First, despite the code being just over two decades old, it was among the first such endeavors of its kind, and it continues to be something of a rarity among cancer care institutions [39].

Second, the code of ethics for MD Anderson Cancer Center was developed with explicit acknowledgment that cancer is, in many respects, a unique disease [39]. This acknowledgment helped to shape one of the primary aims of the code which is "to confront professionals with the need to think seriously about what they perceive as unique problems and thereby arrive at thoughtful solutions rather than to simply react emotionally or selfishly" [39]. By acknowledging the uniqueness of cancer and cancer care from the outset, MD Anderson's code of ethics is well-situated to provide ethical guidance for the institution, especially when they are faced with the ethical challenges that are distinct of cancer care.

Third, the code of ethics for MD Anderson Cancer Center is largely a product of public discourse and community engagement, rather than top-down policy making. Its initial content was formulated by a committee of institutional representatives comprised of individuals from across the institution. Upon the completion of an initial draft, the draft document was presented to the community for comments and a subsequent revision was the focal point of a conference featuring experts from the field of bioethics. The final version took account of all this input, resulting in a document that is responsive to the concerns of the institution and its various constituencies, the field of bioethics, and the surrounding community. Because it incorporates input from the community, MD Anderson Cancer Center's code of ethics can be easily endorsed by that same community and readily utilized as a decision-making tool. Emphasizing community involvement is another way to respond to the ethical challenge of meeting the cancer care needs of a local population while maintaining an international focus.

Finally, the MD Anderson Cancer Center's code of ethics is not simply a code of conduct for its employees. A code of conduct is a set of rules of behavior for a group or organization which individuals must agree to abide by if they wish to participate in that group or organization. A code of ethics, in contrast, is a set of shared values and commitments which provide a standard against which one's actions can be measured. A code of ethics allows for a shared vision of an organization's character. A code of conduct merely informs the individuals within an organization of how they are expected to behave. One can be compliant with a code of conduct simply by following its mandates; one need not endorse any particular value or commitment. Adherence to a code of ethics, in contrast, requires that one invest oneself in a shared vision of both oneself and the organization in which one participates. For example, a health care institution's code of conduct may require that all employees protect confidential patient information in accordance with Health Insurance Portability and Accountability Act (HIPAA) regulations. Contrast this sort of requirement with one of the principles of the MD Anderson Cancer Center's code of ethics which states that "patients justly expect personal information to be confidential, yet their medical records are accessible to all health-care providers. All information must be recorded responsibly. Access also confers a moral obligation. Access must be justified and not harmful to the patients' interests" [39]. Adhering to this sort of expectation involves more than simply being compliant with a particular set of regulations; it also requires that one adopt a particular perspective about the importance of patient confidentiality. In doing so, a code of ethics can help to shape organizational identity as well as individualized instantiations of that identity.

A code of ethics also provides the institution with a decision procedure for ethical decision making because the commitments articulated by an institution's code of ethics can be readily transposed into evaluative standards. For example, another principle of the MD Anderson Cancer Center's code of ethics states that "cancer therapy and research are expensive endeavors demanding conscientious stewardship; however, financial considerations should never dictate the quality of care offered to each patient" [39]. Now, suppose that an institutional decision must be made regarding the provision of uncompensated care, specifically whether uncompensated care should ever be provided. The above principle can be readily transposed into an evaluative standard that can help answer this question. Given their commitment to not allowing financial considerations to dictate the quality of care that is offered to patients, MD Anderson clearly seems to be, and in fact is committed to a policy of providing uncompensated care in at least some cases [40]. In this sense, the institution's commitment to not allowing financial considerations to dictate the quality of the care that is offered to patients gives rise to an evaluative standard that allows the institution to determine what their policy regarding uncompensated care ought to be which is to provide the same quality of care to all patients and to face economic challenges of current and prospective patients in a just manner.

While we have again focused our discussion on a large-scale cancer institution with a strong research focus, it is worth emphasizing that other types of cancer care institutions might also utilize something like a code of ethics as a mechanism for integrating ethical thinking into their organizational identity. For smaller scale institutions, developing a code of ethics may involve a similar version of the same process utilized in the development of MD Anderson's code of ethics. However, an organization of almost any size can identify a set of shared values and commitments and codify them within a publicly available document, and it is this identification and public expression of shared values that is the essential function of a code of ethics. Even if the sort of development process discussed here is not appropriate for a particular cancer care institution, the key features of the MD Anderson Cancer Center's code of ethics, especially its focus on the unique nature of cancer care, its emphasis on shared values and commitments over rules that govern individual conduct and its insistence on community involvement can be appreciated, and where appropriate, duplicated by a cancer care organization of almost any size and scope.

Institutional Decision-Making Process

Thus far, we have claimed that an organization's mission statement and code of ethics are two of the mechanisms one might look to in attempting to incorporate ethical thinking into an institution's organizational identity. However, less formal mechanisms are also important for this purpose. A cancer care institution might, for example, incorporate ethical thinking into its organizational identity by integrating certain ethically grounded evaluative standards into its decision-making process. Institutional decision making requires that financial, legal, political, and social considerations be considered as part of the institutional decision-making process. We propose that there are also certain ethically focused evaluative standards that ought to play a role in the institutional decision making of all cancer care organizations. Let's briefly consider each of these standards and discuss how they can help guide institutional decision makers toward an ethically supportable course of action.

The Consequentialist Standard

Consequentialism is an ethical perspective which states that the ethically required course of action is the one that gives rise to the best overall outcome, from an impersonal point of view [1,41–43]. From an institutional perspective, one should not resort to an entirely consequentialist ethical analysis, since other sorts of considerations also bear on the ethical obligations of institutions. However, we contend that institutions ought to take account of the potential consequences of their decisions, particularly the harms and benefits that may follow from various courses of action. "What are the harms and benefits of this course of action?" is an ethical question that institutions ought to be asking themselves prior to adopting a particular policy or making an institutional decision.

The Deontological Standard

Deontology is an ethical perspective which states that there are ethical obligations, sometimes referred to as duties, which must be met irrespective of the consequences that may follow from adhering to them [1,44,45]. Further, while not all deontologists agree with this, it is generally held that the ethical obligations deontologists recognize as duties give rise to corresponding rights claims. For example, the duty to respect the autonomy of others corresponds to a right of noninterference. If I have a duty to respect your autonomous choices, you thereby have a right to not be interfered with when you make and act on those same choices.

Many prominent approaches to bioethics are, in fact, deontological perspectives. For example, the highly influential, principle-based approach to bioethics articulated by Tom Beauchamp and James Childress is often regarded as largely a deontological approach [6]. What is more, the notion that there are certain ethical obligations that must be met irrespective of the consequences that may follow in a particular case is a significant component of the "common morality" that is the basis of many theoretical approaches to bioethics [6,8]. As such, a cancer care institution's decision-making process ought to incorporate considerations of ethical duty as a relevant component of that organization's approach to institutional decision making. Here, the relevant ethical question is "what are our duties in this case?" or "what rights are at stake, and for whom?" We might also consider how various courses of action might affect the exercise of individuals' rights in the future. Finally, the deontological standard requires that we consider whether a particular course of action or policy amounts to making an exception of ourselves. Duties as well as their corresponding rights are thought to be equally applicable to all moral agents. As such, to fail to adhere to a duty to which we hold others, or violate a right that we claim for ourselves is to violate the deontological standard, since it amounts to making an ethical exception. Here, the relevant ethical questions are "what if everyone did this?" and "what if someone else did this to us?"

The Justice Standard

Justice is a multifaceted ethical concept but at its core, justice has something to do with what is fair. For the most part, the relevant sorts of justice considerations that may come into play within the context of institutional decision making are considerations of distributive justice, which pertain to the fair or equitable distribution of burdens and benefits [1,10–13,33]. For example, institutions may consider whether those affected by a particular institutional decision or policy might either be asked to bear a disproportionate burden or receive an undue benefit as a result of the implementation of that policy or decision. In general, when cancer care institutions make decisions or implement policies, they ought to consider who will bear the burdens, who will reap the benefits, and whether this distribution of burden and benefit is ethically supportable. Here, the relevant question is "Is this a fair distribution of burden and benefits?"

The Common Good Standard

The notion of the common good is not easily characterized but it appears somewhat frequently in both theological and secular approaches to bioethics. For example, the *Ethical and Religious Directives for Catholic Health Care Services* states that "the common good is realized when economic, political, and social conditions ensure protection for the fundamental rights of all individuals and enable all to fulfill their common purpose and reach their common goals" [46]. However, there are also secular interpretations of the common good. For instance, John Rawls characterized the common good in terms of "certain general conditions that are...equally to everyone's advantage" [11]. Martha Nussbaum and Amartya Sen similarly assert that there are certain conditions that are required for individuals to realize all of their capabilities as free agents, which is a requirement for living a good life [47,48]. While they do not adopt this terminology themselves, we might think of this set of conditions as roughly akin to the common good. We might also characterize the common good in terms of human well-being, whereby "the common good" refers to the conditions that make it possible for human beings, generally speaking, to fare well. For our purposes, considerations that pertain to the promotion of the common good are highly relevant for ethically sound institutional decision making. For example, a cancer care institution may be considering closing one of its care centers due to ongoing concerns about profitability. Considerations of the common good may demand that the institution in question evaluate what sort of impact this proposed closure would have on the surrounding community. For instance, would the proposed closure unduly impede the current patient population's ability to access adequate cancer treatment? Would the closure of the facility lead to an unacceptable reduction in quality of life for those who live in the community it serves? These sorts of considerations may also arise when one is applying the consequentialist standard described earlier, since these are clearly questions that pertain to harm and benefit. However, the notion of the common good frames these sorts of considerations in a particular way. Considerations of the common good require that we consider how the potential harms of a particular course of action affect the well-being of or undermine the exercise of capabilities for those who may suffer them.

The Character Standard

Thus far, the ethical standards for institutional decision making that have been addressed have focused on the ethical quality of courses of action. We have identified considerations of consequences, duties, justice, and the common good as relevant ethical standards for institutional decision making. In each case, the structure of our evaluation has been to measure a course of action against some external standard derived from a source that exists outside of the institution itself. The character standard, however, has a different structure. Evaluating courses of action according to the character standard requires that we assess our options according to the institution's own values and commitments. The relevant question, according to the character standard, is whether a particular course of action is consistent with an organization's sense of itself, its own vision of the sort of organization that it wishes to be. And while an institution's mission statement and code of ethics are both mechanisms by which ethical thinking can be incorporated into an organization's identity, a commitment to pursuing courses of action that are consistent with that identity is also an important component of an ethically minded institutional decision-making process. Without a means of putting its ethical vision into practice, the inclusion of ethical thinking into an organization's identity will do little to ensure that ethical thinking is actually taking place at the institutional level.

The Reactive Attitudes Standard

The reactive attitudes are characterized as natural attitudinal reactions to our own Ethics and Org. Identity in Cancer Careperceptions of the goodwill, ill will, or indifference demonstrated by the actions of others [49]. The most familiar examples of the reactive attitudes are praise, which is deployed in response to actions that express respect, concern for others, generosity, or other positive interpersonal sentiments, and blame, which is deployed in response to actions that express malice, disregard, selfishness, or other negative interpersonal sentiments. In general, we deploy the reactive attitudes as a means of expressing that such actions are either acceptable, as in the case of praise, or unacceptable, as in the case of blame. Considering how others might respond to various courses of action is another helpful mechanism for ethical decision making. At the level of institutional decision making, we can utilize the reactive attitudes standard by considering how the surrounding community, or the public more generally, is likely to respond to various courses of action. Alternatively, an organization may wish to consider how certain highly respected individuals are likely to respond to a particular course of action. To put the point into context, if your clergyperson, the President, and your grandmother are all likely to disapprove of a course of action, this may count against an institution's pursuit of that course of action.

We should be careful here about differentiating the reactive attitudes standard from other sorts of considerations related to public perception of an institution. The reactive attitudes standard does not ask, for example, whether a particular course of action is likely to lead people to seek services elsewhere, or discontinue financial support. Rather, this standard is concerned with the attitudinal responses of others, specifically whether a course of action is likely to elicit praise, blame, or some other evaluative attitude.

Assessing which reactive attitudes are likely responses to a particular course of action can help institutions to determine the ethical quality of that course of action.

This list is not intended to be exhaustive of the sorts of evaluative standards that cancer care institutions ought to utilize if they wish to incorporate ethical thinking into their institutional decision-making process. It is also not intended to capture the full spectrum of ethical considerations that may be relevant to a particular institutional decision. Rather, this list is intended to identify the essential elements of an institutional decision-making process that incorporates ethical thinking such that any ethically minded organization will ask these sorts of questions before pursuing a particular course of action or policy measure. Further, unlike our earlier discussion of mission statements and codes of ethics, there is little reason to think that the size and scope of an institution makes much difference regarding the implementation of these evaluative standards into an institution's decision-making process. Even an individual oncologist in private practice can apply these standards to their own organizational policies and decisions.

Leadership Structure

Finally, cancer care institutions can incorporate ethical thinking into their organizational identity by providing a prominent place for ethics within their leadership structure. There are several ways that this might be accomplished. First, an organization can simply invite individuals with a commitment to ethical thinking into leadership positions within their organization. Leaders must do more than simply pay lip service to the importance of ethics; however, ethical leadership is a theoretical construct whereby leaders demonstrate the organization's ethical commitments in their own decisions and behaviors, and empower others to adopt and act on those same values and commitments [50,51]. There is some evidence that leaders who engage in ethical leadership have a positive impact on the decisions and actions of others [51]. Thus, one way that organizations can incorporate ethical thinking into their organizational identity is by inviting ethical leaders into leadership positions within their institution.

Cancer care institutions can also incorporate ethical thinking into their leadership structure by adopting and adhering to ethical standards that support their ethical vision. For example, a large-scale cancer institution with an international scope might look to the Universal Declaration on Bioethics and Human Rights as a source of ethical insight and guidance, rather than something like the Belmont Report, which has been frequently criticized for failing to account for the ethical relevance of differences between patient populations in different geographical locations [52,53]. By adhering to an ethical standard that is intended for an international audience, cancer care institutions can better respond to the international scope of cancer care without undermining their commitment to serving a local population.

Finally, cancer care institutions can incorporate ethical thinking into their leadership structure by establishing mechanisms for ethical oversight and ethical reflection. For example, many cancer care institutions utilize ethics consultation services to evaluate institutional policies, participate in institutional decisions, and help shape organizational identity via the development of mission statements and codes of ethics [54–56]. Some institutions also have formal committees devoted specifically to issues of organizational identity. For example, MD Anderson Cancer Center has recently formed an Integrity and Ethics Advocacy Council for the purpose of critically examining the ethical vision of the institution, revising and updating this vision as needed, and developing an implementation and education strategy to ensure that this vision is disseminated throughout the institution. Ultimately, this is the end goal of any effort to incorporate ethical thinking into an organization's identity—organizations want the individuals that represent them to put the institution's ethical vision into practice. Let's conclude by briefly noting some challenges associated with achieving this goal.

PUTTING AN ORGANIZATION'S ETHICAL VISION INTO PRACTICE: SOME CHALLENGES

Cancer care institutions that seek to incorporate ethical thinking into their organizational identity face a number of challenges when it comes to putting their ethical vision into practice, especially when the goal is to motivate individuals to act in accordance with that vision. First, there is some concern about the motivational efficacy of mission statements, codes of ethics, and other formal mechanisms for expressing organizational identity. Some studies have shown that mission statements are associated with higher firm performance which seems to suggest that individual actors are perhaps motivated by certain aspects of some mission statements [57]. It has also been reported that individuals sometimes perceive themselves as influenced or affected by their organization's mission statement [58]. But this sort of data does not tell us whether mission statements actually motivate the actions and behaviors of individuals, and there is virtually no data regarding whether incorporating ethical thinking into an organization's mission motivates individuals to act in accordance with their organization's ethical vision. This is an area where further research is sorely needed, but for now, the idea that incorporating ethical thinking into one's organizational identity will have an effect on the behavior of individual members of that organization is a theoretical presupposition worth pursuing.

There is another, closely related concern about the strategy of incorporating ethical thinking into an organization's identity as a mechanism for changing the behavior of individuals. There is substantial empirical evidence suggesting that when it comes to decision making in general, and ethical decision making in particular, most people are highly influenced by nonrational factors. By "nonrational factors," we mean to refer to factors that do not give rise to considerations that count in favor of a particular course of action. For example, recent studies have shown that such seemingly irrelevant factors as the smell, temperature, or lighting in a room can have a measurable impact on the choices that individual actors make [59–61]. Further, these sorts of factors have been consistently implicated as salient for our ethical decision-making process [62–66]. Factors that should ideally play no role in our reasoning about ethical issues such as whether we are in a good mood, whether an authority figure is encouraging us to do something, or whether a particular course of action would be in our own self-interest have all proven to be significant factors in the way individuals actually make ethical decisions. What is more, there is some evidence that ethical decision making is not a rational, deliberative process at all, but is instead the result of neural mechanisms of emotion [33,67–69]. This data problematically suggests that even if cancer care institutions are able to incorporate ethical thinking into their organizational identity, there is little reason to think that this will lead to any sort of change in the behavior of individuals in the institution. To put the concern simply, perhaps ethical behavior just is not the sort of thing that is responsive to mission statements or codes of ethics.

This concern, while significant, should not be taken for an argument against incorporating ethical thinking into the identity of one's organization. Rather, this data suggests that any organization that wishes to see its ethical vision put into practice at the level of individuals should not limit itself to simply formulating and disseminating its ethical vision. A comprehensive strategy for putting an organization's ethical vision into practice should include a plan for motivating and inspiring individuals to support that vision by acting in accordance with it. Identifying the best strategy for achieving this goal is well beyond the scope of our discussion here but it represents a potentially fruitful area of further research. For now, the point is simply that while the mechanisms we have been discussing are useful for incorporating ethical thinking into an institution's organizational identity, putting an organization's ethical vision into practice at the level of individual actors requires more than simply adopting these mechanisms.

CONCLUSION

Cancer care institutions face a number of distinct ethical challenges. We have claimed that one way cancer care institutions can begin to meet some of these challenges is by incorporating ethical thinking into their organizational identity. Mission statements, codes of ethics, institutional decision-making procedures, and leadership structure all provide opportunities for cancer care institutions to incorporate ethical thinking into their organizational identity. These mechanisms allow cancer care institutions to shape their organization's character in ways that are directly responsive to the distinct ethical challenges of cancer care, and they often give rise to evaluative standards that allow for ethically sound institutional decision making. This is not to say, of course, that cancer care institutions can ensure that their ethical vision is realized by simply formulating their organizational identity in a manner that grants ethical thinking a prominent place within its structure. Putting an institution's ethical vision into practice requires that institutions also engage with the sorts of considerations that motivate individuals to act ethically, and find ways to promote ethical thinking and behavior within the individuals who represent the institution. As we have noted, determining the best strategies for motivating individuals to act ethically is a significant and important area for further research. In the remaining chapters of this volume, we will see some of the more specific ways that ethical thinking can be pertinent for addressing the ethical challenges of cancer care. However, in each such case, the reader should be mindful of how the ethical vision of one's cancer care organization might either promote or inhibit the sort of ethical thinking being advocated in these more specific cases and situations.

References

[1] Shafer-Landau R. The fundamentals of ethics. Oxford University Press; 2010.
[2] Kuhse H, Singer P. What is bioethics? A historical introduction. 2nd ed. A Companion to Bioethics; 1998. p. 1–11.
[3] Dworkin G. Paternalism. The Monist 1972;64–84.
[4] Kant I, Wood AW, Schneewind JB. Groundwork for the metaphysics of morals. Yale University Press; 2002.
[5] Mill JS, Alexander E. On liberty. Broadview Press; 1999.
[6] Beauchamp TL, Childress JF. Principles of biomedical ethics. Oxford University Press; 2001.
[7] Pellegrino ED. For the patient's good: The restoration of beneficence in health care. Oxford University Press, 1988.
[8] Gert B, Culver CM, Clouser KD. Bioethics: A return to fundamentals. Oxford University Press; 1997.
[9] Ufema JOY. Doing more harm than good. Nursing 2004;34(6):83.
[10] Cohen GA. Equality of what? On welfare, goods and capabilities. Recherches Économiques de Louvain 1990;56:357–82.
[11] Rawls J. A theory of justice. Harvard University Press; 2009.
[12] Roemer JE. Theories of distributive justice. Harvard University Press; 1998.
[13] Sandel MJ. Justice: what's the right thing to do? Macmillan; 2010.
[14] Harvard Health Publications. Cancer treatments may harm the heart. Harvard Heart Letter 2012;22(12):1–7.
[15] Wilt TJ, MacDonald R, Rutks I, Shamliyan TA, Taylor BC, Kane RL. Systematic review: Comparative effectiveness and harms of treatments for clinically localized prostate cancer. Ann Intern Med 2008;148(6):435–48.
[16] Fallowfield LJ. Evolution of breast cancer treatments: Current options and quality-of-life considerations. Eur J Oncol Nurs 2004;8(Suppl. 2):S75–82.
[17] Brons P, Koopman HM, Detmar SB, Abbink F, Engelen V, Raat H, et al. Health-related quality of life after completion of successful treatment for childhood cancer. Pediatr Blood Cancer 2011;56(4):646–53.
[18] Cancer Treatment Centers of America. Cancer Treatment Statistics and Results 2015 [cited 8.10.2015]. Available from: http://www.cancercenter.com/ctca-results/.
[19] Costa Bandeira AK, Azevedo EHM, Vartanian JG, Nishimoto IN, Kowalski LP, Carrara-de Angelis E. Quality of life related to swallowing after tongue cancer treatment. Dysphagia 2008;23(2):183–92.
[20] Gore JL, Kwan L, Lee SP, Reiter RE, Litwin MS. Survivorship beyond convalescence: 48-Month quality-of-life outcomes after treatment for localized prostate cancer. J Natl Cancer Inst 2009;101(12):888–92.
[21] Volkenstein S, Willers J, Noack V, Dazert S, Minovi A. Health-related quality of life after oropharyngeal cancer treatment. Laryngorhinootologie 2015;94(8):509–15.
[22] Center for International Blood and Marrow Transplant. U.S. Patient Survival Report. In: Services HaH, editor. Online.
[23] Drane JF. Competency to give an informed consent: A model for making clinical assessments. JAMA 1984;252(7):925–7.
[24] Macklin R. Bioethics, vulnerability, and protection. Bioethics 2003;17(5–6):472–86.
[25] Roberts LW. Informed consent and the capacity for voluntarism. Am J Psychiatry 2002;156(2):705–12.
[26] Matthews-Juarez P, Weinberg AD. Cultural Competence in Cancer Care: A Health Professional's Passport: Office of Minority Health, US Department of Health & Human Services; 2004.
[27] Kendall S. Being asked not to tell: Nurses' experiences of caring for cancer patients not told their diagnosis. J Clin Nurs 2006;15(9):1149–57.
[28] Shahidi J. Not telling the truth: Circumstances leading to concealment of diagnosis and prognosis from cancer patients. Eur J Cancer Care 2010;19(5):589–93.
[29] Mystakidou K, Parpa E, Tsilika E, Katsouda E, Vlahos L. Cancer information disclosure in different cultural contexts. Supportive Care Cancer 2004;12(3):147–54.
[30] Sabin JE, Cochran D. Confronting trade-offs in health care: Harvard Pilgrim Health Care's Organizational Ethics Program. Health Aff (Millwood) 2007;26(4):1129–34.
[31] Stichler JF. Ethical considerations in healthcare design and construction. HERD 2013;6(4):5–9.
[32] Balmer JM, van Riel CB, Jo Hatch M, Schultz M. Relations between organizational culture, identity and image. EJM 1997;31(5/6):356–65.
[33] Albert S, Ashforth BE, Dutton JE. Organizational identity and identification: Charting new waters and building new bridges. Acad Manage Rev 2000;25(1):13–17.
[34] Albert S, Whetten DA. Organizational identity. Res Organ Behav 1985.

[35] Mayo Clinic. Cancer Care at Mayo Clinic: Our Promise [cited 21.10. 2015]. Available from: http://www.mayoclinic.org/departments-centers/mayo-clinic-cancer-center.
[36] Cancer Treatment Centers of America. About Us: Mission.
[37] Mayo Clinic. The Mayo Clinic Mission and Values [cited 27.10.2015]. Available from: http://www.mayoclinic.org/about-mayo-clinic/mission-values.
[38] MD Anderson Cancer Center. About Us: Mission and Values [cited 21.10.2015]. Available from: http://www.mdanderson.org/about-us/index.html.
[39] Van Eys J, Bowen JM. The common bond: The University of Texas System Cancer Center Code of Ethics. Charles C Thomas Pub Ltd; 1986.
[40] MD Anderson Cancer Center. Uncompensated Care 2014 [cited 26.10.2015]. Available from: http://www.mdanderson.org/about-us/facts-and-history/office-of-health-policy/programs/uncompensated-care.html.
[41] Driver J. Consequentialism. Routledge; 2011.
[42] Railton P. Alienation, consequentialism, and the demands of morality. Philos Public Aff 1984;134–71.
[43] Sinnott-Armstrong W. Consequentialism. Stanford Encyclopedia of Philosophy; 2009.
[44] Darwall SL. Deontology. Wiley-Blackwell; 2003.
[45] McNaughton DA, Rawling JP. Deontology. 2nd ed Principles of Health Care Ethics; 2007. pp. 65–71.
[46] Catholic Church. Ethical and religious directives for Catholic health care services. Washington DC: United States Conference of Catholic Bishops; 2009.
[47] Nussbaum MC. Frontiers of justice: Disability, nationality, species membership. Harvard University Press; 2009.
[48] Sen A. Development as freedom. Oxford University Press; 2001.
[49] Strawson PF. Freedom and resentment. Proc Br Acad 1962;48:1–25.
[50] Brown ME, Treviño LK, Harrison DA. Ethical leadership: A social learning perspective for construct development and testing. Organ Behav Hum Decis Process 2005;97(2):117–34.
[51] Mayer DM, Kuenzi M, Greenbaum R, Bardes M, Salvador R. How low does ethical leadership flow? Test of a trickle-down model. Organ Behav Hum Decis Process 2009;108(1):1–13.
[52] Shore N. Re-conceptualizing the Belmont Report: A community-based participatory research perspective. J Community Pract 2006;14(4):5–26.
[53] Unesco. Universal Declaration on Bioethics and Human Rights: Unesco; 2005 [cited 5.10.2015]. Available from: http://portal.unesco.org/en/ev.php-URL_ID=31058&URL_DO=DO_TOPIC&URL_SECTION=201.html.
[54] Foglia MB, Pearlman RA. Integrating clinical and organizational ethics. Health Prog 2006;87(2):31.
[55] The scope of organizational ethics. In: Khushf G HEC Forum. Springer; 1998.
[56] Silva MC. Organizational and administrative ethics in health care: An ethics gap. Online J Issues Nurs 1998;3(3):4.
[57] Bart CK, Baetz MC. The relationship between mission statements and firm performance: An exploratory study. JMS 1998;35(6):823–53.
[58] Verma HV. Mission statements—a study of intent and influence. J Serv Res 2009;9(2):153.
[59] Leonard TC, Thaler RH, Sunstein CR. Nudge: Improving decisions about health, wealth, and happiness. Constitutional Political Economy 2008;19(4):356–60.
[60] Thaler RH, Sunstein CR, Balz JP. Choice architecture, The Behavioral Foundations of Public Policy. Princeton University Press; 2012.
[61] Vallgårda S. Nudge—A new and better way to improve health? Health Policy 2012;104(2):200–3.
[62] Alfano M, Fairweather A. In: Pritchard D, editor. Situationism and virtue theory. Oxford Bibliographies in Philosophy; 2013.
[63] Isen AM, Levin PF. Effect of feeling good on helping: cookies and kindness. J Pers Soc Psychol 1972; 21(3):384.
[64] Kamtekar R. Situationism and virtue ethics on the content of our character. Ethics 2004;114(3):458–91.
[65] Nelkin DK. Freedom, responsibility and the challenge of situationism. Midwest Studies in Philosophy 2005;29(1):181–206.
[66] Sosa E. Situations against virtues: The situationist attack on virtue theory, In: Philosophy of the social Sciences: Philosophical Theory and Scientific Practice. Cambridge University Press; 2009. p. 274–90.
[67] Greene J, Haidt J. How (and where) does moral judgment work? Trends Cogn Sci 2002;6(12):517–23.
[68] Haidt J. The emotional dog and its rational tail: A social intuitionist approach to moral judgment. Psychol Rev 2001;108(4):814.
[69] Haidt J, Hersh MA. Sexual morality: The cultures and emotions of conservatives and liberals. J Appl Soc Psychol 2001;31(1):191–221.

CHAPTER

2

Experience Matters: A Partnership Between Patient and Physician

Jessica A. Moore, Kathy Denton and Daniel E. Epner

The University of Texas MD Anderson Cancer Center, Houston, TX, United States

INTRODUCTION

Patients make complex medical decisions much differently now than they did just a few decades ago. The internet and social media have brought a world of information to patients' fingertips and patients increasingly want to use that information to help them make well-informed decisions. Patients are no longer comfortable blindly allowing physicians to make decisions on their behalf. Patients now express their autonomy and partner with physicians to make shared decisions, while physicians and other members of the healthcare team approach medical decisions with the patient's perspective in mind. This patient-centered approach empowers patients, but it also imposes new responsibilities upon them. Shared decision making neither relieves physicians of their obligation to guide patients when necessary nor obligates them to agree to patient requests that run counter to their deeply held moral convictions and professional integrity. The patient–physician relationship is now a reciprocal one, with rights and responsibilities on both sides. The old paternalistic model was built almost entirely on a foundation of biomedical knowledge and skills, whereas patient-centered care requires nuanced relational and communication skills in addition to the biomedical knowledge, thereby more completely fulfilling Osler and Peabody's charge to practice both the science and the art of medicine through caring.

When we say experience matters we are referring to several important types of experience: "Patient experience" as a movement, the patient's (life) experience and values as a consideration in shared decisions, and the physician's or other healthcare provider's experience and expertise as a medical professional and values as a fellow journeyman. In this

Ethical Challenges in Oncology.
DOI: http://dx.doi.org/10.1016/B978-0-12-803831-4.00002-6

chapter, we will present clinical scenarios that illustrate ethical challenges that often arise during shared decision making, outline ethical principles that pertain to the scenarios, and describe key strategies and skills that physicians need to partner with the patient and family to develop the best plan of care.

Patient Experience

Patient experience is a field of practice and study that has grown out of the emphasis on shared decision making and patient-centered care resulting from the patient autonomy movement. These three terms represent similar but not identical philosophies and in many ways build upon each other. In this chapter, we will use all three terms interchangeably, acknowledging that they are not completely the same. The most important element of all three terms is "engagement" of the patient and family in medical decisions and the changing landscape of healthcare—patient care, research and quality improvement, policy development, and education of healthcare providers and patients. Equally important are good communication and compassion. These are themes that will weave a common thread through the various subsections of this chapter.

Given the title of the chapter, it is important to have an understanding of the term "patient experience". The widely promulgated and accepted definition of "patient experience" published by the Beryl Institute is: "The sum of all interactions, shaped by an organization's culture that influence patient perceptions, across the continuum of care" [1]. This is not the only definition used among healthcare institutions but it is one that captures the concepts foundational to the expanding field of patient experience and patient-centered care, research, and clinical initiatives. It includes the "key elements [of] an effective framing of the issue" [2]. Table 2.1 depicts these key elements of patient experience based on a review of recent literature. The goal of this review was to (1) identify the key elements, constructs, and themes that were commonly and frequently cited in existing definitions of "patient experience," (2) summarize these findings into what might be considered a common shared definition, and (3) identify important constructs that may be missing from and may enhance existing definition(s). Most definitions and vision statements are consistent with the Institute of Medicine Framework that includes: Compassion, empathy and responsiveness, coordination and education, physical comfort, emotional support, relieving fear and anxiety, and involvement of family and friends.

Patient (and family)-centered care is anchored in a respectful partnership, anticipating and responding to patient and family needs (physical comfort, emotional support, information, cultural, and spiritual needs), and their learning goals [3]. Respecting the patient and engaging them in sharing their healthcare decisions requires genuine caring and intentional listening by the physician and healthcare team members. Good communication begins with listening, not speaking. It takes a deliberate effort by the physician or other healthcare provider to listen to the patient but it goes a long way in improving the patient–physician relationship and the patient experience. Knowing or feeling that you have been heard often equates to a stronger sense of caring and rapport in the patient–physician relationship [4]. We have provided a few tips for improving patient encounters through better listening in Table 2.2.

TABLE 2.1 Themes and Recurring Constructs in Patient Experience Definitions

Current Elements	The Sum of All Interactions...	Shaped by an Organization's Culture...	...That Influence Patient Perceptions...	...Across the Continuum of Care
Expanded description	The orchestrated touch points of people, processes, policies, communications, actions, and environment	The vision, values, people (at all levels and in all parts of the organization), and community engaged and involved with the organization	What is recognized, understood and remembered by patients and support people. Perceptions vary based on individual experiences such as beliefs, values, cultural background	In all facets of the healthcare system, in all encounters, in all settings from nonclinical proactive experiences to long term or hospice; and across the spectrum of services
Supporting themes (for patient experience improvement) and alignment with elements	Integrated nature reinforces that experience from the patient perspective is singular and aligned, not simply a collection of distinct or disparate efforts. It is encompassing of all encounters whether they include quality, safety, or service and these efforts should be coordinated and aligned to support a "one-experience" mindset. [Includes: Beyond survey results, more than satisfaction]		Person-centeredness recognizes that the recipient and deliverer of healthcare experience are at their core human beings. As a component of experience, this reinforces that process or protocol should not trump the broader needs of people engaged (in almost all cases) at any point on the healthcare spectrum. [Includes: Aligned with patient-centered care principles]	
		Patient and Family Partnership (and Engagement) acknowledges that patients, families, and members of their support network are active participants in the care experience and must be engaged as participant owners in their encounters. The voices of these individuals are not only significant in situations of care but also in planning, ongoing operations, and change/improvement efforts. [Includes: Focus on expectations, focus on individualized care]		

The most consistent supporting themes are presented in this graphic, but [the original authors] suggest other practices or concepts may also be proven to support patient experience improvement and performance.
From *Wolf JA, Niederhauser V, Marshburn D, LaVela SL (2014). Defining Patient Experience. Patient Experience Journal, 1(1):3. Available at:* http://pxjournal.org/journal/vol1/iss1/3/

TABLE 2.2 Tips to Improve Patient Encounters Through Listening

1. Avoid interruptions
2. Pay attentions to the ideas, not the delivery
3. Do not multitask
4. Focus on the speaker
5. Listen for the disguised message
6. Listen more than you talk
7. Provide "active listening" feedback
8. Wait to shape your reply until the speaker is finished
9. Watch for nonverbal clues

From *Hirsh L. Patient Experience and the lost Art of Listening. PatientEsperience.com August 4, 2015.* Available at: http://patientexperience.com/patient-experience-lost-art-listening/

LEADERSHIP IN THE FACE OF CLINICAL UNCERTAINTY

Clinical Scenario

Mr. L. is a 66-year-old man with progressive right thigh swelling that was originally attributed to a femoral hernia. MRI revealed a large soft tissue mass invading superficial fascia with three lobulated components, the largest of which was approximately 7 × 10 cm, consistent with soft tissue sarcoma. There were no suspicious regional lymph nodes and staging studies revealed no evidence of distant metastases. Ultrasound-guided core biopsy confirmed high-grade myxoid malignant fibrous histiocytoma. Based on these studies, his clinical stage was III (T2bN0M0, high grade).

The physician met him briefly in medical oncology clinic, where he assessed the patient's general understanding of his illness, conceptually discussed possible treatment options, and arranged to see him back in a few days after discussing his findings in multidisciplinary tumor board. After the physician presented the clinical data, the discussion in tumor board centered on possible treatment options. They began by reviewing National Comprehensive Cancer Network (NCCN) guidelines. The first locus in the guideline for patients with stage II or III soft tissue sarcoma of the extremity, like Mr. L., is to determine whether the tumor is "resectable with acceptable functional outcomes" or "potentially resectable with concern for adverse functional outcomes." Considerable discussion ensued on this point, but the surgeons believed immediate resection of the primary mass would likely be very morbid. The group therefore agreed that primary therapy would be nonsurgical which, according to NCCN guidelines, consists of "preoperative radiation therapy, preoperative chemotherapy, or preoperative chemoradiation."

The Predicament: Informed Decision Making in the Face of Clinical Uncertainty

Everyone at tumor board voiced an opinion regarding which of the three approaches outlined in the algorithm should be offered. Even pathologists and diagnostic radiologists weighed in. Several participants cited studies from the literature, mostly phase I or II, which supported their positions. Nonetheless, available evidence does not clearly favor one approach over the others. There was no clear consensus. Then, someone suggested that the physician should "explain the risks and benefits of each option and let Mr. L. decide."

Obligating the patient to make such a complex medical decision without guidance is reminiscent of the old television game show "Let's Make a Deal." The host of that show, Monty Hall, offered contestants the option of choosing "door number one, door number two, or door number three." A shiny new car stood behind one of the doors, a small prize behind another, and a goat behind the third. Contestants were compelled to make an arbitrary choice. Asking Mr. L. to decide on a treatment course without guidance can represent a subtle form of abandonment rather than respect for his autonomy. How could the physician expect the patient to choose a treatment option if a group of experienced oncologists could not do so? He certainly has the right to decide for himself, but he also has the right to receive advice if he so desires it.

Recommended Action: Be Transparent, Admit Uncertainty

The physician saw Mr. L. back in clinic, where the conversation went something like this:

> P: "Mr. L., we discussed your situation at length yesterday in tumor board. I would like to discuss your options with you today."
> Mr. L.: OK.
> P: "Some medical decisions are what I call 'black and white'. Those are the easy ones, when we know the right answer. However, the decision as to how best to treat your cancer is a difficult one, one I call 'a gray area', since there is more than one approach, sufficient knowledge does not yet exist to be sure what might be best for you."

Some people may confuse this type of transparency with lack of confidence and they may believe admitting uncertainty erodes patients' confidence in physicians. To the contrary, however, this approach usually builds trust [5]. Patients expect physicians to be technically competent and knowledgeable but they do not expect their doctor to be omniscient. They just want the doctor to treat them with respect and honesty. They want to know their physicians care. Admitting uncertainty also engages the patient in the decision-making process and allows physicians to determine how patients wish to decide [6].

Models of Decision Making

There are numerous models of decision making discussed in the literature of the autonomy-based, patient-centered, shared decision-making era. The *four models of the physician–patient relationship* by Emanuel and Emanuel is one of the most commonly quoted in the literature and used in practice, though there are many similarities among the recommended models. Their four models reflect the collaborative and cooperative models of physician guidance that were recommended both before and since the movement has taken hold. Each model has its place in the clinical setting depending on the circumstances. We will quickly outline the Emanuel model here before discussing the important factors in the case further.

In the paternalistic model, the physician is the guardian focused on the patient's health and well-being, not on their preferences or autonomy. Once the physician has determined which treatments are most likely to restore the patient's health or ameliorate pain, selected information that encourages assent may be presented to the patient prior to the initiation of the chosen treatment but the physician has made the treatment decision.

The informative model is quite the opposite. The physician is the technical expert. The patient makes the treatment decision based on the relevant information provided by the physician and the physician executes the selected intervention. The interpretive and deliberative models are situated in-between the paternalistic and informative on the spectrum of decision making and are more consistent with the concept of cooperative shared decision making that is more commonly advocated. The interpretive model casts the physician as the trusted advisor or counselor who helps the patient elucidate and articulate her values in order to determine which intervention among the medically appropriate options described best honors those values through an interpretive process. The physician guides but does not decide. The patient is engaged in a process of self-understanding. In the deliberative model, the physician acts as a teacher empowering the patient in a journey of self-development through a dialogue of deliberation and negotiation, without coercion [7]. In both of these models, the patient makes an informed decision that is implemented by the physician and healthcare team, but it is a well-considered decision resulting from an empowered position of engagement. Quill and Brody call this the "enhanced autonomy" model which "encourages patients and physicians to actively exchange ideas, explicitly negotiate differences, and share power and influence to serve the patient's best interests." This intense collaboration allows the patient to make choices informed by both the medical facts and the physician's experience [8].

Recommended Action: Educate One Another

In Mr. L.'s case, the physician simply stopped to see what the patient would say after raising the issue of uncertainty. Mr. L. started by asking some basic questions, like "What are my options?" The physician responded by defining the terms "chemotherapy" and "radiation," since many patients, even highly educated ones, do not know what these terms mean. This may seem surprising, since nearly all patients are familiar with the idea that "chemotherapy makes your hair fall out" or "people who take chemotherapy cannot be around sick people." Nonetheless, many patients do not truly understand key concepts regarding cancer treatment options. Table 2.3 contains some nontechnical phrases that may help patients understand what chemotherapy and radiation therapy entail.

They then proceeded to discuss potential risks, benefits, and the rationale for each of the three options listed in NCCN guidelines, namely radiation alone, chemotherapy alone, or a combination of the two. The conversation was natural, free-flowing, and reciprocal, a process that has been compared to jazz improvisation [9]. During conversations of this type, providers and patients each have their own agenda. The key to success is integrating these two agendas seamlessly. Table 2.4 lists a few common communication agenda items for providers and patients. Providers should do much more than educate patients about biomedical aspects of the disease and its treatment. They should also understand the illness from the perspective of the patient by determining how the illness is affecting the patients work, family life, friendships, and other important aspects of living. Such an approach helps us to maximize benefit and minimize unavoidable suffering associated with treatment [10].

TABLE 2.3 Key Elements During Conversations Between Provider and Patient

Provider	Patient and Family
Assess patient's understanding of illness and treatment options	Fill in gaps in technical knowledge
Explain biomedical information with clarity by avoiding technical jargon	Determine potential side-effects of treatment options
Assess patient's concerns, preferences, and fears regarding treatment options and illness	Determine prognosis

These are some of the essential elements of successful conversations between provider and patient based on the experienced practice of the senior author.

TABLE 2.4 Examples of Nontechnical Phrases That May Help Patients Understand Treatment Options

Chemotherapy is simply a word that means "medication to treat cancer."
Chemotherapy can be given by many routes, most commonly by vein (IV) or by mouth.
Chemotherapy travels through the body to treat cancer where ever it is, even when it is too small to be visible on CT and other imaging studies.
Different chemotherapy drugs have different side-effects. Many traditional chemotherapy drugs cause the hair to fall out, cause nausea, and block the body's ability to make new blood, although newer targeted drugs have different side-effects.
There are three types of blood cells: Red, white, and platelets. Reds carry oxygen, whites fight infection, and platelets help clot blood. By blocking blood production, chemo can make you feel tired, make you susceptible to infection, or make you susceptible to bleeding.
Radiation can be thought of as a strong X-ray beam that treats cancer in one area of the body.
Radiation damages normal tissues surrounding the tumor.

From *Wolf JA, Niederhauser V, Marshburn D, LaVela SL (2014). Defining Patient Experience. Patient Experience Journal, 1(1):3.* Available at: http://pxjournal.org/journal/vol1/iss1/3/

Note: The most consistent supporting themes are presented in this graphic, but [the original authors] suggest other practices or concepts may also be proven to support patient experience improvement and performance.

Suggested language based on the experienced practice of the senior author that may help patients understand various treatment options.

Recommended Action: Establish the Roles of the Patient and Physician

After Mr. L.'s physician understood his perspective, he paused to ask whether Mr. L. had a preference about which treatment option he wished to pursue. The next step in charting a treatment course in the face of uncertainty is perhaps the most difficult for the provider. The two-part command "explain the risks and benefits of each option and *let him decide*" is not so much incorrect as it is incomplete. It lacks a critical third component, namely a recommendation. The prevalent literature in favor of shared decision making states that the physician has an obligation to make a recommendation and engage in a conversation with the patient where the patient's values will likely become clearer and help guide the "right" decision for the individual patient [7,8,11–13]. This is of course dependent on the patient's age, culture, education, and preferences that may guide the culturally competent physician or other healthcare provider in a nuanced and slightly different course. For the majority of patients in our Western culture today, a recommendation is expected.

The simple phrase "I recommend" is powerful in that it implies several messages:

1. I am confident enough in my knowledge, ability, and experience to take a leadership role in helping you make an important decision.
2. I suggest a particular course of action but the decision is ultimately yours.
3. I am open to discussing advantages and disadvantages of all reasonable approaches.
4. I care about what this illness means to you and how you want to respond to it.

Patients may respond in many ways to unequivocal recommendations made in clear, nontechnical language. Many simply defer to the doctor. Other patients prefer to discuss intricate details of risks and benefits, speak to family members or friends, or consider the information

TABLE 2.5 Key Steps During Conversations With Patients Faced With Clinical Uncertainty (Multiple Treatment Options)

Step	Useful Phrases, Steps, Strategies
Give perspective	The type of tumor you have tends to be difficult to treat. Some people do very well and are cured after treatment but most are not. We will do our best to give you the best possible outcome.
Admit uncertainty	You have three treatment options for your cancer. The options include chemotherapy before surgery, radiation before surgery, or radiation and chemotherapy before surgery. Several studies (clinical trials) have been done to attempt to determine which is the best but we still do not know if one is better than the others.
Assess understanding	What is your understanding of "chemotherapy" and "radiation"? Would you like me to describe them?
Negotiate decision-making style	How would you like to go about deciding which treatment to receive? Do you want me to recommend one? Or would you like to choose yourself after I describe potential risks and benefits of each?
Recommend a course of action if so desired by the patient	"In my clinical opinion ..."
Discuss and follow through on patient's choice	"Do you have a better idea of what you believe would be the best option for you at this time? Tell me more about what went into your decision to pursue this option. What are your priorities?" "Now, let's make a plan for your treatment..."

Suggested language based on the experienced practice of the senior author during conversations with patients faced with clinical uncertainty regarding multiple treatment options.

and make a decision at a later time. Some patients gather more information from the internet and other sources, whereas others do not. Paradoxically, as information becomes increasingly available, patients need more rather than less guidance. Ultimately, most patients who face a complex decision value advice from their physician and other members of the healthcare team. The physician can aid the patient as a partner in the decision-making process by asking the patient what is important to the patient, what goals the patient wants to achieve, and what the patient wants to avoid. The answers to these questions may be significantly impacted by the expected outcomes on the patient's quality of life and in turn will impact the final decision. The final step is to follow through on the choice made by the patient as the ultimate decision maker. The key steps during conversations with patients faced with clinical uncertainty are outlined in Table 2.5.

DISCUSSION

A patient-centered approach such as that described above improves patient experience by reducing, mitigating, and preventing suffering and anxiety. In ethical terms, partnering with the patient in this way to arrive at a shared decision exemplifies beneficence and non-maleficence. Showing respect for the patient, valuing the patient's humanity, and carefully coordinating care among members of the healthcare team greatly enhance quality and reduce avoidable suffering. The Institute of Patient and Family Centered Care asserts that care must go beyond the essential foundation of respect, dignity, and information sharing to engaging the patient (and family) in the decisions through participation and collaboration. Their mantra is "Nothing about me without me. Not to me, not for me, but with me" [14]. This mantra is particularly relevant in the face of complex medical decisions and uncertainty.

In the informed model, unlike the interpretive or deliberative models, information exchange is essentially one way, from physician to patient. This exchange is the very crux of the model, defining the boundaries of the physician's clinical role in decision making. The physician in this model is assumed to be the primary source of information to the patient on medical/scientific issues about the patient's disease and treatment options. To fulfill this role, the physician, at a minimum, needs to give the patient all relevant information from the highest quality research on the benefits and risks of various treatments so that the patient will be enabled to make an informed decision. Beyond information transfer, the physician has no further role in the decision-making process. The remaining tasks of deliberation and decision making are the patient's alone.

Underlying this model are two assumptions. The first is that as long as patients possess current scientific information on treatment benefits and risks, they will be able to make the best decision for themselves. The second is that physicians should not have an investment in the decision-making process or in the decision made [15]. To do so would go beyond the boundaries of an appropriate clinical role because the physician might harm the patient by inadvertently steering her in a certain direction which reflects the physician's own bias or conflicts of interests. Underlying this concern is the assumption that the interests and motivations of the physician and patient may not be the same. This consumer-oriented model emphasizes patient sovereignty and patients' rights to make independent, autonomous choices [8].

The view that physicians have no legitimate role to play in the discussion or recommendation of treatments may be difficult for many physicians to accept since it runs counter to decades of professional medical training and practice in which clinical experience, expertise, and knowledge have been assumed to be the quintessential skills that physicians have to offer. The informed model may meet some patients' needs for autonomy in decision making (for those who value this goal), but it may not meet the needs of the patients who often need our expertise to make an informed decision. In our individual autonomy-focused society, most argue that the patient's needs alone should be served. The physician's need to participate in decision making should always be focused on the patient's benefit, not in the physician's interests or goals. Caution must be taken by the physician and other trusted healthcare providers to avoid coercion or personal bias when engaging in this partnership. We will discuss this further in the next section.

Negotiating as equal partners, however, is not easy for the patient because of the inherent information and power imbalance in the relationship. Physicians, in the usual case, will have superior knowledge of the technical issues involved in treatment decision making and perhaps years of clinical experience with similar types of patients. The physician bears the officially legitimized title of "expert," while the patient may feel particularly vulnerable and frightened during the medical encounter. When education, income, culture, and/or gender differences also exist between the physician and patient, the patient may feel too intimidated to freely and openly express her preferences, let alone negotiate for them with the physician. Creating a safe environment for the patient so that she feels comfortable in exploring information and expressing opinions is probably the highest challenge for physicians who want to practice a shared approach [16]. At the other end of the patient spectrum are those who are well informed about their illness and various treatment options and who have no difficulty expressing preferences. Some of these patients may have already made the treatment decision before entering the physician's office. If the patient's preference is different from the physician's and the physician is not able to change the patient's view, then the process will likely become one of conflict.

In a shared model, both physicians and patients are assumed to have an investment in the treatment decision. The physician can legitimately give a treatment recommendation to patients and try to persuade them to accept the recommendation. However, physicians would also have to concentrate on listening to and understanding why patients might favor a different treatment option.

As Mr. L.'s medical oncologist, the physician in the scenario above assumed the role of coordinating his care and developing a treatment plan. Rather than expecting the patient to make a complex medical decision by choosing from among multiple reasonable options, physicians should instead expect the patient to decide how that decision will be made. Does the patient have preconceived preferences (no chemo, no surgery, etc.)? If so, are those preferences reasonable? Does the patient want to defer to the doctor? Will the decision-making process be shared? The healthcare team needs to establish what decision-making model the patient prefers: Paternalistic, informed, or shared, or some other in between. Healthcare professionals can establish the patient's preferred decision-making model by explicitly asking or by simply exchanging information and determining the patient's preference in the context of the discussion. Lang articulates this concept well when he states, "a 'one communications approach fits all' model will lead to frustrations and problems on [the part of the physician and the patient]. It should be noted that a patient-centered approach does not insist on patient decision making. An approach that is truly patient centered explores the patient's preference for involvement and decision making and respects that preference" [12,17,18].

It is possible that not only can the decision-making approach used in one physician–patient interaction change in the next interaction, it can also change within a single interaction. For example, a physician who starts the consultation with an informed approach may need to switch midstream to a more shared approach if it becomes evident that the patient does not want to make the decision on her own. In this case, what started as an informed approach changes to one in which the physician plays a more active role in making the decision [19,20]. Rather than advocating a particular approach, we emphasize the importance of flexibility in the way that physicians structure the decision-making process so that individual differences in patient preferences can be respected.

THE BATTLE OF AUTONOMIES IN THIS NEW PARTNERSHIP

Clinical Scenario

Mr. P. is a 60-year-old man with stage IV pancreatic cancer, Eastern Cooperative Oncology Group Toxicity and Response Criteria 3 (ECOG 3; capable of only limited self-care, confined to bed or chair more than 50% of waking hours). His cancer is progressive despite extensive chemotherapy and he is suffering from liver and kidney failure. He is no longer a candidate for disease-directed therapy. His medical oncologist recommends transition to a purely palliative strategy. Despite several thorough conversations, the patient insists on more chemotherapy. In today's patient-centered, patient autonomy driven atmosphere, some physicians might choose to offer third line, salvage chemotherapy despite the risks, refer him to a phase I clinical trial, or offer sub-therapeutic doses of single agent chemotherapy that is unlikely to cause much toxicity, as a compromise. In recognition of the shared responsibilities of the physician and the patient to make medical decisions together, within the limits of an appropriate exercise of autonomy on both sides, we recommend a more directive yet compassionate approach. In this scenario, the physician engaged the patient and his wife in a discussion, saying "I wish I could offer more chemotherapy, but I cannot, because I know it will harm you." This was based on a medical determination of the available options and the appropriate limits of such.

The patient ultimately agreed to go to the Acute Palliative Care Unit (APCU). He soon became delirious due to worsening hepatic and renal insufficiency. The patient's wife, who was designated the medical power of attorney (MPoA), noticed he was no longer producing urine and requested hemodialysis. She still hoped for a miracle cure. The PCU-attending physician explained why dialysis would not be helpful and would actually prolong the patient's suffering. Nonetheless, the patient's wife repeatedly insisted on dialysis. The physician considered the options, including consulting nephrology to consider dialysis and hoping they do not offer it, asking the nephrologist to offer dialysis, and drawing blood work as a way of stalling for time and hope the patient's wife changes her mind. The physician remained confident that the best course of action was to continue to be present for the patient and his wife, but continue to not offer dialysis rather than potentially causing additional suffering that would not be outweighed by a medical benefit, taking responsibility for the medical decision.

The Story Continues

The patient developed respiratory failure. His wife asked the physician to revoke the do-not-resuscitate (DNR) order, that the patient agreed to when he chose supportive care only in the APCU, saying: "This is what he would have wanted. He is a fighter." The physician provided presence and support balanced with space and silence to allow her to contemplate the situation. She did not budge. The options before the physician were to revoke the DNR order as the wife requested but in opposition to the patient's autonomous decision when he had capacity, or at the other end of the spectrum, implement a "two-physician DNR order." The physician chose to tell the patient's wife he was going to maintain the DNR status because it would be medically inappropriate to initiate cardiopulmonary resuscitation (CPR) and inconsistent with the patient's previously expressed wishes, thereby relieving the surrogate of the burden of the decision that the patient had already made upon the advice of the medical team. Another option that remained available to the healthcare team and the patient and family in a situation of disagreement was to suggest or facilitate transfer to another institution that might be willing to honor the request.

Physician Autonomy Comes With Responsibility

The scenario above evokes a number of questions. Does a physician have the right to deny a patient treatment that evidence (and extensive clinical experience) shows to be harmful even if the patient insists on receiving that treatment? Preservation of moral and professional integrity would argue in the affirmative. Does a physician have an ethical obligation to deny a patient treatment that evidence shows to be clearly harmful? Some might argue that the principles of beneficence and nonmaleficence would support this claim. In the era of patient autonomy, does the physician's autonomy have a place? We argue that it does. The patient and the physician have equal claims to respect for individual autonomy. "The physician–patient relationship is a moral equation with rights and obligations on both sides and it must be balanced so that physicians and patients act beneficently toward each other while respecting each other's autonomy" [11]. Edmund Pellegrino eloquently wrote in 1994 about this balancing of autonomies in the era of patient autonomy and subsequent patient-centered care. His words, summarized in Table 2.6, stand true today. His perspective on balancing potentially competing autonomies by balancing autonomy and beneficence in the patient encounter are certainly an important and enduring contribution to this ongoing dialogue.

Emanuel and Pearson argue that the defining element of physician autonomy is its place in the patient–physician relationship. Please notice the shift between Emanuel's 1992 description of the physician–patient relationship to the 2012 description of the patient–physician relationship. This alone illustrates the shift in perspective. They continue arguing that the ethical justification for physician autonomy requires the physician to use it to promote the patient's best interests. It is therefore defined as the freedom to determine the conditions of practice and the care delivered aimed at promoting the patient's well-being, through shared decision making with patients who have capacity within the medical standards of professional and technical competence [21].

This focus on the patient's best interests rather than the physician's own is an important foundational precept in medicine and is not new. One might mistakenly understand this expectation and the one stated earlier that the physician must manage his own biases and refrain from coercion to mean that he must be value neutral in his patient encounters. This would be a mistake and would not be consistent with the argument that the physician must engage in a dialogue with the patient, deliberate choices, make recommendations, and share in the decision-making responsibility, even if not in the direct consequences. If values and emotions are removed there is no meaningful discussion and no right or wrong decision, no empathy or compassion. Being nonjudgmental and noncoercive does not mean being free of values and opinions, personal and professional expertise, or experiences beneficial to the deliberation that are essential to the practice of medicine as a

TABLE 2.6 Balancing Beneficence and Autonomy

1) Patient autonomy is a moral right of patients and it is a duty of physicians to respect it.
2) Integrity of conscience and professional judgment are moral rights of physicians. Society and patients have an obligation to respect them.
3) Physician autonomy is limited by a competent patient's or valid surrogate's moral right to refuse proffered treatment. The physician is obliged, however, to help the patient arrive at an autonomous decision by enhancing or empowering the patient's capacity to make authentic, self-governing choices.
4) The patient's autonomy is limited when it becomes a demand for treatment the physician honestly believes is not medically indicated, is injurious to the patient, or is morally repugnant.
5) The physician's autonomy is limited on questions of value, e.g., on questions of the goals or purposes to which medical knowledge may be put for particular individuals or societies.
6) Societies and institutions must establish mechanisms, with only minimal recourse to law, for unilateral discontinuance of the relationship when either patient or physician feels personal integrity is being compromised.
7) The first principle of medical ethics is still beneficence. Beneficence is essential if autonomy is to be authentically expressed and actualized.
In sum, beneficence and autonomy must be mutually reenforcing if the patient's good is to be served, if the physician's ability to serve that good is not to be compromised, and if the physician's moral claim to autonomy and the integrity of the whole enterprise of medical ethics are to be respected.

From *Gallagher CM, Holmes RF. Handling Cases of 'Medical Futility'. HEC Forum 2012;24(2):91–98.* Available at: http://link.springer.com/article/10.1007%2Fs10730-011-9168-3
These steps represent those that have been useful in establishing a framework for working with such difficult situations.

science and an art. It may be difficult for one to know whether his "moral reasoning is personal, professional, both or neither. After all, he is relying on professional codes in assessing the situation but he seems also to be concerned with personal virtue, his patients' welfare, and the importance of being ethically consistent; beliefs that may be grounded in a deeply personal religious or philosophical commitment. This moral reflection will likely affect his practice of medicine" [22]. There will always be some exchange of information aimed at influencing the choice at hand. A balance must be struck. One should not let one's own values, as a healthcare provider or private citizen, unduly influence the potentially vulnerable patient's decision. But physician value neutrality holds patient autonomy too high. When autonomies compete, negotiation may be necessary to come to a shared resolution that does not violate the values/principles of either party. If a resolution of this sort is not possible the physician should suggest that the patient seek care with another physician, as option four above suggests. A sometimes contentious element of a case like Mr. P.'s is the determination of appropriate interventions such as CPR. We discuss this in more detail below.

Limits to Appropriate and Beneficial Therapies

Breakthroughs in cancer care and research have transformed cancer treatment and survival rates in recent decades but advanced cancer remains incurable for many patients despite recent treatment advances. As with many diseases that medicine endeavors to ameliorate and eradicate, there remain limits in the science of treatments for cancer. At a certain point, life-sustaining treatment may no longer be possible or appropriate. When treatment is requested and continued under these circumstances, it is often referred to as being futile.

Medical futility is defined in many varying ways by different authors and ethicists but is "commonly understood as treatment that would not provide for any meaningful benefit for the patient" [23–27]. As such, many ethicists and authors prefer to use the terms "medically nonbeneficial" or "medically inappropriate" to refer to these types of therapies. These phrases are slightly more accepted by both healthcare professionals and patients and families because they are gentler language that may be easier to define. Though the application of the labels may be just as difficult to determine they seem to be slightly less ambiguous in definition and feel less value laden. A number of issues that are reported along with medically inappropriate or nonbeneficial therapies include withdrawing or withholding life-sustaining procedures; questions about whether the level of treatment is appropriate, particularly in terms of whether to shift the goals of care from curative to palliative or regarding the patient's resuscitation status; and issues of quality of life and pain control. In addition to determining which of the many different aspects of a patient's medical care are appropriate and beneficial, which are no longer so, and when this shift from appropriate to inappropriate occurs, other confounding factors such as cultural or spiritual differences, family dynamics, finances, benefits to others, and value-laden judgments that affect decision-making and determinations of quality of life make discerning "futility" very challenging.

When a physician has determined that an intervention is medically inappropriate because it is not effective in accomplishing what it is designed to do, does not stop the deterioration of the patient's condition toward death, the harms of the intervention outweigh the benefits, or the intervention only prolongs the dying process or even prevents natural death from the underlying causes, he may face opposition from the patient, family, or surrogate decision maker. This opposition may result from a lack of understanding of the medical circumstances or the gravity of the situation, a difference in cultural or religious values that affects the expectations for medical care and thereby affects medical decisions, or a misunderstanding of the patient's or family member's decision-making role and authority in determining the appropriateness of interventions. When faced with situations where there is disagreement regarding the appropriateness of an intervention, particularly life-sustaining or life-prolonging interventions, the physician and other healthcare professionals must communicate accurate and pertinent information to the patient and or surrogate decision maker in a manner that is both clear and compassionate. It may be helpful to explain which interventions are still effective and will continue and which interventions will be withheld or withdrawn due to ineffectiveness along with the assurance that comfort measures will always be provided. The medical facts must then be weighed against the values and wishes of the patient and caregivers to determine the qualitative benefits and risks, typically in a shared decision-making fashion; the quantitative benefits and risk having already been considered in the medical determination, exercising physician autonomy in an appropriate manner [28]. Finally or concurrently, discussing and applying the jointly defined goals of care to the treatment plan for the patient is important to the decision-making process and may help alleviate disagreements. Gallagher and Holmes have provided a list of steps to consider when talking with patients, families, or surrogate decision makers requesting nonbeneficial treatments that can be found in Table 2.7.

In addition to the recommendations found in Table 2.7, Gallagher and Holmes suggest the following methods when implementing those steps, or preparing to do so: (1) Utilize

TABLE 2.7 Steps to Consider When Talking to Patients, Families, or Surrogate Decision Makers Requesting Nonbeneficial Treatments

1. Clarify goals of care	• To establish a common ground for all parties. • Allows the physicians to outline their perception of the medical possibilities. • Helps the family to discuss their understanding of the situation and express how the patient would wish to be treated. • Shifts in goals of care and what to expect should be clearly expressed.
2. Assess whether all reasonable options have been attempted	• Clinicians can both assess whether they have exhausted all of the medically appropriate options and reassure the family or surrogate decision maker that they have indeed offered what they can. • Provide comfort to both parties by ensuring that all options are considered and appropriate measures are taken.
3. Do not offer options that are not medically appropriate	• Offering treatments that have a high risk to benefit ratio, a low likelihood of achieving notable results, or treatments that the physician or medical team do not feel comfortable in offering only add to the decision maker's confusion in an already challenging situation.
4. Establish guidelines and limits for interventions in place	• Setting limits and key benchmarks for interventions already in place is important for all parties. • Physicians and caregivers should know why the intervention is in place, what role it serves in the patient's care, and the duration that it will be in place. • [Establish] how it will be utilized, whether it will be altered and why, and when or if that intervention will be removed.
5. Seek to address emotional needs of the [family] caregiver	• Empirical assessments can be useful in attempting to defuse emotionally charged situations. • Offer support [of] the emotional needs of those involved. [This] can help to resolve other conflicts between family members or between the family or surrogate decision maker and providers. • Using terminology such as "futility" often gives rise to family discomfort and the assumption that the physician and clinical team have "given up" on the one the family loves. • It is often helpful to use language such as beneficial and nonbeneficial treatment in place of the professional language of medically futility.
6. Use interdisciplinary team resources	• Utilize the expertise of chaplains, social workers, and others with psychosocial training whenever possible. Often family members relate better to these professionals as they use different language and ask them to comment on different aspects of the patient such as spirituality and thinking processes.

From *Pellegrino ED. Patient and Physician Autonomy: Conflicting Rights and Obligations in the Physician-Patient Relationship, 10 J. Contemp. Health L. & Pol'y 47, 1994:10(1).* Available at: http://scholarship.law.edu/jchlp/vol10/iss1/8
A list of several precepts that [according to the original author] need to be built into the current reexamination of the foundations of professional ethics.

interdisciplinary patient care conferences/family meetings to encourage and facilitate good communication between all parties; (2) review the steps with a colleague for rehearsing, brainstorming, and further support; and (3) request an ethics consult for a neutral, outside perspective on a challenging situation [23].

One of the more common life supporting measures that is often deemed to be "medically inappropriate" or "nonbeneficial" is CPR. This situation often requires a respectful yet appropriately directive discussion of the medical recommendation or decision to implement a DNR physician's order, employing clear and consistent communication. CPR is often considered to be a low-yield intervention in cancer patients. In patients with metastatic disease, the procedure is thought to be ineffective and therefore inappropriate. DNR orders are physician's orders that are often put in place ideally with the agreement of the patient or the patient's surrogate decision maker/family in the setting of a terminal illness or injury. The goal is usually to avoid prolonging the dying process; causing additional harm, injury, or suffering; or implementing measures that are not expected to benefit the patient or restore the patient to previous health and are therefore considered medically inappropriate.

Most state codes and institutional policies regarding the ethical consideration of foregoing life-sustaining measures consider patient autonomy to be one of the cornerstones upon which these decisions are made. But this principle must be balanced with other competing values and principles. Cantor et al. discuss "four reasonable justifications [identified by Howard Brody] for physicians' decisions to withhold [medically inappropriate] treatments." The first of these is based on the principle of beneficence and nonmaleficence. "[T]he goals of medicine are to heal patients and to reduce suffering; to offer treatments that will not achieve these goals subverts the purpose of medicine." The second justification is based upon the first principle of the American Medical Association Principles of Medical Ethics. "[P]hysicians are bound to high standards of scientific competence; offering ineffective treatments deviates from professional standards." The third of Brody's justifications is also founded in the American Medical Association (AMA) Principles of Medical Ethics in the first and second principles. "[I]f physicians offer treatments that are ineffective, they risk becoming 'quacks' and losing public confidence." The final statement asserts that, "physicians are justified in risking harm to patients only when there is a reasonable chance of benefit; forcing a physician to inflict harmful procedures on patients makes them 'agents of harm, not benefit.'" This is again based on the principles of beneficence and nonmaleficence and a balance of benefit to burden. It is also a matter of professional integrity. Cantor et al. and Brody assert that, "the right of a patient to demand a treatment that is futile is limited by the need for physicians to provide care that meets high ethical, clinical, and scientific standards" [29,30]. But ethical and legal standards are not always congruent.

Communication

The language that we use to communicate is important to how the message is received. For example, some consider the term DNR to be both vague and threatening [31]. Chen and Youngner suggest that "while DNR universally means that the patient will not receive cardiopulmonary resuscitation in the event of cardiac or respiratory arrest, DNR orders rarely specify what medical care should be provided to DNR patients before they experience an arrest, thus leaving matters open to individual interpretation" [32]. This can lead to misunderstandings and disagreements about decisions that are in the patient's best interest or that honor the patient's values and goals and may lead to unnecessary or inappropriate interventions, such as CPR and or intubation, that could cause harm and prolong suffering. Discussion and documentation about the care plan is necessary to avoid confusion for patients, families, nurses, and other care providers. It is therefore important to discuss the plan for continued care leading up to the possible arrest, including a clarification of goals and appropriate interventions in light of the changed medical prognosis and accounting for the patient's values and best interests. It is also beneficial to discuss not just what will not be done in the event of a cardiopulmonary arrest, but also what will be done up to and during that arrest to address the patient's other medical needs and comfort until the time of death or improvement in condition that allows for discharge. DNR orders alone do not necessarily entail or imply other limits on care and should not affect other aspects of care [33,34].

Discussions about DNR, or beginning, withdrawing, or withholding any other medical intervention, should include an explanation of the meanings of terms in language that is understandable to the listener and should not use jargon. The healthcare provider giving the information should not assume understanding but should ask the listener to repeat what they have understood the provider to say. When the healthcare provider makes treatment recommendations, the provider must take responsibility for the determination of what is a medically appropriate intervention but must also seek a minimum of assent, if not consent, from the appropriate decision maker through a process of shared decision making [35]. There is a delicate balance to removing the burden of responsibility for a decision while encouraging participation. Likewise, the healthcare providers must balance autonomies (the patient's and their own) with beneficence and nonmaleficence in regard to the patient's needs.

Difficult decisions such as withholding CPR or withdrawing other life sustaining interventions may require more than one conversation. It is often helpful to establish a timeframe for when the next meeting will take place. It is often beneficial to include other members of the healthcare team in these meetings, such as the nurse and social worker or chaplain, who can provide ongoing support and continued conversation and clarification to the patient and family after and between meetings. If there is disagreement with the recommendation to withhold CPR, or other interventions deemed to be nonbeneficial, consultation with patient advocacy, social work, pastoral care, palliative care, or ethics, when these services are available, can be helpful during the conflict resolution process. Having an established rapport with these individuals may be advantageous [33,34,36].

Good communication is sensitive to cultural, generational, and educational differences. Attention to these factors may determine who the decision-maker will be. Discussions may take place with a surrogate decision maker either because the patient is unable to participate or prefers not to participate. In either circumstance, it is important to identify the appropriate person as decision-maker, clarify the surrogate's role, and establish an accurate understanding and acceptance of the patient's condition and prognosis, before discussing the decisions that need to be made. Clarifying the surrogates role includes explaining that the surrogate is asked to use substituted judgment to inform shared decisions made in the patient's stead, grounded in the patient's known wishes or best interests, based on family values or cultural norms and analysis of the benefits and burdens.

Breaking bad news is an important part of the healthcare professional's job and requires both experience and expertise. We have an ethical and legal obligation to tell patients (or their decision makers) the truth—but the manner of doing so is an important predictor of the outcome of the meeting. Conversations that "break bad news" include two parts that often occur simultaneously: Divulging of information to patient or decision maker and therapeutic dialogue by which you listen to, hear, and respond to reactions to the information [37]. Robert Buckman is considered an authority on "delivering bad news." He suggests the following format for communicating "bad news": (1) Start off well, the setting is important; (2) find out how much the decision maker knows; (3) find out how much the decision maker wants to know; (4) share the information (aligning and educating); (5) respond to feelings and reactions; and (6) make a plan and follow through [37,38]. This is very similar to the model we suggested in an earlier section of this chapter.

Good communication skills by the physician can encourage the patient to feel they are a partner in the decision. Collaborative, effective shared decision making will leave the patient feeling informed, supported, and open to express their preferences, and at the same time, the patient (or surrogate) will not feel burdened with the full responsibility of making the final care decision for the next step in their treatment and care.

CONCLUSION

Over the last 45 years, there has been a shift in medical decision making from the paternalistic approach of decisions made by physicians toward an exercise of patient autonomy. This occurred as a reaction to the concerns that the rights of patients to refuse unwanted treatment had been neglected for too long [11]. In the past, the physician accepted the responsibility to make recommendations based on the physician's experience. The shift toward patient autonomy moves the decision-making responsibility to the patient. The goal of the shift toward shared decision making is to empower patients to select the most appropriate treatment or clinical plan that best aligns with their personal values and needs aided by the expert advice of the physician. This move toward patient engagement in decision making is aimed at increasing patient satisfaction because the lack of communication to patients about their treatment options is the most common cause of patient dissatisfaction [39,40]. Patient-centered care with shared decision making incorporates the desired informational components about the disease and the treatment options and it addresses the patient's other needs from a whole-person perspective.

Most patients want to be engaged and participate in the decision making with the physician where the physician presents the options, the statistics, and odds for a successful outcome [41]. Patients want to consider and should be given the physician's expert opinion regarding the options. A collaborative conversation where the physician and patient discuss options and outcomes is the desired approach to patient decision making; not two monologues, one of the physician to the patient and one of the patient to the physician. It should be a dialogue in which the physician and the patient identify and appreciate the medical possibilities and uncertainties and discuss the patient's preferences regarding quality of life-related to the treatment options. The key is that neither the physician nor the patient is the decision maker in isolation of one another. The patient needs to feel comfortable rejecting the physician's recommendations if they do not align with other aspects of the decision and the physician needs to feel comfortable allowing the patients to make decisions from among the medically appropriate options that may go against her recommendations. Shared decision making requires more than a reciting of the physician's expertise. Shared decision making requires the physician to communicate with the patient on a personal level. The patient's values, experiences, and goals need to be part of the discussion and be included in the decision-making process. While physicians must respect the patients as the ultimate decision maker, they must also provide to the patient the information he or she needs to make the decision, recognizing not all patients want to be the final decision maker so he or she might be depending on the physician to make the final care recommendations.

Shared decision making requires time by the physician to communicate clearly with the patient the rational and possible outcomes behind each of the treatment options. Physicians also need to assess the type and amount of information a patient wishes to hear. Not all patients want to know all the information about the clinical options and the biomedical aspects of their disease. Patients want practical information that helps them manage their disease and that answers their practical questions [42]. Patients want the physician to provide information that will help they be able to make a decision based on not only the statistics of the treatment but they want the physician to discuss the possible outcomes based on his or her expertise and experiences. Consistent and understandable communication between the physician and the patient is a key component in shared decision making when medical uncertainty exists. Physicians should establish trust and strive for transparency when communicating to patients about uncertain treatment options. The information should be easy for the patient to understand [43]. To achieve patient-centered shared decision making when medical uncertainty exists physicians should share medical expertise fully, listen to patient's preferences and goals, and use collaborative communication techniques with the patient to talk through the options and to reach mutual understanding, realizing the patient has the final decision regarding next steps in their care [8].

We have discussed why experience matters and in what ways experience is linked to the patient–physician relationship: Patient experience as a movement, the patient's (life) experience and values as a consideration in the shared decision, and the physician's experience and expertise as a medical professional and as a fellow journeyman. The reciprocal relationship created by the shared decision-making model obligates both parties to understand their roles and responsibilities and balance-competing autonomies to benefit the patient's well-being. Employing the strategies and skills offered here may help to more completely fulfill the charge to practice both the science and the art of medicine.

References

[1] The Beryl Institute. Defining Patient Experience Bedford, Texas; 2015. Available from: ⟨http://www.theberylinstitute.org/?page = DefiningPatientExp⟩; [accessed 10.13.2015].
[2] Wolf JA, Niederhauser V, Marshburn D, LaVela SL. Defining patient experience. JPE 2014;1(1):7–19.

[3] Balik B, Conway J, Zipperer L, Watson J. Achieving an exceptional patient and family experience of inpatient hospital care. Cambridge Mass. Institute for Healthcare Improvement; 2011.
[4] Hirsch L. Patient Experience and The Lost Art of Listening 2015. August 4, 2015. Available from: ⟨http://patientexperience.com/patient-experience-lost-art-listening/⟩; [accessed 10.13.2015].
[5] Groopman J. The uncertainty of the expert. How doctors think. Boston: Houghton Mifflin Company; 2007. p. 307.
[6] Parascandola M, Hawkins JS, Danis M. Patient autonomy and the challenge of clinical uncertainty. Kennedy Inst Ethics J 2002;12(3):245–64.
[7] Emanuel EJ, Emanuel LL. Four models of the physician-patient relationship. JAMA 1992;267(16):2221–6.
[8] Quill TE, Brody H. Physician recommendations and patient autonomy: Finding a balance between physician power and patient choice. Ann Intern Med 1996;125(9):763–9.
[9] Haidet P. Jazz and the 'art' of medicine: Improvisation in the medical encounter. Ann Fam Med 2007;5 (2):164–9.
[10] Dempsey C. Making the connection: Reducing suffering with compassionate connected Care. In: Ganey P, editor. Quality improvement and evidence based practice. Institute for Innovation; 2014.
[11] Pellegrino E. Patient and physician autonomy: Conflicting rights and obligations in the physician-patient relationship. J Contemp Health Law Policy 1994;10:47.
[12] McCullough LB. The professional medical ethics model of decision making under conditions of clinical uncertainty. Med Care Res Rev 2013;70(1 Suppl):141S–58S.
[13] Gurmankin AD, Baron J, Hershey JC, Ubel PA. The role of physicians' recommendations in medical treatment decisions. Med Decis Making 2002;22(3):262–71.
[14] IPFCC. In: Patient and family centered care: Core concepts, Bethesda MD: Institute for Patient- and Family-Centered Care; 1992. p. 1–2.
[15] Eddy DM. Anatomy of a decision. JAMA 1990;263(3):441–3.
[16] Guadagnoli E, Ward P. Patient participation in decision-making. Soc Sci Med 1998;47(3):329–39.
[17] Lang F. The evolving roles of patient and physician. Arch Fam Med 2000;9(1):65–7.
[18] Charles C, Gafni A, Whelan T. Shared decision-making in the medical encounter: What does it mean? (or it takes at least two to tango). Soc Sci Med 1997;44(5):681–92.
[19] Charles C, Gafni A, Whelan T. Decision-making in the physician–patient encounter: revisiting the shared treatment decision-making model. Soc Sci Med 1999;49(5):651–61.
[20] Epstein RM, Alper BS, Quill TE. Communicating evidence for participatory decision making. JAMA 2004;291 (19):2359–66.
[21] Emanuel EJ, Pearson SD. Physician autonomy and health care reform. JAMA 2012;307(4):367–8.
[22] Beckwith FJ, Peppin JF. Physician value neutrality: a critique. J Law Med Ethics 2000;28(1):67–77.
[23] Gallagher CM, Holmes RF. Handling cases of 'medical futility'. HEC Forum 2012;24(2):91–8.
[24] Callahan D. Medical futility, medical necessity: The-problem-without-a-name. Hastings Cent Rep 1991;21 (4):30–5.
[25] Lantos JD, et al. The Illusion of futility in clinical practice. Am J Med 1989;87:81–4.
[26] Youngner S. Who defines futility? JAMA 1988;260:2094–5.
[27] Truog R, Brett A, Frader J. The problem with futility. N Engl J Med 1992;326:1560–4.
[28] Schneiderman L, Jecker N, Jonsen A. Medical futility: Its meaning and ethical implications. Ann Intern Med 1990;112:949–54.
[29] Cantor M, Braddock III C, Derse A, et al. Do-not-resuscitate orders and medical futility. Arch Intern Med 2003;163:2689–94.
[30] Brody H. Medical futility: A useful concept? In: Zucker MB ZH, editor. Medical futility and the evaluation of life-sustaining treatment. Cambridge, Mass: Cambridge Press; 1997. p. 1–14.
[31] Venneman S, Narnor-Harris P, Perish M, Hamilton M. "Allow natural death" versus "do not resuscitate": Three words that can change a life. J Med Ethics 2008;34:2–6.
[32] Chen Y, Youngner S. "Allow natural death" is not equivalent to "do not resuscitate": A response. J Med Ethics 2008;34:887–8.
[33] Lang F, Quill T. Making decisions with families at the end of life. Am Fam Physician 2004;70(4):719–23.
[34] Sulmasy D, Sood J, Ury W. The quality of care plans for patients with do-not-resuscitate orders. Arch Intern Med 2004;164:1573–8.

[35] Curtis J, Burt R. Point: The ethics of unilateral "do not resuscitate" orders. Chest 2007;132(3):748–51.
[36] Manthous C. Counterpoint: Is it ethical to order "do not resuscitate" without patient consent? Chest 2007;132(3):751–4.
[37] Buckman R. How to break bad news: A guide for health care professionals. Baltimore: The Johns Hopkins University Press; 1992.
[38] Baile WF, Buckman R, Lenzi R, Glober G, Beale EA, Kudelka AP. SPIKES—a six-step protocol for delivering bad news: Application to the patient with cancer. Oncologist 2000;5(4):302–11.
[39] Coulter A. Engaging patient in their healthcare: How is the UK doing relative to other countries? Oxford, United Kingdom: Picker Institute Europe; 2006.
[40] Elwyn G, Laitner S, Coulter A, Walker E, Watson P, Thomson R. Implementing shared decision making in the NHS. BMJ 2010;341.
[41] Barry M, Edgman-Levitan S. Shared decision making—the pinnacle of patient-centered care. N Engl J Med 2012;366:780–1.
[42] Durif-Bruckert C, Roux P, Morelle M, Mignotte H, Faure C, Moumjid-Ferdjaoui N. Shared decision-making in medical encounters regarding breast cancer treatment: The contribution of methodological triangulation. Eur J Cancer Care 2015;24(4):461–72.
[43] Braddock CH. Supporting shared decision making when clinical evidence is low. Med Care Res Rev 2013;70(1 Suppl):129S–40S.

CHAPTER

3

Patient Experience and End-of-Life Care: A Discussion and Analysis of Four Patients

Angelique Wong and Donna S. Zhukovsky

The University of Texas MD Anderson Cancer Center, Houston, TX, United States

INTRODUCTION

A driver of the patient experience, person-centered, family-oriented care is recognized nationally as a key component of end-of-life (EOL) care. A recent Institute of Medicine (IOM) report, "Dying in America," emphasizes the role of patient-centered, family-oriented care in harmonizing medical care with psychosocial and spiritual support for individuals of all ages as they approach the EOL [1]. In this chapter, we will identify some common clinical situations in EOL care and describe ethical challenges inherent in the care of these patients. We will then apply selected bioethical principles, grounded by a patient-centered, family-oriented care approach, to offer potential strategies for optimizing care outcomes. This chapter is not meant to be exhaustive in nature but rather to enhance awareness of issues that may arise and potential approaches for their prevention or resolution. Patient examples used to illustrate challenges and potential strategies to optimize care outcomes have been modified to prevent identification of patients and their family members.

PERSON-CENTERED, FAMILY-ORIENTED CARE, AND THE PATIENT EXPERIENCE

Early definitions of person-centered family-oriented care were based on the endorsement of deep respect for patients as unique living beings and the obligations to provide care for them on their own terms [2]. In 2001, in the report "Crossing the Quality Chasm," the IOM recognized patient centeredness as one of the six domains requisite for improving

Ethical Challenges in Oncology.
DOI: http://dx.doi.org/10.1016/B978-0-12-803831-4.00003-8

the United States' (US) health-care system. Patient-centered care was defined as providing care that is respectful of and responsive to individual patient preferences, needs, and values and ensuring that patients' values guide all clinical decisions [3]. Embracing the concept of person-centered family-oriented care, the specialty of pediatrics developed a definition based on the understanding that the family is the child's primary source of strength and support and that the perspectives and information provided by families, children and young adults are essential components of high-quality decision-making. They further concluded that when patients and families are recognized as full partners of the health-care team, they shape all areas of health care including direct patient care, education, policy making, program development, implementation and evaluation, and health-care facility design. Family is broadly defined and may evolve over time. The six core principles of patient-centered family-oriented care are detailed in a policy statement put forth by the American Academy of Pediatrics [4]. Notably, patient-centered, family-oriented care is not restricted to pediatrics; it is a tenet of palliative care for people of all ages that takes into account culture, traditions, values, beliefs, and language and that evolves with patient and family needs [5].

Person-centered family-oriented care is often used interchangeably with the term the patient experience, but actually is one of the core components of the patient experience. Seemingly self-evident, there is no consensus on the definition of the patient experience [6]. A definition in common use is one developed by The Beryl Institute, self-described as a global community of practice dedicated to improving the patient experience through collaboration and shared knowledge. Their definition references "the sum of all interactions, shaped by an organization's culture that influences patient perceptions, across the continuum of care" [7]. The President of The Beryl Institute, with colleagues from other organizations, conducted a literature review with narrative synthesis. They concluded that there was no consistent definition of the patient experience, but that the Beryl Institute definition captured the common themes identified in their analysis: the sum of all interactions, the influence of organizational culture, patient perceptions and the importance of considering experiences across the continuum of care. The authors identified three additional consistent themes that they believed would enhance the Institute's definition of the patient experience: (1) active patient and family partnership and engagement, (2) the integral need for person-centeredness, and (3) acknowledgment of the broad and integrated nature of the experience overall [6]. For individuals with advanced medical illness, there are clear implications of the integration of person-centered family-oriented care along the continuum of care for providers of all types, not just those who specialize in palliative care. Providing authentic care of this type has the potential to proactively prevent some of the ethical challenges that arise in these complex circumstances and minimize harm from those that do occur.

BIOETHICAL PRINCIPLES AND END-OF-LIFE CARE

Ethical analysis provides guidance on how to resolve challenging clinical situations. As individuals approach the EOL, conflictual situations may be especially prevalent given the tumultuous emotions they, their families, and their health-care providers often experience, the diversity of backgrounds influencing personal values of those involved and the existential questions that arise. Adding to these challenges is that involved stakeholders may have different values and goals that compete with one another, making it difficult to know how to frame the issues. Taboda provides a useful discussion of the main approaches in contemporary bioethics and the ethical principles upon which they are based. Two of the more influential approaches are principlism and personalism. The former, the most commonly used analytical approach in the Anglo-Saxon world, is based on the principles of respect for autonomy, nonmaleficence, beneficence, and justice. Personalism, more commonly used in continental Europe and parts of Central America, is based on the primary principles of respect for human life and death, totality (also known as therapeutic principle), freedom and responsibility, and sociability and subsidiarity. Taboda finds personalism to be more congruent with the philosophy and goals of palliative medicine than she does principlism. Regardless of the paradigm used, she recommends a systematic approach to the ethical analysis of clinical cases, starting with identifying the ethical problem, formulating the right question and referring to the ethical value or principle involved. Once this is accomplished, one can move on to collect and analyze the relevant clinical information before exploring the patient's values and preferences with regard to medical decision-making, after establishing his or her competence for medical decision-making in both the cognitive and affective domains. After formulating a solution and alternatives, one can then consider the best way to implement the proposed solution and reflect on lessons learned [8]. This systematic approach incorporates active involvement of patients, surrogate decision-makers, families and the health-care team in problem-solving, as is part of person-centered family-oriented care. How this is done will influence the patient experience for better or for worse.

SOME CHALLENGING CLINICAL SITUATIONS IN END-OF-LIFE CARE

When Disclosure and Decision-Making Preferences Differ Among Clinicians, Patient and Family

The patient is a 75-year-old woman from Saudi Arabia who comes to the palliative-care clinic accompanied by her adult son and daughter. She is currently receiving immunotherapy for refractory lung cancer metastatic to bone, which has progressed after receiving multiple different treatments in several different countries.

She and her children, who are very involved in her care, have come to the US for yet another opinion about her cancer care. The patient understands some English, but looks to her son and daughter to translate major portions of the conversation. The patient's major complaint is that she cannot walk due to overwhelming weakness. She appears sad which you interpret as speaking volumes about her concern regarding the cause of her fatigue. Importantly, you note that she has been wheelchair-bound since her initial presentation months ago and as such, she is unlikely to regain much function or strength. After initial discussion with the patient and her family, the son asks to speak with you alone, to which the patient agrees. He requests that you not discuss the extent of her disease nor her prognosis with his mother, for fear that such information would cause her to give up. He would like his mother to believe that her cancer is localized and stable.

Nondisclosure is a controversial topic and is not the norm in the US, where truth telling plays a prominent role in biomedical ethics [9,10]. However, major cross-cultural differences in truth-telling attitudes and practices exist among Western and nonWestern countries. In particular, disclosure remains a controversial topic among Middle Eastern countries [11] and often does not take place, invalidating the informed consent process which is prioritized in Western medicine.

As the palliative-care physician, you are concerned that without full disclosure, the patient may continue to blame herself for her inability to improve her strength and function. You fear that this may further worsen her mood and detract from her ability to focus on other goals that may be important to her at this point in her illness such as returning home to Saudi Arabia to be with her husband and the rest of her children. You also fear that without full knowledge of her situation, she may not be able to fully assess the risks and benefits of continued treatment while in a foreign country. Further complicating the issue of different perspectives in truth telling is the language barrier. You do not wish to burden the son with translation, despite his willingness to do so and would like to be confident that translations are an accurate representation of what you say to the patient. Thus, you conduct all encounters with the help of an onsite trained professional medical Arabic translator.

The approach to the ethical conflict presented in this scenario is not straightforward. The answer is neither what the physician thinks is best in regards to full disclosure and truth telling, nor what the son is requesting in terms of nondisclosure. We must also take into account the patient's preferred form of decision-making. In the US, the physician–patient relationship has evolved from one of paternalisms to that of shared decision-making between the two [12]. However, in many cultures, patients prefer to include family members in the decision-making process [13–15] and favor a family-centered model of care [16]. A survey done in Saudi Arabia found that 65% of the patients interviewed endorsed that family should be informed of the diagnosis of incurable cancer, and then family should decide whether or not to inform patient of the diagnosis [17]. Regardless of prevailing cultural norms related to disclosure and truth telling, many individuals personally wish to be actively involved in their own health-care decisions [15], often in contrast to their relatives' opinions [18].

Not wanting to sever the relationship between yourself and the patient's son or that between the patient and her son, you decide not to push your views and perceived obligations. In hopes of better understanding the family's view, you choose to first explore further the son's request for nondisclosure. You learn that as the eldest of eight children, he feels a sense of duty and responsibility to protect his mother and worries that further bad news may further cripple his ill mother, causing intense anxiety and depression. He also does not want to let down his other siblings who are waiting for him to return his mother home in good health.

After acknowledging his struggles to obtain the best care possible for his mother, you ask if he and his mother have discussed how she may be feeling at present or her thoughts on how to proceed. You then express that you, too, wish for her to receive good care and gently explain your obligation as her physician to tell her the truth regarding her diagnosis and prognosis so that she can make informed decisions regarding her care. You also acknowledge the importance of respecting his mother's preferences and their family traditions. He remains hesitant with regards to full disclosure, but you and he negotiate a plan with which he feels comfortable. The son agrees that you may explore the patient's knowledge and awareness of her illness without telling her more than she discloses. He also agrees to have you ask her if she wants to be included in decision-making.

Is withholding the truth harming the patient? Is telling the truth harmful to her well-being? Is she able to make the best decisions for her care without the full truth? Does she want to know the truth? Or, does she already know the extent of her diagnosis and prognosis and perhaps is an active party in creating an illusion of nondisclosure in order to protect her son from this knowledge? Does she want her son to make medical decisions for her? You must consider all these in analyzing the situation, as each patient is unique, regardless of their cultural origin.

Beauchamp and Childress define personal autonomy as "self-rule that is free from both controlling interference by others and from limitations, such as inadequate understanding, that prevent meaningful choice" [19]. Respecting autonomy does not imply "truth-dumping." Respecting autonomy takes into account the patient's preferences for receiving information, communication, and decision-making style and role of the family [20–22].

You decide that the best approach is to ask the patient if she wants to be involved in decision-making and whom she wants involved. Returning to the room with the medical interpreter and son, you gently explore the patient's understanding of her illness and her goals for care. She explains that she has received several different treatments for lung cancer but has not become stronger. She defers further questions about her illness to her son and tells you that she relies on him to receive medical information and make medical decisions on her behalf. She adds that her family is very important to her, that all her children are included in the decision-making process since they are more educated than she and that she is only told about the decisions. She trusts their decisions for her. Her only goal at present is to be able to walk some so as to not to burden her children.

You respect her decision to defer medical decision-making to her children and her unspoken decision to not address prognosis as a form of patient autonomy, along with her request to focus on more immediate goals of pain relief and physical therapy. However, you recognize that this is not true informed consent, as valued in our medical system and wonder how to balance her values with our own. As guidance, The University of Texas MD Anderson Cancer Center's informed consent policy recognizes that factors other than capacity for medical decision-making have the potential to impact the patient's willingness to engage in the informed consent process, among them emotional and psychosocial readiness and cultural considerations. In such situations, one may consider Sulmasy and Synder's approach to surrogate decision-making, based on substituted interest and best judgment as one way to represent what the patient would have decided if she had participated in full disclosure [23].

McCabe et al. suggest a step-wise approach to requests for nondisclosure so that patients and families can gradually adapt to information sharing that meets with their cultural traditions but also allows the physician to meet his/her professional obligations [24]. For this patient, obtaining a better understanding of the patient's communication style, her preference for receiving information, and the role of her family involvement in decision-making helped build a trusting relationship so that a step-wise approach to disclosure and truth telling may be explored and approached again in the future, if appropriate.

When Patients and Families Demand Care Outside the Usual System

The patient was a 67-year-old CEO of an oil and gas company with pancreatic carcinoma metastatic to liver and peritoneum. Prior to his cancer diagnosis, he had no health problems. A high-powered business executive, he continued to work throughout his illness, despite pursuing aggressive chemotherapy in the face of inexorably progressive disease and a known chronic small bowel obstruction. After 18 months of treatment with conventional and experimental therapies, his oncologists informed him that he was not a candidate for additional chemotherapy and recommended hospice care. Shortly thereafter, he was admitted to the acute palliative-care unit with a 2-week history of abdominal pain, nausea, vomiting, and failure to thrive. His wife is insistent that he is seen only by another member of the palliative-care team who is not on service, stating that she will call her friend who is a senior physician in the hospital's administration, if she cannot have Dr. Q be her husband's physician.

Is it ethically appropriate to acquiesce to the wife's demand for change of physician? Using the principlist approach, we can consider the situation from the principles of respect for autonomy, nonmaleficence, beneficence, and justice. To collect the clinically relevant data, we must first decide who we are addressing—is it the wife, the patient, the different doctors involved, involved members of the health-care team, or other patients on the unit? Initially, the situation seems to be limited to the patient, his wife, and the involved physicians, with most agreeing that the patient is the primary focus. After establishing that Dr. Q is willing to assume the care of the patient, the team social worker meets with the wife to understand her reason for requesting the change, as she has yet to meet the palliative-care unit physician, Dr. P. The patient's spouse reports that she values relationships and has a great deal of trust in Dr. Q; she does not want to meet a new physician with whom she has not established a relationship. Next, the social worker meets with the patient separately to see if the request for change in physician is consistent with his wishes. You also clarify with him that he likes to make his own decisions and does not want his wife to speak for him (respect for autonomy). The interdisciplinary team, including both physicians, then meets to see how the change might impact the patient's care and the team. For example, would Dr. Q have the same availability as Dr. P to see the patient should a problem arise (nonmaleficence)? What is the potential benefit of the change of the patient and his wife—would they feel that they were listened to and heard? Would this provide more confidence in their medical team and make the difficult decisions confronting them easier to handle (beneficence)? How will this affect the care of the other patients on the unit and also, Dr. Q's other patients, as he will be caring for this complicated patient in addition to his usual responsibilities (justice)?

The team decides that on balance, the situation would be manageable if Dr. Q assumed the patient's care, with the benefits to the patient and family outweighing the additional work for Dr. Q, anticipated to be modest (beneficence). However, the situation rapidly escalates, with the patient's spouse who is focused on feeding her husband an organic, plant-based diet, despite his obvious discomfort with oral intake so that he can heal, get more chemotherapy, and return to work. She repeatedly declines pain medication on his behalf, stating that his nausea, vomiting, and fatigue are due to constipation from opioids. She also declines interventions for her husband and herself from the team social worker, counselor, and chaplain. In contrast, the patient welcomes all members of the palliative-care-unit-based interdisciplinary team into his room and asks that he receives whatever the team recommends to control his symptoms so that he may be expeditiously discharged home with hospice care.

A family meeting is held to clarify goals of care, including the patient, his wife, four adult children, Drs. P and Q, and the interdisciplinary team. The patient and his family members all indicate their understanding that no curative treatment was available, that oral intake would exacerbate his abdominal pain given the bowel obstruction and that nutritionally, oral or artificial nutrition would not extend his life, given his grave prognosis. At the conclusion of the meeting, the agreed upon plan as directed by the patient was discharge home with hospice care as soon as good symptom control was established (respect for patient autonomy). His wife agreed not to feed him against his will (respect for autonomy, nonmaleficence).

However, his wife continued to feed her husband and decline pain medicine on his behalf, which he would request as soon as she left the room. Each day, Dr. Q, the nursing staff, and the interdisciplinary team spent considerable time in the patient's room on rounds, hearing the wife's concerns. She would demand that the nursing staff page Dr. Q at all times of the day and night, wanting to discuss her husband's situation further. The palliative-care unit medical team, nursing staff, and interdisciplinary team once again found themselves in an ethical quandary. They felt that the patient's spouse was impairing the quality of her husband's care by declining medications for symptom relief and by finding perceived obstacles to his discharge. Team members also expressed their frustrations that they had been unsuccessful in relieving the wife's distress and that her demands limited the quality of care of other patients on the palliative-care unit. Many also recognized signs of burn out developing in themselves.

On reflection, the team's question in this evolving situation changed from "Is it ethically appropriate to acquiesce to the wife's demand for change of physician?" to "Is it possible to optimize care for the patient (beneficence and nonmaleficence) without interfering with the care of other patients (justice) and team members' wellbeing" (nonmaleficence)? Knowing now that the wife had a history of a difficult relationship with her husband and multiple other family members and that she had not become less distressed, despite the efforts of the team, the group decided to establish ground rules with the wife in order to minimize disruption to unit processes and patient care. She was told that Dr. Q would not round on her husband until after rounds were completed by Dr. P and the interdisciplinary team for the other patients on the floor and that requests for physician pages would not be honored unless the nursing staff felt that they were medically justified. Furthermore, all urgent medical needs would be attended by Dr. P, since he was based on the unit and could meet the patient's needs in a more timely fashion. With this plan in place, a plan that prioritized justice and nonmaleficence, team care for patients on the palliative-care unit began to run more smoothly and shortly thereafter, the patient was discharged home with hospice care and subcutaneous fluid administration, meeting both his and his wife's needs.

This case, while extreme in some of its features, represents some of the challenges often seen in patient care at the EOL and the multiple and often competing ethical issues that can arise. The wife's behavior impacted not only the patient's care but also care of the other patients in the palliative-care unit. How do you interrupt that behavior that is negatively impacting patient care? Is it possible or even appropriate to disrupt what appears to be a long-standing dynamic of the couple? Do we simply accept the behavior for the benefit of temporary peace? Is it fair to the patient himself or the other patients to accept the wife's behavior and demands? Which issue is prioritized, the supporting principles to consider, and how far to assess the consequences (i.e., patient, family, health-care providers, organizational resources) may change over the care continuum. How these circumstances are approached has clear-cut implications for the provision of patient-centered, family-oriented care and the patient experience and team well-being.

When the Value Driving Patient or Family Requests for Care is Not Recognized

The patient was a 77-year-old long widowed Mandarin speaking retired high school and college mathematics teacher from Shanghai with recurrent rectal carcinoma who had immigrated to the US 15 years prior to his diagnosis. All three of his adult children had been born in Shanghai, although his two sons had received their college education in the US, with the elder son returning to Shanghai and the younger remaining in the US. The patient initially underwent preoperative chemotherapy followed by resection of tumor, before deciding against postoperative chemotherapy that was recommended to reduce the likelihood of relapse.

He did well for 4 years until he was found to have painful bone metastases. He received palliative radiation to sites of bone involvement with relief of pain and agreed with "do not resuscitate" status, at the advice of his sons during that admission. On discharge home to the care of his 51-year-old daughter who spoke only Mandarin, the younger son declined hospice care. Shortly thereafter, he was readmitted with urinary retention. During each of these admissions, his younger son requested that the patient be discharged rapidly, so that he could more easily feed his father home-cooked Chinese food. However, once again the patient is admitted to the hospital, for the third time in 5 weeks. He has a 1-week history of anorexia, continued weight loss, and confusion. His older son, a US educated PhD scholar, arrives from China the next day and respectfully but assertively requests that his father be fed intravenously with total parenteral nutrition so that he does not starve to death.

In order to clarify the patient's and family's understanding of the patient's prognosis and goals of care, you invite the patient and his children to a family meeting. Through the Mandarin translator, the patient tells you that he does not need to participate because he knows that he will die from the cancer and his children can make any medical decisions needed for him. His sons did not ask the daughter, who has been the patient's main caregiver, to attend because they do not want to upset her with the news that their father is dying. On meeting with the sons and the interdisciplinary team, you confirm your understanding established from previous conversations that both know that their father will likely die from cancer in the near future. However, they still resist taking him home with hospice because they believe he will starve to death if he does not receive intravenous feedings and this cannot be done on hospice care. You are confused because they clearly state that they know he is dying of the cancer and because the younger son has consistently told you that his father is happier at home than in the hospital.

Is it acceptable to provide artificial nutrition and hydration for those who so desire even though from a medical perspective it would not prolong the patient's life and would keep the patient in the hospital without the potential of medical benefit when he would rather be a home? Many times clinicians continue treatments with limited prospect of benefit to minimize psychological distress for patients and caregivers. This clinical situation illustrates many of the complexities that may complicate care when clinicians care for patients of cultural backgrounds different from their own. Furthermore, the family itself demonstrates different degrees of acculturation as reflected by differences in disclosure of prognostic information to various family members. Regarding disclosure, the patient and his two sons all confirm their knowledge of the terminal nature of his disease, consistent with disclosure as practiced in the US, yet they do not want to inform the patient's daughter because in their culture, it is often considered rude or disrespectful to talk about death which may even bring bad luck. As you explore the situation further, you learn that in some Asian Chinese traditions, words that sound like death, such as the number four and clock, and wearing the color white are also avoided, as they may bring bad luck [25].

As the patient has demonstrated his understanding of the terminal nature of his disease and his preference to be at home, you are surprised that he has delegated decision-making authority to his sons and appears content to remain in the hospital. Concerned about patient autonomy, you review decision-making preferences and learn that many Chinese prioritize family decision-making over their own autonomy [25]. You are still stymied by the sons' insistence on artificial nutrition and their reluctance to bring their father home with hospice care, so you and the team sit down with them once again, in an effort to see how you can best meet their and the patient's needs without providing what is seemingly futile care from your perspective. After going around and around about the lack of benefit of parenteral nutrition in their father's situation, the younger son says, "but he must die with his stomach full!" You ask him to explain that to you and learn that a good death means dying surrounded by family members with a full stomach. You also learn that if nonfamily members are perceived as being the main caregivers, that some in the Chinese community will perceive that children have not fulfilled their obligation of filial piety, which will bring shame to their parents' reputation, portraying the patient and his wife as poor parents [25,26]. With that knowledge in hand, you offer a trial of subcutaneous fluids that includes sugar (dextrose) and vitamins for nutrition and explain that the hospice team's role is to support the family in caring for the patient, not to replace them. The sons indicate their agreement with that plan. The patient is discharged home with a hospice provider experienced with caring for patients of Chinese origin. Three weeks later, you receive a card from the older son on behalf of his family, informing you that his father died comfortably with his stomach full, surrounded by those who loved him. He thanks you for your willingness to stick with them and the respect that the team showed them in their efforts to understand what was important to their family.

Reflecting on this case scenario, we see that the issue as initially framed, a request for medically futile artificial nutrition, was not what was actually underlying the situation. The patient and his sons understood his diagnosis and prognosis and that medically, nutrition would not benefit him. With respectful listening to understand what was at the root of their request, you learned that the driving value was "dying with one's stomach full" while being cared for by family. In their seminal article, 'Negotiating Cross-Cultural Issues at the end of Live, "You Got to Go Where He Lives," Kagawa-Singer and Blackhall eloquently discuss the impact of patient and clinician culture on patient care. Recognizing the wide variation of individual beliefs and values among persons from the same ethnic population, they state that stereotyping individuals from a particular background or failing to take culture into account may both lead to harm. From a practical perspective, the authors offer six commonly encountered issues encountered in EOL care, the possible consequences of not attending to these issues and potential strategies with which to address them. Topics discussed include responses to inequities in care, communication/language barriers, religion and spirituality, truth telling, family involvement in decision-making, and hospice care [27].

When Support of Patient Autonomy Is Counterintuitive to Beneficence and Nonmaleficence

The patient was a 34-year-old landscape business owner in the prime of life when he was diagnosed with stage IV Burkitt's lymphoma. Married 18 months, he had a 12-month-old son and a wife whom he adored and a fulfilling career. Despite state-of-the-art treatment, he was frequently admitted to the hospital with complications of his treatment and his disease. This time, 6 months after his initial diagnosis, he is directed to the emergency room by his primary oncologist due to a 3-day history of confusion and agitation. He is found to have a serious infection and to be in renal failure, thus, is admitted and placed in the intensive care unit where he is initiated on dialysis and started on broad spectrum antibiotics for suspected sepsis. Three days later, when his mental status has not improved, the intensive care unit physician and the inpatient lymphoma attending who is "on-service" talk with his wife, recommending Do Not Resuscitate status and that dialysis be discontinued. Acknowledging his limited life expectancy, they also recommend Supportive Care consultation for symptom management and possible transfer to the Acute Palliative Care Unit for EOL care. As the surrogate decision-maker for her husband and because the patient remains delirious and unable to participate in medical decision-making, the spouse agrees to all those recommendations and the patient is transferred to the Acute Palliative Care Unit.

Over the weekend, the patient becomes more alert and communicative. He is surprised to find himself under the care of the palliative-care team but indicates that he knows his time is limited, he is tired of battling his disease, and that he would like to be at home with his wife and son for his final days when his symptoms are better controlled. He is open to receiving hospice care. As he begins to feel better mentally and physically, his wife asks the palliative care attending physician to consult with the lymphoma team to see if any more lymphoma treatment would be reasonable. She states "we know that he is going to die from this disease, but given his improvement, we are cautiously optimistic that he may benefit from treatment and live a little longer." The attending physician confirms this request with the patient (respect for autonomy) and contacts the lymphoma physician. Dr. N comes to visit the patient and agrees to investigate experimental clinical trial options on his behalf. The next evening, the palliative-care unit attending receives an email from her lymphoma colleague that no conventional or experimental trial options exist for the patient. Dr. N states that he will visit the patient and his wife the following evening to share this information. However, after meeting with Dr. N, the patient is transferred back to the lymphoma service where he receives chemotherapy. He rapidly becomes hypotensive and somnolent. The lymphoma team requests transfer back to the Acute Palliative Care Unit. After discussion with the wife, who is distraught about encouraging her husband to receive this last course of chemotherapy, the patient is transferred back to the Acute Palliative Care Unit. The next day, he dies without regaining consciousness, surrounded by his wife, parents, and son.

As the palliative-care unit physician for this patient, you question if there is more you could have done for this patient and his wife to prevent the multiple transitions that took place during the final days of his course and the additional suffering he and his wife experienced. You also question if you contributed to the patient's and wife's unrealistic hope by contacting the lymphoma team for further patient evaluation, when initially you perceived yourself to be supporting patient autonomy. In view of his poor prognosis, if you had been more paternalistic with the family, could you have prevented some of the additional suffering the patient and his wife endured?

This situation illustrates the challenges frequently experienced in EOL care, when sound bioethical principles compete with one another. Contextually, the patient and his wife had been drivers of his medical care since the time of diagnosis, making informed decisions after seeking copious information. Was it inappropriate to support the same type of decision-making (respect for patient autonomy), given his overwhelmingly poor prognosis at this point in his disease trajectory (beneficence and nonmaleficence)? Unbeknownst to the palliative-care interdisciplinary team at the time of request for follow-up with the lymphoma team, was the patient's desire to please his father, a life-long pattern. This information was elicited by the team chaplain after the decision about seemingly futile chemotherapy had been made. The father, who had not seen his son since the patient's diagnosis, had pleaded with his son to try whatever was available so that they could have more time together. Reflecting on this knowledge may yield a different balance of respect for autonomy versus beneficence and nonmaleficence. Decisions occur within the context of social relationships and as such, emotions are part of the equation. Stated otherwise, autonomy is influenced by the social nature of individuals and the impact of individual choices and actions of others and is not focused exclusively on reason [28]. For this patient, the beneficence of pleasing his father may have superseded the importance of nonmaleficence for himself and his wife. Clearly, understanding as much of the social milieu as possible will aid the treating team in supporting decisions that are most congruent with the patient's values and goals. Forearmed with this knowledge, the team might have been able to intercede with the patient's father in a way that would have better supported the well-being of the patient–family unit as a whole.

ETHICAL CHALLENGES INHERENT IN END-OF-LIFE CARE AND STRATEGIES TO FACILITATE OPTIMAL OUTCOMES

The challenges to seamless patient-centered, family-oriented care as patients approach the EOL are numerous. Some of the more common challenges are noted in Table 3.1. While some of the examples in the different categories are the same, it is identification of the underlying challenge and associated values that is crucial, in order to proactively solve the problem. Failure to correctly identify the challenge may lead to incorrect assumptions and clearly has implications for the patient, family, clinicians, and the health-care system. From a patient and family perspective, these may include dissatisfaction with patient care, lack of trust in the provider, caregiver burden, psychological morbidity, risk for complicated bereavement in survivors [29], and increased financial stressors. From the perspective of the health-care team, these issues may result in impaired quality of patient care, increased clinician stress, suboptimal team function, and ineffective resource utilization.

Central to most of these challenges is the need for expert, culturally sensitive communication on the part of the health-care team in order to better understand the patient's and family's needs and how best to meet them. Use of expert communication skills, together with other strategies as noted in Table 3.2, can often prevent potentially conflictual situations and deescalate existing emotionally charged scenarios in a manner that more

TABLE 3.1 Common Challenges in Delivering End-of-Life Care

Challenge	Examples
Demands for care outside the usual structure (i.e., "special care")	Total parenteral nutrition (TPN) and blood product support for patients receiving hospice care as hospice care does not typically include TPN nor blood products; care from provider who is not part of usual care team
Differing goals or priorities of involved stakeholders	Discharge when patient no longer meets acute care criteria (hospital system) versus longer hospitalization due to patient and caregiver distress (patient or family)
Differences in disclosure preferences among involved stakeholders	Family request for nondisclosure of diagnosis and/or prognosis to patient versus Western medical clinician norm of disclosure to patient
Differences in preferred style of decision-making	Patient versus family versus community
Medical futility	Requests for ongoing chemotherapy despite medical recommendation that it is no longer beneficial. Individuals with widespread metastatic disease who request TPN despite severe cachexia and hypoalbuminemia
Literacy	Frequent emergency room presentations and hospitalizations for pain control because unable to read prescription bottles or medication instructions
When the value driving requests for care are not recognized	Patient requests for artificial nutrition without anticipated medical benefit versus cultural norm of "dying with one's stomach full"
Unrecognized nonmedical benefits underlying requests for medical interventions	Patient concurs with CPR despite personal preference for withholding CPR at the end of life because s/he places higher priority on family decision-making over personal choice

TABLE 3.2 Strategies That Promote Patient-Centered, Family-Oriented Care at the End of Life

Expert communication	
Hearing from the "other"	• Respectful listening (patient, family, or colleague) • Observe for nonverbal cues or "unspoken" messages • Openness to different ways of doing things (cultural sensitivity) • Do not superimpose personal agenda • Avoid bias from one's own values • Do not formulate a response while listening
Clear communication	• Use easy, succinct language • Deliver information in short bursts • Use professional medical translators when language differences exist (not family members) • Check for understanding in individual's own words • Attend to emotion: acknowledge and validate, pause delivery if emotions heightened • Repeat conversations or key points, as needed (allows for consolidation of information) • Ask for clarification of questions or statements • Reframe questions or statements to ensure understanding
Boundaries and limit setting	• Appropriate to situation • Nonrigid
Caring	
Stay the course	• Presence • Nonabandonment • Empathy
Show empathy	• Verbal communications • Touch, as appropriate
Align patient/family goals with the team's care goals	
Work from areas of agreement	• Consider different ways of saying the same thing • Allow patient and family members to share their thoughts, values, and expertise/experiences • Use patient's and family's words to reflect on what they have said • Negotiate what is desired with what is possible
Involve interdisciplinary team	• Listen to each team member and allow them to share their thoughts, values, and expertise • Consider health-care provider and team well being • Align team goals with goals of patient/family goals, as possible
Consider family meeting with all key stakeholders	• Clarify if patient would like to attend or defers to family • Ascertain with patient who to invite and who to exclude (ideally those most affected by and most likely to be effective in patient care) • Clarify patient and family priorities for discussion prior to meeting • Navigate truth telling and disclosure among individuals with different preferences • Summarize meeting outcome, action plan • Follow up to convey caring and address outstanding items

(*Continued*)

TABLE 3.2 (Continued)

Proactive advance care planning	
Start early in provider–patient relationship	• Outpatient setting preferred • Build rapport • Invite patient to discuss understanding of current condition and treatment options • Ask patient to bring family member most involved in his/her care, or medical power of attorney, if established • Consider as a process, not "one-stop shopping"
Establish key information	• Explore patient's understanding of disease process and treatment options • Identify patient's communication preferences (i.e., detailed or big picture, to whom) • Identify preferred involvement of patient and/or family in decision-making • Explore patient's preferences, values, fears, concerns, and hopes for the future • Ask what it means to the patient to "live well" (what it is that gives their life meaning) • Assess readiness for advance care planning and proceed accordingly
Prepare the patient and surrogate for decision-making	• Provide information tailored to patient's current condition; fill in gaps in medical knowledge and clarify misperceptions • Discuss options for care • Clarify role of surrogate decision-makers • Provide education regarding types of advance directives and their role as a means of supporting patient preferences • Encourage discussion of preferences with family and other people important to the patient and with involved clinicians
Discuss specific preferences for life sustaining therapies for patients with advanced disease	• Focus discussion around their concept of "living well" as a starting point, not on foregoing different interventions per se • Encourage patient to be specific in their wishes regarding CPR, mechanical ventilation, tube feeding, dialysis, comfort-care options and place of death, if known • Include surrogates and other key stakeholders in discussion of these preferences

Communicate preferences (transparency)	• To key stakeholders, including family, significant others and the health-care team
	  ◦ Verbally
	  ◦ In the form of advance directives that are made available

Review regularly	• Review preferences annually, with changes in disease status and at transition points (i.e., hospitalization, intensive care unit admission, referral to hospice, nursing home discharge) • Communicate changes and update advance directives, as appropriate

effectively promote patient-centered, family-oriented care. Much of this can be integrated in to the patient's care under the guise of proactive advance care planning, focusing on the patient's values and goals, over the course of the illness trajectory [30,31].

It is important to be mindful that these are not easy discussions for any of the involved parties. A recent metasynthesis of qualitative studies focused on oncologists' perspectives on breaking bad news to patients details the many challenges they encounter. Disclosure was not only limited to diagnosis but also included discussion of prognosis, relapse, transition to palliative care, and transition to EOL. Universal challenges included balancing how much information to disclose with sustaining hope and managing strong emotions of the oncologist, as well as the patient. External factors that shaped the physician–patient encounter included family relationships as barriers or facilitators, the influence of culture on the family's role in disclosure and decision-making as compared to the prevailing Western medical model of patient autonomy and systemic and institutional factors. Among the latter were socioeconomic constraints impacting available care options, insufficient training to conduct these sensitive conversations and time constraints to the actual process [32].

Given the multicultural society in which we live, it is also important to recognize the influence of culture on communication and EOL care preferences. One definition of culture is the beliefs, values, and behaviors that are shared within a group, such as a religious group or a nation. Culture includes language, customs, and beliefs about roles and relationships [33]. An individual's ethnicity is defined by one's sense of identity as a member of a cultural group, embedded within the multicultural whole and the sociohistorical lens through which it is viewed. Accordingly, ethnicity is a dynamic construct that must be evaluated within the context of the individual's circumstances at the time. Consequently, not all cultures are visible and ethnicity is very much a personalized construct. Clinician and patient/family differences in preferred styles of disclosure and decision-making are two areas in which conflicts often arise when cultures and ethnicities of involved stakeholders differ. In one study of individuals older than 64 years who self-identified as being African-American, European-American, Korean-American, or Mexican-American, Korean-Americans and Mexican-Americans were more likely than the others to believe that patients should not be told of a metastatic cancer diagnosis or a terminal prognosis nor should they be the ones to make decisions about the use of life-sustaining technology. Instead, these groups favored disclosure to family members and family-centered models of decision-making [16]. Inquiring with whom medical information should be discussed and how they and their family like to make decisions are proactive ways for clinicians to minimize distress in these areas. However, for some, even this level of discussion may be harmful to individuals with certain values and beliefs. For example, traditional Navajo values and ways of thinking prohibit discussion of any negative information. This belief contrasts with Western bioethics which prioritizes individual autonomy and self-determination, with the attendant need for disclosure of diagnosis, prognosis, and potential risks and harms when considering treatment options and the possibility of death when discussing advance care planning. [34]. Kagawa-Singer provides a 7-level model of cultural assessment, with examples of questions that can be used to evaluate each area, as a means of affording more culturally competent care [35].

CONCLUSIONS

As patients approach the EOL, ethical challenges are laced with heightened emotions, often in the context of differing cultural backgrounds, belief systems, values and goals among involved stakeholders. Guided by the constructs of patient-centered, family-oriented care, these sensitive EOL issues must be considered and can be expertly navigated using bioethical principles. Key to optimal outcomes is formulating the correct question, in order to select the most helpful ethical principles for its solution. Overarching all is the need for clear, careful, empathic, and culturally sensitive communication. Future research is needed to explore how best to evaluate disclosure, decision–making, and communication preferences, as well as effective proactive advance care planning strategies, and their impact on patient, clinician, health-care team, and institutional outcomes.

LIST OF ACRONYMS AND ABBREVIATIONS

End of Life EOL
Institute of Medicine IOM
United States US

References

[1] Pizzo PA, Walker DM, Bomba PA, Bruera E, Fahey CJ, Hinds PS, et al. Summary. IOM (Institute of Medicine). Dying in America: improving quality and honoring individual preferences near the end of life. Washington, DC: The National Academies Press; 2015. p. 1–20.
[2] Epstein RM, Street RL. The values and value of patient-centered care. Ann Fam Med 2011;9(2):100–3.
[3] Committee on Quality of Health Care in America. Executive Summary. IOM (Institute of Medicine). Crossing the quality chasm: a new health system for the 21st century. Washington, DC: National Academies Press; 2001. p. 1–22.
[4] Committee on Hospital Care and Institute for Patient-and-Family-Centered Care. Patient- and family-centered care and the pediatrician's role. Pediatrics 2012;129(2):394–404.
[5] Introduction. IOM (Institute of Medicine). Dying in America: improving quality and honoring individual preferences near the end of life. Washington, DC: The National Academies Press; 2015. p. 21–43.
[6] Wolf JA, Niederhauser V, Marshburn D, LaVela SL. Defining patient experience. Patient Experience J 2014;1 (1):6–19.
[7] The Beryl Institute. ⟨http://www.theberylinstitute.org⟩; 2017 [accessed 18.04.2017].
[8] Taboada P. Bioethical principles in palliative care. In: Bruera E, Higginson I, von Gunten CF, Morita T, editors. Textbook of palliative medicine and supportive care. 2nd ed Boca Raton, FL: CRC Press; 2015. p. 105–18.
[9] Bruera E, Neumann CM, Mazzocato C, Stiefel F, Sala R. Attitudes and beliefs of palliative care physicians regarding communication with terminally ill cancer patients. Palliat Med 2000;14(4):287–98.
[10] Novack DH, Plumer R, Smith RL, Ochitill H, Morrow GR, Bennett JM. Changes in physicians' attitudes toward telling the cancer patient. JAMA 1979;241(9):897–900.
[11] Khalil RB. Attitudes, beliefs and perceptions regarding truth disclosure of cancer-related information in the Middle East: a review. Palliat Support Care 2013;11(01):69–78.
[12] Reiser SJ. Words as scalpels: transmitting evidence in the clinical dialogue. Ann Intern Med 1980;92 (6):837–42.
[13] Noguera A, Yennurajalingam S, Torres-Vigil I, Parsons HA, Duarte ER, Palma A, et al. Decisional control preferences, disclosure of information preferences, and satisfaction among Hispanic patients with advanced cancer. J Pain Symptom Manage 2014;47(5):896–905.
[14] Yennurajalingam S, Noguera A, Parsons HA, Torres-Vigil I, Duarte ER, Palma A, et al. Family caregiver preferences for patient decisional control among Hispanics in the United States and Latin America. Palliat Med 2013;27(7):692–8.
[15] Yennurajalingam S, Parsons HA, Duarte ER, Palma A, Bunge S, Palmer JL, et al. Decisional control preferences of Hispanic patients with advanced cancer from the United States and Latin America. J Pain Symptom Manage 2013;46(3):376–85.
[16] Blackhall LJ, Murphy ST, Frank G, Michel V, Azen S. Ethnicity and attitudes toward patient autonomy. JAMA 1995;274(10):820–5.

[17] Mobeireek AF, Al-Kassimi F, Al-Zahrani K, Al-Shimemeri A, al-Damegh S, Al-Amoudi O, et al. Information disclosure and decision-making: the Middle East versus the Far East and the West. J Med Ethics 2008;34 (4):225–9.
[18] Zekri JM, Karim SM, Bassi S, Bin Sadiq BM, Fawzy EE, Nauf YI. Breaking bad news: comparison of perspectives of Middle Eastern cancer patients and their relatives. J Clin Oncol 2013;31, abstr 9568.
[19] Beauchamp TL, Childress JF. Respect for autonomy. Principles of biomedical ethics. 5th ed. New York: Oxford University Press, Inc; 2001. p. 57–112.
[20] Hallenbeck J, Arnold R. A request for nondisclosure: don't tell mother. J Clin Oncol 2007;25(31):5030–4.
[21] Jotkowitz AB, Glick S, Porath A. A physician charter on medical professionalism: a challenge for medical education. Eur J Intern Med 2004;15(1):5–9.
[22] Jotkowitz A, Glick S, Gezundheit B. Truth-telling in a culturally diverse world. Cancer Invest 2006;24 (8):786–9.
[23] Sulmasy DP, Snyder L. Substituted interests and best judgments: an integrated model of surrogate decision making. JAMA 2010;304(17):1946–7.
[24] McCabe MS, Wood WA, Goldberg RM. When the family requests withholding the diagnosis: who owns the truth? J Oncol Pract 2010;6(2):94–6.
[25] Yeo G, Hikoyeda N. Cultural issues in end-of-life decision making among Asians and Pacific Islanders in the United States. In: Braun KL, Pietsch JH, Blanchette PL, editors. Cultural issues in end-of-life decision making. Thousand Oaks, CA: Sage; 2000. p. 101–25.
[26] Blank RH. End-of-life decision making across cultures. J Law Med Ethics 2011;39(2):201–14.
[27] Kagawa-Singer M, Blackhall LJ. Negotiating cross-cultural issues at the end of life "You got to go where he lives." JAMA 2001;286(23):2993–3002.
[28] Beauchamp TL, Childress JF. Moral norms. Principles of biomedical ethics. 5th ed New York: Oxford University Press, Inc; 2001. p. 1–23.
[29] Wright AA, Zhang B, Ray A, Mack JW, Trice E, Balboni T, et al. Associations between end-of-life discussions, patient mental health, medical care near death, and caregiver bereavement adjustment. JAMA 2008;300 (14):1665–73.
[30] Sudore RL, Stewart AL, Knight SJ, McMahan RD, Feuz M, Miao Y, et al. Development and validation of a questionnaire to detect behavior change in multiple advance care planning behaviors. PLoS ONE 2013;8(9): e72465.
[31] Pizzo PA, Walker DM, Bomba PA, Bruera E, Fahey CJ, Hinds PS, et al. Clinician–Patient communication and advance care planning, IOM (Institute of Medicine). *Dying in America: improving quality and honoring individual preferences near the end of life*, 3. Washington, DC: National Academies Press; 2015. p. 117–219.
[32] Bousque G, Orri M, Winterman S, Brugiere C, Verneuil L, Revah-Levy A. Breaking bad news in oncology: a metasynthesis. J Clin Oncol 2015;33(22):2437–43.
[33] National Cancer Institute. Dictionary: Culture. National Cancer Institute. ⟨http://www.cancer.gov/⟩; 2017 [accessed 18.04.2017].
[34] Carrese JA, Rhodes LA. Western bioethics on the Navajo reservation. Benefit or harm? JAMA 1995;274 (10):826–9.
[35] Kagawa-Singer M, Valdez Dadia A, Yu MC, Surbone A. Cancer, culture, and health disparities time to chart a new course? CA Cancer J Clin 2010;60(1):12–39.

CHAPTER

4

Ethical Issues for Children With Cancer

Kevin Madden, Jessica A. Moore and Patrick Zweidler-McKay

The University of Texas MD Anderson Cancer Center, Houston, TX, United States

THE ETHICAL TREATMENT OF CHILDREN IN RESEARCH STUDIES

Children and adolescents are considered a vulnerable population when it comes to participating in research. This concept has led to additional regulations and oversight which are required to perform research on children. The United States Code of Federal Regulations Title 45, Part 46, Subpart D, specifically addresses additional protections for children involved in research [1]. Beyond the general principles of ethical research practices outlined in the Nuremberg Code, the Belmont Report, and the Declaration of Helsinki, which apply equally to children, several key concepts are of particular importance when children participate in research, namely the direct benefit to the child, the proportionality of risks and benefits, parental permission, and assent of the child. These concepts will be addressed briefly with the intent of further discussion and exploration as we expand and evolve the participation of children in research.

Direct Benefit to Minor Patients

One of the critical concepts to evaluate for all research studies is the potential direct benefit for the research participant. If a trial has the prospect of directly benefitting the participant, with alleviation of symptoms, or prolongation of life, etc., the ethical and regulatory guidance suggests that additional "risks" may be involved (see the "Proportionality of Risks and Benefits" section). Indeed, the Code of Federal Regulations allows research in children if the weighing of benefit to risks "is at least as favorable to the subjects as that presented by available alternative approaches" (Section 405). Unfortunately, many early phase cancer trials are not very likely to provide direct benefit. For example, an early phase trial (Phase I) where the potential of a drug is truly unknown, based on decades of experience, participation on such trials would not likely directly benefit the patient [2].

Ethical Challenges in Oncology.
DOI: http://dx.doi.org/10.1016/B978-0-12-803831-4.00004-X

Despite very limited therapeutic success in Phase I trials, it is argued by some that any chance at response "holds out the prospect of direct benefit." Alternatively, researchers may learn something about the drug as a treatment for a specific disease, and this would potentially benefit other patients with the same cancer as the patient, but without direct benefit to the individual child. This is a critical distinction as the Code of Federal Regulations only allows participation of children on a trial without direct benefit if the risk represents "a minor increase over minimal risk" or if the research may produce "vitally important scientific knowledge about the disorder or condition" (Section 406), which are not easier to define than the direct benefit of a trial. The federal regulations define minimal risk as "the probability and magnitude of harm or discomfort anticipated in the research are not greater in and of themselves than those ordinarily encountered in daily life or during the performance of routine physical or psychological examinations or tests" (Section 102). However, "a minor increase over minimal risk" must be defined by local institutional review boards (IRB), and thus there is the potential for great variation in the application of these federal regulations [3]. Indeed, most IRBs would judge any therapeutic trial, especially phase I dose-escalation studies as significantly greater than minimal risk. The "daily life" clause has brought up a debate between absolute versus relative standard of minimal risk. Briefly, should minimal risk be a fixed definition for all children (common examples of minimal risk are an exam, a blood draw, a chest X-ray), or should the definition of minimal risk be relative to the daily risks faced by the children to be enrolled on the study [4]. Some would argue that the lives of children with relapsed cancer likely include multiple daily risks and procedures which would not typically occur for other children, so they may be able to be exposed to greater risks. However, this argument for a relative standard was rejected by the National Human Research Protections Advisory Committee in favor of a standard "indexed" to the experiences of a "normal, healthy, average child" with the goal of maintaining equity of exposure to risk, regardless of the child's circumstance [5]. Thus, it can be difficult to satisfy the Code of Federal Regulations requirements for treatment studies in children with cancer.

Proportionality of Risks and Benefits

The other half to this limitation of risks is that the risks and potential benefits must be proportional, i.e., in balance. This is based on the fundamental medical principles of beneficence and nonmaleficence, where there must be an intent to benefit the patient, and simultaneously to do no harm [6]. This requirement is even more stringent in research involving children and is difficult to assess when the potential benefit is unknown and the risks are unknown. One concern is the overestimation of this potential benefit, termed therapeutic misconception. Pediatric oncologists are often ambiguous about whether a clinical trial is research or treatment [7] and may truly believe that trial participation is in the child's best interest despite the conflicting goals of the research and the child's best interest [6]. And parents often assume that clinical trials are designed to benefit their children, whereas their true purpose is to answer a therapeutic question and contribute to knowledge [8]. Common clinical trial methods such as randomization are not designed to benefit the individual child but rather provide the researcher a more robust analysis of the differences between treatment options. Parents often do not know that they are agreeing to a research protocol and do not fully understand the concept of randomization [9], which contributes to therapeutic misconception. Another method used in Phase I studies is dose escalation to a maximum tolerated dose; although one may argue that higher doses may have more potential for therapeutic benefit, they are even more likely to have additional toxicities. In addition, the common practice of requesting research samples such as tumor biopsies and blood for pharmacokinetics and pharmacodynamics does not benefit the individual child. Tumor samples, which are not needed for the routine care of such a patient, can be painful and add risk of anesthesia, infection, etc. Parents often do not fully understand that these procedures are voluntary, and that there are nonexperimental treatment alternatives, citing confidence in the medical team as the principal factor in their decision to participate, rather than a true balancing of risks and benefits [9]. When the likelihood of direct benefit is low and the risks are high, the proportionality of risks and benefits is not equal and IRBs may question the ethical or regulatory soundness of the study. Another ethical challenge is clinical equipoise, where treatment options are felt to be equal, with no advantage to any treatment arm. Indeed, the concept of randomization depends on the principle of equipoise, where it would be unethical to assign patients to a treatment that was known to be inferior to another available treatment. Oncologists and parents struggle with this and often believe that the new drug, or new regimen, will be better than the standard-of-care treatment. This bias toward benefit contributes to exposure of many children to potential risk, and efforts to restore equipoise are essential.

Parental Permission

Informed consent is a manifestation of the ethical principle of respect for autonomy, where an individual has the right to all of the pertinent information regarding participation in a research study, and may make the sole decision about whether or not to participate in that study. The informed consent process includes these discussions and the ultimate act of coming to an agreement of participation, which often includes a signature(s) confirming participation on the study as well as an understanding of the goals, potential benefits, and risks involved. This becomes more complex when parents are given the responsibility of making medical decisions on behalf of their child, especially for Phase I trials where understanding the risks and benefits is challenging [10]. Due to parental preference to protect their child from hearing difficult information or a perceived lack of interest of the child, these informed consent discussions often do not involve the child directly or completely, and parents are left to make informed parental permission decisions without full participation of the child. This issue is directly addressed in several regulations, including the federal regulations, which require the assent of the child.

Child Assent

As children are not fully mature in their decision-making processes and are under the protection of their parents or a guardian, the parents or a proxy decision maker must provide informed permission for the child to participate in research. However, the understanding and agreement of the child is ethically desirable and can take the form of coconsent (both proxy and child consent) or more commonly permission from the proxy and assent from the child. Assent is "a child's affirmative agreement to participate in research" and stipulates that mere failure to object cannot, without affirmative agreement, be taken as assent. Thus, many guidelines and regulations incorporate the requirement for child assent with the goal of involving the child in the discussions and decisions about participation in research. Since cognitive capacity increases with age, IRB requirements for assent often change throughout childhood, based on cognitive and emotional maturity and psychological state. Although assessment of each child's ability to comprehend is ideal, the "rule of sevens" is often used as a guide [11]. In the rule of sevens, children 0–7 years are presumed to not have the ability to decide for themselves, children 7–14 years are presumed not to have this ability unless they are assessed to be able to make such decisions, and 14–21 years are assumed to be able to make these decisions (age of informed consent is decided by local government, often 18 years, or emancipated minors). However, it is important to remember that cognitive and emotional maturity vary significantly between children of the same age, and for patients with neurocognitive issues, such as Down syndrome, the cognitive age, and chronological age may be quite different. Based on this, some have recommended a more personalized approach of assessing the capability of assent, though they acknowledge the subjectivity and difficulty in enforcing this approach [12].

REFUSAL OF CARE

As with research, parents generally have the responsibility and authority to make medical decisions and accept or refuse medical treatment for their children. There are many reasons why a parent or guardian, or even an adolescent patient, may refuse medical treatment including religious or cultural reasons, a weighing of recommended and alternative options against personal/familial values, a past experience with side effects deemed unacceptable, or possibly a misunderstanding or lack of understanding of the benefit or necessity for the recommended therapy. Respect for parent's decision-making authority is an important principle, but it must be weighed against the best interests of the minor child. There are some exceptions to parental authority in the case of medical neglect/abuse and or life-threatening situations that the courts have decided with varied outcomes.

When a parent refuses a recommended therapy, it is essential that the physician and other members of the interdisciplinary health-care team inquire into the reasoning behind that decision. This allows for a greater understanding of differences influenced by culturally motivated or faith-based decision-making and exploration of medically appropriate alternatives that are sensitive to those differences. This exploration also facilitates clarification of misunderstandings and opportunities to fill in knowledge gaps that may lead to an agreed upon treatment plan that will satisfy the needs of the patient. Appealing to the courts and or reporting to protective services should be reserved for very serious and rare cases when attempts at establishing a shared resolution through interdisciplinary discussion, mediation, and negotiation, potentially aided by ethics consultation, have reached an impasse. In most cases, good communication, compassion, and patience will lead to a mutually agreed upon medically beneficial solution. In the end, the health-care team is called to balance the respect for persons exercised through the autonomous choices of the parents and patient with the beneficence/nonmaleficence of *medical* best interests formulated with the aid of experience and expertise of the medical professional. This equation must take into account the *overall* best interests and hopefully result in a resolution that incorporates considerations of physical, psychosocial, emotional, and spiritual benefit and health.

Religious Objections

A refusal of treatment or some aspect of recommended effective therapy by a pediatric patient or his/her parents may be in direct opposition to what is available and the standard of care in the medical community. A pediatric health-care professional may be faced with the decision of whether to accept the parents' refusal of treatment. It is important to remember that informed consent, parental permission, and pediatric assent are a process. Permission to treat must be given or refused with full knowledge of the benefits, risks, and burdens of what is being offered and the alternatives to the proposed therapy.

Culture and/or faith can play a large role in how patients or their legal representative make medical decisions. Because culture is a combination of attitude, behavior, words, beliefs, perceptions, and values, we must take the time to understand these factors about our patients when dealing with their decision-making. To understand a patient's values in making a decision, the clinician must be culturally competent, have the desire to inquire and learn, and adapt his or her presentation. This will provide the necessary foundational information about the patient based on the new knowledge that is acquired to engage in a respectful conversation of how that culture may affect medical decision-making. Because there is typically a spectrum of practice in any community, it is dangerous to assume that all people of the same or similar faith or cultural tradition believe and practice in the exact same way. Obtaining information from this patient about her belief and how it frames her decisions is important. She may be following the teachings of a faith or her belief may be based on familial experience in which someone received the same therapy and his or her situation did not turn out well [13]. It may also be helpful to engage the assistance of a cultural or religious liaison, such as a chaplain, medical translator, or community representative for health care, to assist with the communication and mutual understanding. This may also help put particular tenets of the faith or culture into a larger perspective and avoid cherry-picking or misapplication [14].

Culture and faith likewise can shape the values of health-care providers, thereby affecting their decision-making processes and actions. It is particularly difficult to know how to react to a treatment refusal when a patient expresses a cultural or faith system that diverges from the clinicians. When the clinician attempts to give information to a patient with some understanding of his or her values, accepting the patient's refusal and remaining compassionate and respectful while addressing any symptom that can be managed under the limitations placed by the patient or parents may be necessary, even if you do not agree with the decision being made or the reasons for that decision [13].

It is beyond the scope of this chapter to be able to discuss the specific beliefs of the many different faith traditions or cultures that may at some point in a person's life affect medical decision-making. Most cultures and faith traditions have modern interpretations of what the teachings of the faith or cultural expectations allow in terms of certain types of medical treatment, such as customs or prohibitions, dietary practices, importance of family, gender and modesty, the cause of illness, beliefs regarding death and dying, beliefs regarding birth and naming, beliefs around diagnosis/prognosis, beliefs regarding autonomy/truth-telling and informed consent, and the use of traditional medicine and folk-medicine as complementary or alternative methods. These may translate into decisions about abortion or artificial fertilization; maintaining, withholding, or withdrawing life-sustaining measures, including ventilators, artificial nutrition and hydration, dialysis, and blood transfusions; organ donation; certain medications; and surgery.

The most common cases in the courts and the literature are Christian Scientists and Jehovah's Witnesses, particularly in pediatric medical cases. But religious objections to certain treatment options are by no means limited to these two communities, nor do certain objections always equate to rejection of any and all treatment. Health-care professionals should become familiar with the cultural and faith traditions of the patients that they treat and care for most often so that they can engage in culturally competent and respectful conversations about medical care for those patients. It is often possible to discover a mutually agreed upon solution that facilitates good medical care that benefits the whole person. Because the literature is so full of cases from these two specific traditions, we commend the reader to seek out more education regarding the teachings of these and others that may affect the medical treatment and care of their patients [15–22].

Legal Remedies

Court decisions regarding health care have varied over the years as evidenced in Table 4.1 [23]. There have been decisive rulings regarding medical neglect such as the Section 504 of the Rehabilitation Act of 1973 and the 1984 and 1985 amendments to the Child Abuse Prevention and Treatment Act, but even the interpretations of these laws have evolved and depend on the treatment and its effects and outcomes [24,25].

The 1985 federal regulations governing the treatment of severely handicapped infants, based on the 1984 amendments to the Child Abuse Prevention and Treatment Act, require that, except under certain specified conditions, all newborns receive maximal life-prolonging treatment. States can refuse the federal funds in question and thus be under no obligation to comply with the regulations. There was significant controversy about these regulations that lasted for years to even decades. In 1988, Kopelman et al. sent questionnaires to the 1007 members of the Perinatal Pediatrics Section of the American Academy of Pediatrics (AAP) to determine their views on the Baby Doe regulations and whether the regulations had affected their practices; 49% responded. Of the respondents, 76% believed that the current regulations were not necessary to protect the rights of handicapped infants; 66% believed that the regulations interfered with parents' right to determine what course of action was in the best interests of their children; and 60% believed that the regulations did not allow adequate consideration of infants' suffering. The responding neonatologists' concerns about the Baby Doe regulations were similar to those

TABLE 4.1 Various Results of Legal Remedies

Case Name	Case Description	Outcome
Jacobson v. Commonwealth of Massachusetts 197 U.S. 11, 37–39 (1905)	During an increasing epidemic of smallpox in Massachusetts, a seemingly healthy man is prosecuted for refusing to receive the smallpox vaccine and comply with the compulsory vaccination law mandating that all must receive the vaccine, meet the exception, or pay the fine for refusing the vaccine	Court holds that the statute was constitutional and the state was allowed to use its police powers to pass regulations that protect the welfare and health of its citizens, so the compulsory vaccination law was allowed to stand. The man was mandated to pay the fine for refusing the vaccine
Zucht v. King 260 U.S. 174, 176–177 (1922)	A child was excluded from public school because she refused to be vaccinated	Court allows the San Antonio city ordinance to stand and refuse children that were not vaccinated from attending their public schools
Prince v. Massachusetts 321 US 158, 170 (Mass. 1944)	A Jehovah's Witness woman is convicted for violating child labor laws by letting her niece sell pamphlets about the tenets of their faith on the street with her	Court affirms the conviction because parental rights can be limited when it is in the public interest, and religious freedom does not prevent the State from this limitation
Application of President and Directors of Georgetown College, Inc. 331 F. 2d 1000 (D.C. Cir. 1964)	A Jehovah's Witness loses two-thirds of her blood from a ruptured ulcer and is refusing life-saving blood transfusions because of her religious convictions.	Court granted the emergency writ because they act on the side of life, life-or-death situations don't allow much time, and if they refused to act, then found out they should've acted, it would've been too late
Jehovah's Witnesses in State of Washington v. King County Hospital Unit No. 1 278 F Supp 488 (WD Wash 1967), affirmed per curiam 390 US 598 (1968)	Jehovah's Witnesses in Washington brought a class action suit against various hospitals and doctors to prevent them from performing blood transfusions on them in the future because they had proven to do so in the past against expressed objections	Court affirms the holding in *Prince* and allows the state courts to continue to declare minors to be dependent for the purpose of giving blood transfusions to the minors against expressed objections by their parents
Custody of a Minor 379 NE2d 1053 (MA 1978)	Court requires a child's parents to allow the child to receive chemotherapy for leukemia	Court affirms removing custody from parents if they refused to allow the chemotherapy treatment on their child because the state has an interest in the welfare of children whose lives are being threatened by their parents refusing medical care for them
***Matter of Hofbauer** 393 NE2d 1109 (NY App. 1979) 47 N.Y.2d 658 1979	Parents were reported for neglect for pursuing alternative therapy for the treatment of Hodgkin's rather than standard therapy	Appellate Court upheld the parents' right to choose alternative therapy prescribed by a licensed doctor and dismissed neglect petitions

(*Continued*)

TABLE 4.1 (Continued)

Case Name	Case Description	Outcome
Walker v. Superior Court of Sacramento County 763 P2d 852, 860 (Calif 1988), cert denied, 491 US 905 (1989)	Mother is charged with involuntary manslaughter and felony child-endangerment after her 4-year-old daughter dies from acute meningitis because she provided spiritual treatment in lieu of traditional medical care	Court affirms the conviction claiming that the "spiritual treatment" exemption in the misdemeanor child neglect statute was not a defense to prevent being prosecuted for only using prayer treatment when the child is under life-threatening circumstances
In the Matter of Elisha McCauley 409 Mass. 134, 136, 565 N.E.2d 411 (1991), citing 375 Mass. 733, 379 N.E.2d 1053 (1978)	Hospital wanted authorization from the court to give a blood transfusion to an 8-year-old girl with suspected leukemia after her Jehovah's Witness parents refused	Court affirms the authorization given to the hospital to administer the blood transfusion and treat the girl for leukemia against her parents' wishes
****In re E. G.** 515 NE 2d 286, 549 NE 2d 322, Ill 1990	17-year-old Jehovah's Witness with leukemia, and her parents, refused blood transfusion	Trial judge determined patient to be mature enough to make this claim on her own and refuse as an adult. Decision was upheld on appeal
***Newmark v. Williams** 588 A. 2d 1108 (Del. 1991)	Child Protective Services sought temporary custody of a child needing chemotherapy because the child's Christian Scientists parents refused the treatment	Court reverses the temporary custody because the child was not being neglected when the parents refused a radical form of chemotherapy that only had a 40 % chance of being successful
State v. McKown 475 N.W.2d 63 (Minn 1991), cert denied, 112 S Ct 882 (1992)	Parents were indicted for second-degree manslaughter after their 11-year-old child with diabetic ketoacidosis died as a result of only Christian Science spiritual treatments being used	Court affirms its prior decision that the "spiritual treatment" exception in the child neglect statute does not protect parents from all criminal prosecution, but it dismisses the indictment because statute failed to give fair notice, which violates the parents' right to due process
Care and Protection of Beth 412 Mass. 188, 195 & n. 11, 587 N.E. 2d 1377 (1992)	Department of Social Services petitioned the court for substituted judgment to decide what further medical treatment should be given to an infant in an irreversible coma following a car accident	Court decides that enough evidence is shown to allow a do-not-resuscitate order in the event of respiratory or cardiac arrest because the infant is in an almost "brain-dead" condition that would be terminal
Matter of Hughes 259 N.J. Super. 193, 611 A. 2d 1148 (1992)	A 39-year-old Jehovah's Witness patient undergoing a hysterectomy signed forms expressing the desire to not receive any blood or blood products during her treatment. After emergency situation arises after surgery, the trial court allows the appointment of her husband as the decision maker, and he authorizes the necessary blood transfusions	The appellate court affirms this decision because surrogate decision makers should not withhold life-saving treatment unless the decision maker is "manifestly satisfied" that the patient would make the exact same decision in the particular situation

(Continued)

TABLE 4.1 (Continued)

Case Name	Case Description	Outcome
Banks v. Medical University of South Carolina 444 S.E.2d 519 (S.C. 1994)	Mother of an 8-year-old child, who died from pulmonary emboli, brought a wrongful death suit against the physicians and medical university for providing inadequate care and administering blood transfusions without parental consent	Court finds merit in some of the mother's claims, but rules that the physicians were not liable for administering the blood transfusions because parents do not have the right to withhold necessary medical treatment
***In re Rena** 46 Mass. App. Ct. 335, 705 N. E.2d 1155 (1999)	Trial court allows the hospital the authority to administer blood transfusions to a 17-year-old Jehovah's Witness patient with a lacerated spleen if a life-threatening traumatic event occurred. Parents and patient had previously made clear that they would object to the use of blood products	Court decides that the trial court erred in allowing this authorization because the court relied on the representations of the patient's parents, did not determine the patient's maturity to make informed decisions, and did not hear the patient's testimony about her preferences
In re K.I. 735 A.2d 448 (D.C. 1999)	Medical guardian was granted the request to issue a "do not resuscitate" order with respect to a neglected child who was born prematurely, who suffered from serious medical problems, and who was in comatose state. The biological mother appealed	Court held that the trial court correctly applied the best interests of the child standard, rather than substituted judgment standard, with respect to a "do not resuscitate" order, and the child was properly assessed as a neglected child with the medical guardian being the appropriate decision maker over the parents
***In re Maxin** JU124198 (Court of Common Pleas, Stark County, OH 2002)	Parents were reported to Department of Job and Family Services for neglect and court petitioned to mandate standard chemo therapy; Holistic physician reported to State Medical Board	Court upheld parents' right to choose alternative therapy prescribed by a licensed doctor and monitoring for relapse; charges against licensed family medicine holistic physician determined to not be warranted
***Virginia v. Cherrix 2007**	The Commonwealth of Virginia sued to force 16-year-old boy to undergo further conventional medical treatment for relapsed Hodgkin's disease. The *patient* rejected any further use of chemotherapy or radiation because of the side effects. His parents supported his choice and were subsequently accused by the state of medical neglect of their child. The lower court decided against the parents	The decision was overturned on appeal and the parties reached a compromise in a consent decree, in which Cherrix would receive treatment from a board-certified specialist of Cherrix's choice for alternative therapies. The case resulted in a new law that increased the rights of patients aged 14–17 years in Virginia to refuse medical treatment

(*Continued*)

TABLE 4.1 (Continued)

Case Name	Case Description	Outcome
State v. Neumann 2013 WI 58, 348 Wis. 2d 455, 832 N.W.2d 560 *cert. denied*, 134 S. Ct. 544, 187 L. Ed. 2d 389 (2013)	Parents were convicted with second-degree reckless homicide after their 11-year-old dies from diabetic ketoacidosis resulting from only receiving spiritual treatment	Court holds that the "spiritual treatment" exception in the child abuse statute does not give immunity from other criminal prosecution. It also holds that the due process fair notice requirement is met when the second-degree reckless homicide statute, criminal child abuse statute, and "spiritual treatment" exception are read together
In re Cassandra C. 316 Conn. 476, 112 A.3d 158 (2015)	Mother and 17-year-old cancer patient appeal the decision to allow temporary custody to Department of Children and Families after the refusal to undergo chemotherapy	Court does not find that the patient meets the "mature minor" determination and allows the department to continue to make all medical decisions for the patient

**Decided in parents' favor*
***Decided in favor of minor patient*
We would like to extend a special thanks to Whitney Morgan for her assistance researching the current status of the Baby Doe regulation and other cases found in the table.

expressed by the United States Supreme Court in rejecting an earlier set of Baby Doe regulations. The authors contended that, for that reason, the Baby Doe regulations of 1985 should be reevaluated [24]. A number of modifications were made in the second version of the rule. The requirement to treat handicapped newborns was modified so that clearly futile therapies "which would merely prolong an infant's process of dying" were exempted. Furthermore, the new rule stated that "reasonable medical judgment" would be respected regarding the choice of a treatment plan. The Department of Health and Human Services (DHHS) also encouraged hospitals to establish infant care review committees to review treatment decisions for handicapped newborns [26]. In June of 1986, the American Medical Association, the American Hospital Association, and the American College of Obstetricians and Gynecologists challenged DHHS's final rule in *Bowen v. American Hospital Association et al.* [27]. The United States Supreme Court held that Section 504 of the Rehabilitation Act [28] does not apply to situations where the parents of a handicapped infant refuse to consent to life-saving treatment of their child.

In Report D (A-91), the AMA Council on Judicial and Ethical Affairs concluded that when a patient has never been competent, decisions regarding life-sustaining treatment should be guided by the best interest standard. The best interest standard requires a weighing of the benefits and burdens of treatment options including nontreatment as objectively as possible. Factors that should be considered when making decisions about life-sustaining or life-saving treatment for a seriously ill newborn include (1) the chance the therapy will succeed, (2) the risks involved with treatment and nontreatment, (3) the degree to which the therapy if successful will extend life, (4) the pain and discomfort associated with the therapy, and (5) the anticipated quality of life for the newborn with and without treatment. Appropriate considerations about an infant's quality of life do not include any form of assessment of the patient's social worth or value to other persons, such as the infant's parents. An infant's quality of life must be evaluated from the infant's perspective. Individual decision makers must avoid making decisions for the infant based on what they would want for themselves. This task can be more difficult than it may appear at first [26].

The current Baby Doe Amendment [29] became effective as of January 3, 2012 and was last approved on August 7, 2015. There have been few changes made from the initial Baby Doe rule under Section 504. Enforcement of the current regulation is still based on the threat of loss of federal funding in public hospitals. Current text of the Statute reads: 42 U.S.C. § 5106a (West).

(C) an assurance that the State has in place procedures for responding to the reporting of medical neglect (including instances of withholding of medically indicated treatment from infants with disabilities who have life-threatening conditions), procedures or programs, or both (within the State child protective services system), to provide

1. coordination and consultation with individuals designated by and within appropriate health-care facilities;
2. prompt notification by individuals designated by and within appropriate health-care facilities of cases of suspected medical neglect (including instances of withholding of medically indicated treatment from infants with disabilities who have life-threatening conditions); and
3. authority, under State law, for the State child protective services system to pursue any legal remedies, including the authority to initiate legal proceedings in a court of competent jurisdiction, as may be necessary to prevent the withholding of medically indicated treatment from infants with disabilities who have life-threatening conditions.

Decisions are less consistent when the condition is not immediately life-threatening, the treatment course is a prolonged one, or treatment offers limited benefits or has significant side effects. Chemotherapy is one example of treatment decisions that are not assured in the courts [30,31].

Physician's Responsibility and Role

The role of the physician is no longer as simple as the expert sole decision maker for a patient in need of medical treatment. In the age of patient and family autonomy trending back to a moderate position of shared decision-making, the physician's role is rather, more often than not, a partner with medical expertise and experience working in cooperation with the parents of a minor child who are together making decisions on behalf of or with, depending on age, the patient. Often respect for autonomy and beneficence may appear to come in conflict when the physician and the parent do not agree on the best course forward or the most appropriate treatment plan for the patient for various reasons. Likewise, difficult situations such as these might cause concerns regarding professional integrity and ethical/legal obligations to the patient, cultural competence and sensitivity, or irrevocably damaging a physician/patient relationship that might further endanger the ongoing and future health of the minor patient due to lost trust in the physician or the health-care system.

The AAP and the International Society of Pediatric Oncology (SIOP) assert that the physician has an obligation to seek legal remedies to disagreements that put a child's life or well-being in danger of serious harm, when a treatment to prevent such exists [32–34]. Health-care providers bare a significant weight in protecting children from medical neglect and are obligated by professional standards and state laws to report the situation to the authorities [14]. The AAP in particular and to some extent the SIOP take a strong stance against religious exemption laws that endanger the health or life of minor children who are not able to exercise their own autonomy due to lack of capacity or legal competence. They are not only in favor of legal accountability of the parents for such actions but also recommend community education programs to combat or prevent religious objections or decisions for other reasons that might be considered medical neglect or child endangerment. These strong positions are provided along with the guidance to be flexible and sensitive toward families who hold religious beliefs that may lead them to refuse or limit appropriate medical treatment [31,33,34]. See Table 4.2. Additional information and resources can be found on the Children's Health Care is a Legal Duty website, a national organization for information on current laws, exemptions, and descriptions of religious doctrines [25].

This expectation to seek state involvement and or legal recourse is a foundational precept in pediatric medicine owing to the understanding and position that "medical caretakers have an ethical and legal duty to advocate for the best interests of the child when parental decisions are potentially dangerous to the child's health, imprudent, neglectful, or abusive." These decisions should be challenged through respectful discussion, ethics consultation, and finally reporting to and intervention by state child protective agencies and the courts [36]. It is essential that the responsible physician do so with great caution and consideration because this action can have profound effects on the patient's care, the ongoing relationship with the patient and her parents, and the emotional state of each of the individuals involved as well as the other members of the family and the health-care team.

Parents' Responsibility and Role

Parents have the responsibility and authority to make decisions including medical decisions such as accepting or refusing medical treatment for their children. They are generally considered the best judge of what is in their child's best interests and are permitted to make decisions in concordance with their values. That being said, they are not allowed to sacrifice the lives of their children, who they are supposed to protect, for religious or any other reason before the child is legally able to make that decision for themselves [23,25,31,37]. There are many reasons why a parent or guardian, or even an adolescent patient, may refuse medical treatment including physical discomfort, fear of side effects or disfigurement, misunderstanding or uncertainty regarding the implications of the disease or the treatment, inadequate information or poor communication, desire to pursue alternative therapies due to cultural or religious beliefs or mistrust in the medical system, financial or other socioeconomic reasons, lack of education about or access to resources, anxiety and emotional turmoil, or simply becoming tired and depressed by ongoing therapies [25,34].

TABLE 4.2 Professional Guidelines Regarding Physicians' Obligations to Protect Minors

American Academy of Pediatrics—1997 Committee on Bioethics [33]	American Academy of Pediatrics—2013 Committee on Bioethics [31]	Childhood Cancer International (formerly ICCCPO) SIOP Working Committee on Psychosocial Issues in Pediatric Oncology [35]
The AAP calls for all those entrusted with the care of children to	Pediatricians, pediatric medical subspecialists, and pediatric surgical specialists should	
1. Show sensitivity to and flexibility toward the religious beliefs and practices of families	1. Respect families and their religious or spiritual beliefs and collaborate with them to develop treatment plans to promote their children's health	1. Facilitate open, honest, and thorough level of communication with all patients, especially adolescents and their families
2. Support legislation that ensures that all parents who deny their children medical care likely to prevent death or substantial harm or suffering are held legally accountable	2. Report suspected cases of medical neglect to state child protective services agencies, regardless of whether the parents' decision is based on religious beliefs	2. Identify factors predictive of a tendency toward noncompliance even at the time of diagnosis
3. Support the repeal of religious exemption laws	3. (and the AAP and its chapters) Work to repeal religious exemptions to child abuse and neglect laws and to prevent public payment for religious or spiritual healing practices	3. Addressing reasons of personal finances: The health-care team can help prevent noncompliance in such families by (a) providing or assisting with housing near the hospital, (b) Helping to find assistance with the needs of the other family members (babysitting, cooking meals), (c) giving financial help for transportation, (d) keeping flexible hours for medical visits and appointments, (e) insisting on and implementing continuous communication between all family members and the entire health-care team, and (f) facilitating support groups
4. Work with other child advocacy organizations and agencies and religious institutions to develop coordinated and concerted public and professional action to educate state officials, health-care professionals, and the public about parents' legal obligations to obtain necessary medical care for their children		4. Requests for an otherwise medically valid treatment which the physician has already ruled out as not being appropriate for that particular patient: Listen to the parents' request with respect and provide a reasoned exposition of the plan formulated to achieve the best therapy and care. Parents and patients should be allowed and even encouraged to seek a second opinion whenever they have doubts about the appropriateness of the therapy being offered

(Continued)

TABLE 4.2 (Continued)

American Academy of Pediatrics—1997 Committee on Bioethics [33]	American Academy of Pediatrics—2013 Committee on Bioethics [31]	Childhood Cancer International (formerly ICCCPO) SIOP Working Committee on Psychosocial Issues in Pediatric Oncology [35]
		5. Remain nonpatronizing and open to discussion when alternative medical treatments are proposed by parents or patients. When the physician cannot dissuade parents [from pursuing nonbeneficial or potentially harmful alternative methods], judicial intervention may be called for. Careful attention should be paid to the parents' expressed religious or cultural value system. Often an alternative [or complementary] treatment does not significantly interfere with the prescribed treatment and is in itself medically and financially harmless. Physicians should in such cases encourage the parents to ***supplement*** conventional therapies with their own cultural and/or religious approaches
		6. Assure the parents and patients that no matter what else happens, the medical staff will take care of the child until the end. Even during the palliative phase when cure is no longer possible, physical and emotional support remain as important as ever
		7. When communication breaks down and it becomes a matter of life or death for the child, the medical team may need to pursue judicial intervention. When parents adamantly refuse treatment or completely replace the medically sound measures with ineffectual alternative medicine, it may be necessary to resort to legal proceedings to override the parents' at once, and thus insure that the best interests of the child are served
		**It is important in such an event to discuss the reasons for this decision

(Continued)

TABLE 4.2 (Continued)

American Academy of Pediatrics—1997 Committee on Bioethics [33]	American Academy of Pediatrics—2013 Committee on Bioethics [31]	Childhood Cancer International (formerly ICCCPO) SIOP Working Committee on Psychosocial Issues in Pediatric Oncology [35]
		with the child at his developmental level of understandingIt is also important to help the child to retain trust in the parents, by assuring all the family members that the judicial intervention is specific and limited only to the medical treatment
Reproduced with permission from PEDIATRICS Vol. 99 No. 2 Pages 279–281	Reproduced with permission from PEDIATRICS Vol. 132, No. 5, Pages 962–965	Reproduced with permission from Med Pediatr Oncol. 2002;38(2):114-7 John Wiley and Sons, Publishers

***This statement is not part of the original cited text from which this column was adapted.*

As a society, we permit parents to make choices that may appear to stray from the medical recommendation when the indications for treatment are less defined and the prognosis is uncertain or poor. When the treatment is clear and the prognosis is good or excellent, we expect parents to follow the medically and ethically sound recommendations of the health-care professionals because we assume that is what is in the patient's best interests and that the parents and physicians would agree on this point. But this expectation takes a lot for granted. There must be agreement on the concept of good or excellent prognosis. Is this determination of best interest simply a medical calculation or does it take into account psychosocial, emotional, and spiritual well-being and best interests? Does it consider quality of life? By whose determination of quality? Does it consider cultural and spiritual expectations and prohibitions of the adherent community member and their consequences? Does the equation include the financial burden of the treatment? Does this estimation of best interests evaluate the effect on the integrity of the family or the consequences for the community in which the child lives, who may be caring for him and financing his care? Many would argue that a determination of medical neglect does not and should not consider anything more than the physical effect on the individual patient [30], but real situations are rarely so simple. These are difficult decisions to make and there are many important factors that must be considered. The AAP reminds health-care providers to consider the value systems of patients and families beyond prognosis alone, maintaining the essential role of the parents in shared decision-making balanced with the best interests and safety of the child [14,30,33].

Patient's Responsibility and Role

For years, there has been a growing trend toward allowing adolescents more autonomy in health-care decision-making. This trend began in large part due to what might be referred to as an epidemic of untreated venereal diseases in young people not seeking medical care because most physicians would not treat without parental permission. Laws were enacted to allow treatment of minors for venereal disease, then drug and alcohol abuse [38]. The ability of minors to legally consent to health services including sexual and reproductive health, mental health services, and drug and alcohol abuse treatment has continued to expand. This expansion is due to recognition that many teens may not seek treatment if required to involve their parents in decision-making and permission. The Guttmacher Institute keeps an up-to-date listing of state minor consent laws [39]. In addition to consent by minors for treatment of specific issues, many states have more general "mature minor' laws that recognize the developing maturity of adolescents to participate in or solely direct their medical care, along with other aspects of their lives, as if they were adults. The determination of such usually requires that the minor be at least 14 years old and has demonstrated a level of understanding and medical decision-making that approximates that of an adult, the treatment is of benefit to the minor and does not present a great risk and is within established medical protocols. Some states allow a physician to make this determination but most require a judicial determination [36,40]. Furthermore, several states have provisions for emancipated minors to consent to care as adults. The emancipated minor is one who is married, in the military, or is living at home or separately but is self-supporting. Many states consider an unmarried minor mother emancipated [36,38].

In the case of a minor refusing medical treatment, the situation is less defined [41]. Most physicians recognize the importance of including the minor patient in an age appropriate manner in decisions that affect the minor's health but still defer to the parents as the legal decision maker. The research of Talati, Lang, and Ross confirmed their hypothesis and the general trend that "'pediatricians' decisions whether to respect treatment refusals for minor patients are multifactorial. When prognosis is good, best interest dominates. When prognosis is poor, parental authority is more important in younger minors, and minor autonomy is more important in older minors" [41]. One important shift to note is the passing of "Abraham's Law" in Virginia that states "that parents of a child at least 14 years old with a life-threatening condition could refuse medically recommended treatment (i.e., it would not be considered neglect) provided 1. the parents and child made the decision jointly 2. the child is sufficiently mature to have an informed opinion on the treatment 3. other treatments have been considered, and 4. they believe in good faith that their choice is in the child's best interest." Although the court has some discretion to determine that a child is not sufficiently mature to understand, the law makes it possible for a 14-year old with a life-threatening illness to choose not to accept the recommended treatment or to pursue other options [43]. Some might argue, as does Ross, that respecting the autonomy of an adolescent to make medical decisions such as refusing treatment before they have fully developed their decision-making skills, goals, and values, we are denying that minor the right to fully develop into a mature, autonomous, independent adult and that we deny the obligation to protect patients from harms due to their immaturity or lack of perspective [43]. Most pediatric ethicists argue that the wishes of older children regarding their medical care should be taken seriously and honored. If disagreement between parties exists, attempts to resolve the conflict through respectful discussions should occur. It may be beneficial to request assistance from other members of the interdisciplinary team such as patient advocacy, a child-life specialist, psychologist, or social worker, chaplaincy or the ethics consultation service [36].

In general, pediatric health-care providers do recognize the developing maturity of minors and encourage the participation of children with the developmental ability to understand what is happening to them in decisions about their care. As their decision-making capacity matures, their voice should be heard and weighed and they should be given an opportunity to assent to (or refuse-dissent from) treatment. The elements of assent are well known and clearly outlined by the American Academy of Pediatrics Committee on Bioethics [44]. In situations in which the patient will have to receive medical care or investigational procedures despite his or her objection, the patient should be told that fact and should not be deceived. The child should be informed of the procedures to be performed and the reasoning behind that decision, unless there is a compelling reason not to do so. Good communication is an essential and continuing process that often leads to greater adherence to medical recommendations.

Involving children in decisions that affect their own welfare offers positive benefits to those children. Among other things, it allows children to obtain needed practice and skills in making responsible decisions, contributes to their perception that they have some control and influence over their lives, and results in a greater sense of self-esteem and competence, while reducing anxiety and fear of the unknown. Factors such as the patient's emotional and psychological readiness and willingness to engage in the informed consent process in addition to age and maturity should be weighed when determining the patient's ability to make independent or semi-independent decisions. One should also consider capacity in relation to the complexity of the decision and developmental regression from disease or hospitalization versus contextual experience from the same, as well as variability across individuals, families, and cultures. Not every 16–19-year old is equally ready to make decisions and some 12–14-year olds may be mature enough to have a fully informed opinion that should be honored.

ETHICAL ISSUES AT THE END OF LIFE FOR CHILDREN WITH CANCER

Approximately 20% of children with cancer eventually die of their disease or complications of their treatment, making the provision of palliative care to children with advanced cancer an expected standard of care [45]. In order to truly understand the scope of ethical issues that children and their families face at the end of life, it is crucial to understand where the majority of children with cancer will die.

Recent research has examined all deaths in children ages 0–19 years of age from 1989 to 2003. During that time period, approximately 20% of children died at home [46]. However, for children and adolescents with cancer, there was an increasing trend for death to occur at home, rising from 28% in 1989 to 41% by 2003. Despite this move toward a home death, the children with cancer may die in a hospital or at home, so health-care professionals need to be aware of the variety of ethical issues that children and families face at the end of life in either circumstance.

Prognostic Disclosure

Health-care providers often do not provide full prognostic disclosure due to the fear of taking away parental hope. It is often assumed that "hope" is synonymous with "cure," and while every parent desires his or her child to be cured, usually a more nuanced and sophisticated cognitive process is occurring that is unspoken. Research supports the idea that parents who receive a greater number of elements of prognostic disclosure are more likely to report communication-related hope, even when the likelihood of a cure was low [47]. Parents will often reframe hope as the desire that their child not suffer unnecessarily. This is seen in another recent study that demonstrated that higher levels of parental hopeful patterns of thinking are significantly associated with increased odds of limiting medical or surgical interventions [48]. Parents' most fundamental hopes are rooted in the desire that their child is cared for and does not suffer and possessing more information about prognosis almost uniformly allows parents to make the most informed decision they can for their child.

"Don't tell my child they are dying!"

One of the most difficult decisions parents of seriously ill children face is whether they should talk with their child about their poor prognosis and impending death. Pediatrics spans an enormous range of developmental capacities to comprehend suffering, loss, and death (Table 4.3), and conversations must take into account their developmental understanding of death.

The varying ability of children to conceptualize death combined with themes commonly seen in adult populations about prognostic disclosure—cultural norms about death and dying, expected rapidity/trajectory of decline, adaptive or maladaptive coping strategies used by the caregiver/parents—only serve to make the decision about disclosure in pediatrics even more complex.

Health-care professionals tend to comply with parents' well-intentioned, but potentially harmful, requests for nondisclosure regarding the child's prognosis for many reasons. Often this decision is rooted in respect for parental decision-making authority. However, nondisclosure is usually not beneficial because children are likely aware that they are dying, despite lack of concrete, direct communication. Unfortunately, without direct communication and concrete answers children often fear and assume the worst but find it difficult to communicate this fear, particularly to the family, because they too want to protect others from "the truth."

Parents who make the courageous and loving decision to inform their child of their prognosis and inevitable death often seek guidance about how to speak with their child. It should therefore be within the skill set of a Pediatric Oncology or Pediatric Palliative Care physician to assist and guide parents. In general, published guidelines stress the

TABLE 4.3 Typical Developmental Understanding of Death

Infants to 2 years of age: Children do not have cognitive capacity to understand death
2–6 years of age: Death is a reversible phenomenon, which children do not personalize
7–12 years of age: Death is final, personal, and unpredictable
Teenagers: Death is final and universal but distanced

importance of open and honest communication with the child [49,50] but the idea that information and prognosis should be foisted upon a child in one fell swoop does not account for the developmental capacity of a child to absorb, process, and comprehend information. It is for that reason the information should always be "titrated" to the child. It is paramount to consider the child's developmental understanding of their illness and to compassionately ask the parents what the child already understands about their illness and prognosis. When presented within an empathetic framework, parents often provide valuable insight about how best to share the information with the child. It is also important to provide "anticipatory guidance" about what worries and fears the child may have. Children are often concerned about their changing physical appearance, the loss of their physical or cognitive abilities, concern about being forgotten after they die, experiencing inadequately treated pain, and how their death will inflict further emotional distress in those they will leave behind, specifically their parents, siblings, and friends [51].

For families insistent on not informing their child in a developmentally appropriate manner of the severity of their illness, it is often helpful and informative to tell the parents that despite their wishes to protect their child from this information, children are often aware of how critically ill they are. Children are exquisitely sensitive to external cues such as interpreting the body-language of physicians, the shifting tones of conversations and changes in communication patterns (i.e., abruptly changing from conversations in the room including the child to conversations outside of the room without the child) and the overall mood of their parents. When you combine this with children's keen ability to discern whether they feel healthy or ill, it should not be surprising that even without direct verbal communication children often are aware that they are dying. As documented in the seminal work "The Private Worlds of Dying Children," almost all children universally progress through a series of stages that reflect cognitive understanding of their illness and its implications [52]:

1. I am seriously ill.
2. I am seriously ill and will get better.
3. I am always ill and will get better.
4. I am always ill and will never get better.
5. I am dying.

For most families, reinforcing the idea that many other families have made the loving and caring decision to speak in a developmentally appropriate way with the child about death is reassuring. It may also be helpful for them to know that research shows that no parent who talked with their child about their death regretted doing so; conversely 27% of those parents who did not speak frankly about imminent death regretted not having done so. Furthermore, among those who did not talk with their children about death, 47% of parents who sensed their child was aware of imminent death felt regret after their child died [53]. While most guidance and counseling for disclosure is framed in the context of how honesty and forthrightness is beneficial for the child, this study highlights the potential harm nondisclosure may have on parents after the child's death.

In summary, the best way to prevent the harm that can come from withholding information is education—including the child in age- and developmentally appropriate discussions about health-care decision making from the beginning is highly recommended.

Advance Care Planning

Far too often, however, good decision-making regarding quality end-of-life care is undermined and precluded by the lack of dialogue around advanced care planning. Sadly, health-care providers often identify the barriers to these conversations squarely within the domain of the parents rather than as failures of communication that are within their control. A survey of pediatric oncology and critical care physicians and nurse practitioners found that the top three barriers to advanced care planning in seriously ill children are as follows [54]:

1. The parents are "unrealistic."
2. The parents "do not understand the poor prognosis."
3. The parents "are not ready" to have a conversation.

While 92% of health-care providers surveyed felt that advanced care planning should take place at the time of diagnosis or during a period of stability, 60% report that it actually takes place during an acute illness or when death is clearly imminent [54]. Educational interventions that improve provider communication skills may help better advanced care planning and quality end-of-life care for children.

Facilitating conversations with parents about the end of life is crucial for many reasons. These complex, emotional, and difficult discussions have been shown to improve parental satisfaction with the care their child received [55,56], helps parents feel more prepared for their child's death [57], and improves bereavement outcomes [57–59].

Faith and cultural traditions as well as prior personal or familial experience can influence specific medical interventions, such as initiating, withholding, or withdrawing life-sustaining measures. Decisions to withhold or withdraw life-sustaining treatments are often difficult to make, particularly in the pediatric setting. This difficulty may be eased slightly with the benefit of an ongoing long-term relationship with the patient and his or her family and time to weigh the decision to limit life support based on medical circumstances, if considered in advance. However, these decisions undeniably are an integrated part of medical activity. Physicians are under no ethical obligation to provide or maintain treatments they judge to be of no benefit to patients, but making that judgment can be difficult in the first few moments of a life-threatening emergency. Once the patient has been stabilized, assessment of his or her medical condition, underlying disease, and cause of acute deterioration may lead to the determination that withdrawal of treatment is an appropriate option.

In Western bioethics, withholding and withdrawing nonbeneficial treatments hold equal weight in the abstract because either of them, when appropriately applied, allows death to occur naturally owing to the underlying condition. In practice, however, they can feel very different. Some physicians may feel that withholding an intervention is more appropriate than withdrawing one in progress because withdrawal could be interpreted as participating in or hastening the patient's death. Others believe that a stronger argument exists for initiating treatment in an emergency situation and withdrawing it if appropriate when more information is available and can be weighed carefully.

Often, assessment of beneficial treatment yields different outcomes according to the medical team and patient or surrogate/guardian. This may result from different estimations of acceptable quality of life or definitions of benefit and burden or harm. Quality of life and burden of treatment are socially defined concepts that are determined by the patient, and in this case the patient's parents. The patient's/family's preferences for or against treatment based on these indicators should be respected within the bounds of medically and ethically appropriate options. The pediatric health-care professional respects the principle of nonmaleficence by always seeking to maximize the benefits of treatment and minimize the risk of harm.

Euthanasia, Physician-Assisted Suicide, and Palliative Sedation

It is almost unbearable for a parent to bear witness to the death of their child. In fact, in one study approximately 1 out of every 3 parents would have considered hastening death for their child under certain circumstances, with uncontrolled pain being the most common reason [60]. Hastening death in a child would universally be viewed as euthanasia, which is defined as the "compassion-motivated, deliberate, rapid and painless termination of the life of someone afflicted with an incurable and progressive disease" [61].

The United States federal government, all 50 states and the District of Columbia prohibit euthanasia under general homicide laws. The distinction between physician-assisted suicide or aid in dying and euthanasia is based in who administers the fatal agent, the patient or the physician. The federal government does not have laws pertaining to assisted suicide, and in general these cases are handled at the state level [62,63]. With respect to adults, currently four states—Oregon (1997) [64], Washington (2009) [65], Vermont (2013) [66], and California (2016) [67]—have legalized physician-assisted suicide via legislation or public referendum while one state, Montana (2009) [68], permits, or more correctly does not prohibit, physician-assisted suicide via court ruling. Several states have attempted to pass similar laws or appeal relevant court cases to the State Supreme Court. To this point, children have never played a role in any legal process involving assisted suicide. It is important to distinguish the prohibited acts of both euthanasia and physician-assisted suicide in pediatric patients from medically and ethically appropriate decisions to withhold or withdraw medical interventions that are not expected to be or are no longer beneficial (medically, emotionally, or psychologically) to the patient or effective in reversing their condition. Doing so may have a secondary effect of allowing natural death to come faster than if those interventions were to continue in a state that often would be considered prolonging death or dying.

Palliative sedation, on the other hand, is an ethical intervention used to treat refractory symptoms [69] such as pain, dyspnea and delirium, and is distinct from euthanasia. The difference between a "difficult" symptom and a "refractory" symptom can be confusing but a clear definition [70] that is generally accepted by the medical community is provided in Table 4.4.

TABLE 4.4 Attributes of a Refractory Symptom

1. Intensive efforts short of sedation fail to provide relief
2. Additional invasive or noninvasive treatments are incapable of providing relief
3. Additional therapies are associated with excessive or unacceptable morbidity, or are unlikely to provide relief within a reasonable time frame

It is important to highlight that palliative sedation therapy is distinct from physician-assisted suicide and euthanasia. Palliative sedation has the intent to provide symptom relief, is a proportionate therapeutic intervention and the death of the patient is not the intended goal (as is the case with assisted suicide and euthanasia). Legally, the United States Supreme Court has upheld a physician's ability to provide treatment for refractory symptoms in terminally ill patients even when the patient may die during the process of delivering such care [69].

Death may occur during palliative sedation, but there have been numerous studies documenting that palliative sedation does not hasten death [71]. From an ethical standpoint, the Doctrine of Double Effect provides the ethical foundation to justify the practice. Originally developed by Catholic theologians, the Doctrine of Double Effect has been adopted by the legal, bioethical, and medical communities. It states that when an action has two possible outcomes—one inherently good, the other inherently bad—that the action is justifiable when certain criteria are met [72]:

- The act in itself, apart from the circumstances and consequences, must not be morally wrong.
- The agent must not intend the bad effect as a means to be sought; they are merely foreseen and tolerated.
- The bad effect must not be a means to the good effect; and the good effect must not be achievable without the indirectly intended foreseen bad effect.
- The good effect must be equal to or greater than the bad effect and the good effect must follow the action at least as immediately as does the bad effect.

Palliative sedation therapy should always be explicitly discussed with all staff members, including the nursing staff, pharmacy staff, physicians as well as the patient (if developmentally appropriate) and family members before starting.

Withholding Nutrition and Hydration

Forgoing or discontinuing artificial nutrition and hydration (ANH) is supported by many medical professional associations including the American Medical Association [73] and AAP [74], but it is not without strong, differing opinions [75,76]. It is paramount that pediatric providers be aware of the deeply rooted, universal, and sacred importance of providing food and liquids to seriously ill children. For parents, providing nutrition—and its perceived benefits—to their child is one of the last things that is within their control; as such, it can often be a tenacious issue between health-care providers who want to discontinue ANH and families who want it continued.

The ethical dilemma of continuing or discontinuing ANH can often be resolved by straightforward, open, and clear communication between the health-care providers and families. When explained that ANH can cause suffering in the form of fluid overload, electrolyte abnormalities, dyspnea, or clearly prolonging the dying process [77], families often understand and acknowledge that they do not want their child to suffer unnecessarily and will agree with the recommendation to discontinue ANH.

CONCLUSION

Ethical dilemmas are borne out of conflicts of values and principles, the resolution of which reasonable people may disagree. This may manifest in competing patient/guardian and physician autonomies; competing principles of autonomy versus beneficence, which ideally must be balanced; differing goals of care; or conflicting definitions of beneficence.

Ethical dilemmas encountered during the course of treatment for a pediatric patient involve additional layers of complexity. One such layer is the need for parents or surrogate decision makers to make medical decisions on the minor's behalf, due to the patients' developmental immaturity. Determining the best interests of a patient who cannot participate or cannot fully participate in the decision without biasing that determination toward the decision maker(s) own wishes can be quite complicated. Likewise, eliciting the patient's own definition of the good/best interests when they often lack the understanding of a mature adult can be difficult. Some of the more common ethical issues that provide additional challenges beyond the obvious efforts to treat a child diagnosed with cancer include consent to participate in research studies, better described as patient assent and parental permission; refusal of care by the minor patient or their legal decision maker; and emotional end-of-life decision-making in young patients.

When time allows, exploration of factors that affect medical decision making and the resultant spectrum of practice may aid shared decision-making or elucidate the reasons behind peculiar or conflicting decisions. When potentially inflammatory disagreement regarding initiating, continuing, withholding, or withdrawing medical treatments exists, it should be challenged through respectful discussion, ethics consultation, and finally, as a last resort, reporting to and intervention by state child protective agencies and the courts. It is essential that the responsible physician do so with great caution and consideration for the profound effects on the patient's care, the ongoing relationship with the patient and his/her parents, and the emotional state of each of the individuals involved.

Good communication is essential in every health-care encounter particularly when a parent or guardian must make health-care decisions at the end of life on behalf of a small child or guide and share in the decisions of an adolescent/young adult. Possessing more information about prognosis almost uniformly allows parents to make the most informed decision they can for their child. Parents' most fundamental hopes are rooted in the desire that their child is cared for and does not suffer. Honest, compassionate communication can facilitate decisions that do not destroy hope yet honor the well-being of the young patient.

References

[1] The United States Code of Federal Regulations: US Government Publishing Office. Available from: <http://www.hhs.gov/ohrp/humansubjects/guidance/45cfr46.html#subpartd>; 2009 [21.9.2015].

[2] Schechter T, Grant R. The complexity of consenting to clinical research in phase I pediatric cancer studies. Paediatr Drugs 2015;17(1):77–81.

[3] Shah S, Whittle A, Wilfond B, Gensler G, Wendler D. How do institutional review boards apply the federal risk and benefit standards for pediatric research? JAMA 2004;291(4):476–82.

[4] Westra AE, Wit JM, Sukhai RN, de Beaufort ID. Regulating "higher risk, no direct benefit" studies in minors. Am J Bioethics: AJOB 2011;11(6):29–31.
[5] Clarifying Specific Portion of 45 CFR 46 Subpart D that Governs Children's Research.
[6] de Vries MC, Houtlosser M, Wit JM, Engberts DP, Bresters D, Kaspers GJ, et al. Ethical issues at the interface of clinical care and research practice in pediatric oncology: a narrative review of parents' and physicians' experiences. BMC Med Ethics 2011;12:18.
[7] Dekking SA, van der Graaf R, Kars MC, Beishuizen A, de Vries MC, van Delden JJ. Balancing research interests and patient interests: a qualitative study into the intertwinement of care and research in paediatric oncology. Pediatr Blood Cancer 2015;62(5):816–22.
[8] Appelbaum PS, Roth LH, Lidz CW, Benson P, Winslade W. False hopes and best data: consent to research and the therapeutic misconception. Hastings Center Rep 1987;17(2):20–4.
[9] Chappuy H, Baruchel A, Leverger G, Oudot C, Brethon B, Haouy S, et al. Parental comprehension and satisfaction in informed consent in paediatric clinical trials: a prospective study on childhood leukaemia. Arch Dis Childh 2010;95(10):800–4.
[10] Baker JN, Leek AC, Salas HS, Drotar D, Noll R, Rheingold SR, et al. Suggestions from adolescents, young adults, and parents for improving informed consent in phase 1 pediatric oncology trials. Cancer 2013;119 (23):4154–61.
[11] Wendler DS. Assent in paediatric research: theoretical and practical considerations. J Med Ethics 2006;32 (4):229–34.
[12] Waligora M, Dranseika V, Piasecki J. Child's assent in research: age threshold or personalisation? BMC Med Ethics 2014;15:44.
[13] Gallagher C, Moore J, Farroni J. Cancer care ethics in the emergency center. In: Mansullo EF, Gonzalez CE, Escalante CP, Yeung S-CJ, editor. Oncologic emergencies. MD Anderson Cancer Care Series. New York: Springer; 2016.
[14] Orr RN, William E, Perkins, Ronald M. Faith-based decisions: parents who refuse appropriate care for their children. AMA J Ethics (Virtual Mentor) [Internet] 2003; 5(August):[3 p.]. Available from: http://journalofethics.ama-assn.org/2003/08/ccas1-0308.html [accessed 7.5.2015].
[15] Network for the Advancement of Transfusion Alternatives. Blood transfusions, Jehovah's witnesses and the American patients rights movement: network for the advancement of transfusion alternatives. Available from: <http://www.nataonline.com/np/405/blood-transfusions-jehovahs-witnesses-and-american-patients-rights-movement>; 2012 [accessed 9.9.2015].
[16] Galanti G-A. Caring for patients from different cultures. 4th ed. Philadelphia: University of Pennsylvania Press; 2008.
[17] Tanenbaum Center for Interrelious Understanding. The medical manual for religio-cultural competence: caring for religiously diverse Populations. First ed. New York: Tanenbaum Center for Interrelious Understanding; 2009.
[18] Remmers PA, Speer AJ. Clinical strategies in the medical care of Jehovah's witnesses. Am J Med 2006;119 (12):1013–18.
[19] Swan R. Faith healing, Christian science, and the medical care of children. N Engl J Med 1983;309 (26):1639–41.
[20] Relman AS. Christian science and the care of children. N Engl J Med 1983;309(26):1639.
[21] Talbot NA. The position of the Christian science church. N Engl J Med 1983;309(26):1641–4.
[22] P.B. Perry, S. Sorajjakool, R. Yelland, K. McMillan. Health care and religious beliefs. Loma Linda, CA: Loma Linda University Medical Center. Available from: <http://lomalindahealth.org/media/medical-center/departments/employee-wholeness/healthcare-religious-beliefs.pdf>.
[23] Patsner B. Faith versus medicine: when a parent refuses a child's medical care. In: UoHL Center, editor. Health law perspectives. Houston, TX: Health Law & Policy Institute; 2009.
[24] Kopelman LM, Irons TG, Kopelman AE. Neonatologists judge the Baby Doe regulations. N Engl J Med 1988;318(11):677–83.
[25] Linnard-Palmer L, Kools S. Parents' refusal of medical treatment based on religious and/or cultural beliefs: the law, ethical principles, and clinical implications. J Pediatr Nurs 2004;19(5):351–6.
[26] American Medical Association Council on Ethical and Judicial Affairs, Treatment decisions for seriously ill newborns. Contract No.: CEJA Report I-A-92; 1992.

[27] Bowen v. American Hospital Assn., 84 U.S. 1529; 1986.
[28] 29 U.S.C. § 701–797.
[29] Chapter 67 – Child Abuse Prevention and Treatment and Adoption Reform: Grants to States for programs relating to investigation and prosecution of child abuse and neglect cases. Sect. 5106c.; 2012.
[30] Antommaria AHM, Collura CA, Antiel RM, Lantos JD. Two infants, same prognosis, different parental preferences. Pediatrics 2015;135(5):918–23.
[31] American Academy of Pediatrics Committee on Bioethics. Conflicts between religious or spiritual beliefs and pediatric care: informed refusal, exemptions, and public funding. Pediatrics 2013;132(5):962–5.
[32] Hord JD, Rehman W, Hannon P, Anderson-Shaw L, Schmidt ML. Do parents have the right to refuse standard treatment for their child with favorable-prognosis cancer? Ethical and legal concerns. J Clin Oncol: Off J Am Soc Clin Oncol 2006;24(34):5454–6.
[33] American Academy of Pediatrics Committee on Bioethics. Religious objections to medical care. Pediatrics 1997;99(2):279–81.
[34] Spinetta JJ, Masera G, Eden T, Oppenheim D, Martins AG, van Dongen-Melman J, et al. Refusal, non-compliance, and abandonment of treatment in children and adolescents with cancer. A report of the SIOP Working Committee on Psychosocial Issues in Pediatric Oncology. Med Pediatr Oncol 2002;38 (2):114–17.
[35] SIOP Working Committee on Psychosocial Issues in Pediatric Oncology, Guidelines for refusal, non-compliance, and abandonment of treatment in children and adolescents. Available from: http://www.childhoodcancerinternational.org/guidelines-for-refusal-non-compliance-and-abandonment-of-treatment-in-children-and-adolescents/; 2002 [updated 11.3.2011; cited 8.9.2015].
[36] Diekema D. Parental decision making. Washington: University of Washington School of Medicine. Available from: <http://depts.washington.edu/bioethx/topics/parent.html>; 2014 [cited 9.9.2015].
[37] Prince v. Massachusetts. US; 1944. p. 158, 70.
[38] Holder AR. Childhood malignancies and decision making. Yale J Biol Med 1992;65(2):99–104.
[39] Guttmacher Institute, An overview of minors' consent law. Washington, DC: Guttmacher Institute. Available from: <http://www.guttmacher.org/statecenter/spibs/spib_OMCL.pdf>; 2015 [cited 9.9.2015].
[40] US Legal.com, Right to refuse lifesaving treatment—special considerations regarding children: US Legal, Inc. Available from: <http://death.uslegal.com/right-to-refuse-lifesaving-treatment/special-considerations-regarding-children/#sthash.tZcalUXg.dpuf>; 2014 [7.5.2015].
[41] Talati ED, Lang CW, Ross LF. Reactions of pediatricians to refusals of medical treatment for minors. J Adolesc Health 2010;47(2):126–32.
[42] Deleted in review.
[43] Mercurio MR. An adolescent's refusal of medical treatment: implications of the Abraham Cheerix Case. Pediatrics 2007;120(6):1357–8.
[44] American Academy of Pediatrics Committee on Bioethics. Informed consent, parental permission, and assent in pediatric practice. Pediatrics 1995;95(2):314–17.
[45] SEER Cancer Statistics Review, 1975–2012 Bethesda, MD: National Cancer Institute; 2015 [20.8.2015].
[46] Feudtner C, Feinstein JA, Satchell M, Zhao H, Kang TI. Shifting place of death among children with complex chronic conditions in the United States, 1989–2003. JAMA 2007;297(24):2725–32.
[47] Mack JW, Wolfe J, Cook EF, Grier HE, Cleary PD, Weeks JC. Hope and prognostic disclosure. J Clin Oncol: Off J Am Soc Clin Oncol 2007;25(35):5636–42.
[48] Feudtner C, Carroll KW, Hexem KR, Silberman J, Kang TI, Kazak AE. Parental hopeful patterns of thinking, emotions, and pediatric palliative care decision making: a prospective cohort study. Arch Pediatr Adolesc Med 2010;164(9):831–9.
[49] Freyer DR. Care of the dying adolescent: special considerations. Pediatrics 2004;113(2):381–8.
[50] Goldberg A, Frader J. Holding on and letting go: ethical issues regarding the care of children with cancer. Cancer Treat Res 2008;140:173–94.
[51] Hinds PS, Schum L, Baker JN, Wolfe J. Key factors affecting dying children and their families. J Palliative Med 2005;8(Suppl 1):S70–8.
[52] Bluebond-Langner M. The private worlds of dying children. Princeton, NJ: Princeton University Press; 1978, xv, 282 p.

[53] Kreicbergs U, Valdimarsdottir U, Onelov E, Henter JI, Steineck G. Talking about death with children who have severe malignant disease. N Engl J Med 2004;351(12):1175–86.
[54] Durall A, Zurakowski D, Wolfe J. Barriers to conducting advance care discussions for children with life--threatening conditions. Pediatrics 2012;129(4):e975–82.
[55] Mack JW, Wolfe J. Early integration of pediatric palliative care: for some children, palliative care starts at diagnosis. Curr Opin Pediatr 2006;18(1):10–14.
[56] Mack JW, Hilden JM, Watterson J, Moore C, Turner B, Grier HE, et al. Parent and physician perspectives on quality of care at the end of life in children with cancer. J Clin Oncol: Off J Am Soc Clin Oncol 2005;23 (36):9155–61.
[57] Dussel V, Kreicbergs U, Hilden JM, Watterson J, Moore C, Turner BG, et al. Looking beyond where children die: determinants and effects of planning a child's location of death. J Pain Sympt Manage 2009;37(1):33–43.
[58] Meert KL, Thurston CS, Sarnaik AP. End-of-life decision-making and satisfaction with care: parental perspectives. Pediatr Crit Care Med 2000;1(2):179–85.
[59] Meert KL, Thurston CS, Thomas R. Parental coping and bereavement outcome after the death of a child in the pediatric intensive care unit. Pediatr Crit Care Med 2001;2(4):324–8.
[60] Dussel V, Joffe S, Hilden JM, Watterson-Schaeffer J, Weeks JC, Wolfe J. Considerations about hastening death among parents of children who die of cancer. Arch Pediatr Adolesc Med 2010;164(3):231–7.
[61] Roy DJ, Rapin C-H. Regarding euthanasia. Eur J Palliative Care 1994;1(1):1–4.
[62] Washington v. Glucksberg. US: Supreme Court; 1997. p. 702.
[63] Vacco v. Quill. US: Supreme Court; 1997. p. 793.
[64] Or. Rev. Stat. §§ 127.800–.897; 1994.
[65] Rev. Code Wash. §§ 70.245.010–.903; 2008.
[66] Act 39: An act relating to patient choice and control at end of life (Bill S.77 "End of Life Choices"); 2013.
[67] End of Life Option Act (ABX2–15); 2015.
[68] Baxter v. Montana, 2009 MT 449. Mont. LEXIS 695; 2009.
[69] Burt RA. The Supreme Court speaks—not assisted suicide but a constitutional right to palliative care. N Engl J Med 1997;337(17):1234–6.
[70] Cherny NI, Portenoy RK. Sedation in the management of refractory symptoms: guidelines for evaluation and treatment. J Palliative Care 1994;10(2):31–8.
[71] Maltoni M, Pittureri C, Scarpi E, Piccinini L, Martini F, Turci P, et al. Palliative sedation therapy does not hasten death: results from a prospective multicenter study. Ann Oncol 2009;20(7):1163–9.
[72] Reich WT. Encyclopedia of bioethics. New York: Free Press; 1978.
[73] American Medical Association Council on Ethical and Judicial Affairs, Decisions near the end of life. JAMA 1992;267(16):2229–33.
[74] American Academy of Pediatrics Committee on Bioethics. Guidelines on foregoing life-sustaining medical treatment. Pediatrics 1994;93(3):532–6.
[75] Solomon MZ, Sellers DE, Heller KS, Dokken DL, Levetown M, Rushton C, et al. New and lingering controversies in pediatric end-of-life care. Pediatrics 2005;116(4):872–83.
[76] Feltman DM, Du H, Leuthner SR. Survey of neonatologists' attitudes toward limiting life-sustaining treatments in the neonatal intensive care unit. J Perinatol 2012;32(11):886–92.
[77] Diekema DS, Botkin JR, Committee on Bioethics,. Clinical report—forgoing medically provided nutrition and hydration in children. Pediatrics 2009;124(2):813–22.

CHAPTER

5

Ethical Issues in Cancer Survivorship

Maria Alma Rodriguez, Paula Lewis-Patterson and Guadalupe R. Palos

The University of Texas MD Anderson Cancer Center, Houston, TX, United States

INTRODUCTION

Oncology health care providers will encounter ethical challenges when caring for patients in our complex medical world. Cancer survivorship, a distinct phase of the cancer experience, is governed by the same ethical challenges as other facets of clinical care. Thus, questions of ethical decision-making present themselves in the dynamic process of survivorship care. The following statement provides a brief but thorough definition and approach to ethical decision-making in caring for cancer survivors:

> Medical ethics has been defined as a process for identifying the principles at stake, evaluating relative obligations of the clinician's towards potentially competing outcomes or interests, and providing moral justification for choosing one path or the other [1].

Major advances in the detection and treatment of cancer have contributed to the growing number of cancer survivors, many of whom return to a healthy and productive life. To date, there are approximately 14 million cancer survivors living in the United States with a projected increase to 18 million by 2022. This trend is due to the growth of an aging population with a higher incidence of malignancies and, thus, increase in cancer survivors [2]. The survivor population ranges from being an individual with no evidence of disease to a person who will to live with the disease for the remainder of their life [3].

The 2005 Institute of Medicine's (IOM) seminal report, *From Cancer Patient to Cancer Survivor: Lost in Transition*, brought attention to the critical need for cancer survivorship care [4]. This report became the roadmap for the delivery of comprehensive and multidisciplinary survivorship care. The IOM report also defined key components of survivorship care, including surveillance for new/recurrent cancers, management of late or long-term effects, risk reduction/prevention, and monitoring psychosocial functioning. A core tenet of survivorship care introduced in the report was that coordination of care between

Ethical Challenges in Oncology.
DOI: http://dx.doi.org/10.1016/B978-0-12-803831-4.00005-1

primary care providers and oncologists would contribute to a better quality of life for survivors, as well as improved health.

Since the release of the IOM report, empirical evidence related to survivorship care has grown exponentially with literature on models of care, clinical practice guidelines, health promotion interventions, and several other topics. In clinical practice, providers have integrated much of this evidence into their practice with cancer survivors. Although, this growth in published evidence has contributed to better care of cancer survivors, there is a gap in the availability of information related to ethical challenges encountered in survivor–provider interactions. Several trends in the United States society such as the Affordable Care Act, substantial increase in the cost for cancer care, and growing demand for patient-centered communication will increase the need to focus on the ethical nuances of survivorship [5–8]. Health care providers in any type of clinical settings or practice will need to prepare themselves to address ethical challenges when caring for cancer survivors.

The following is a composite case which reflects ethical dilemmas often encountered in clinical practice with cancer survivors.

Vignette 1: I Can Beat This Cancer

Mr. R is a 72-year-old male patient who presents with weight loss, increased hoarseness and difficulty in swallowing, which has been present over the past year. He was diagnosed and treated successfully for head and neck cancer in your facility ten years ago. This patient lives in a small city with limited health care resources, including access to providers specializing in medical oncology. Although, he lives quite a distance from the clinic, he faithfully returned to his primary oncologist for annual follow-up visits the first 5 years. At the fifth visit, he informed the clinic staff he would not be returning since he had been cured of his cancer, reached his five-year anniversary of no cancer, and had difficulty with transportation to the clinic. Records were sent to his primary care physician, who specialized in internal medicine and he was assured that he could return if any new problems arose.

Five years later, the patient arrives alone and without notice to the clinic asking for an appointment. He states he wants to see his original oncologist, who is still on staff, and an appointment is made. His work-up indicates the presence of a nodule on his left vocal cord, which comes back positive for cancer and will require extensive surgery. His previous surgeon is no longer with the practice and he is referred to a new surgeon. He does not like the surgeon and informs you he is going back home to look for his own doctor to do the surgery. He is confident that he can beat this cancer like he did with his first diagnosis. What are your obligations to communicate the risks associated with the complex surgery needed and the benefits in having a surgeon who specializes in this field?

This vignette of a patient's experience raises multiple issues unique to the survivorship experience. First, it illustrates that a cancer survivor's experience can last a number of years as in this patient's case. He had lived cancer-free for ten years with a good quality of life. Although he initially returned for regular follow-up care, it is not clear what type of follow-up cancer care was obtained once he stopped returning to the oncology clinic. This issue reflects the *temporality of the survivorship trajectory*. Second, the patient has moved through the various stages of survivorship. His experience began with the *acute phase,* which included his initial diagnosis and treatment, and then he entered the *intermediate stage*, when he finished his curative treatment and lived in remission for a number of years. After 5 years, it appeared he had entered into the *long-term phase* of survivorship, where he had passed the highest risk of cancer (i.e., his 5-year anniversary) and could focus on maintaining his health [3]. This patient had entered the long-term phase, but with his new symptoms, he re-entered the cancer trajectory. A third issue that confounds the survivorship experience is the level of risk a person has to a relapse or secondary effect related to their diagnosis or treatment. This characteristic calls for *risk stratification* and is critical in determining follow-up care for the survivor in terms of surveillance, risk reduction, late effects management, and monitoring psychosocial functioning [9]. Because this patient was previously diagnosed with a head and neck cancer, his outcomes in treating this second cancer would be based on his previous and current risks. The conflict evident in this vignette revolves shared decision-making to achieve an equitable balance between the wishes of the patient (i.e., autonomy) and the concerns of the oncology team regarding the patient's decision to beat his cancer (i.e., risk versus benefit).

These unique characteristics of the cancer survivors' journey raise ethical issues for the health care provider related to communication, self-determination, autonomy, and truth-telling. Smith and Bodurtha identified four ethical tenets fundamental to oncology care: (1) General principles of ethics, (2) conflict of interest, (3) risk management, and (4) patient decision-making at the bedside [10]. In addition, these authors propose beneficence, respect, and justice are three basic principles that must be used to guide oncology providers in responding to ethical dilemmas. The complexity of these issues and their impact on an individual's life reflect the types of ethical issues encountered when caring for cancer survivors [4,11,12].

In this chapter, we will provide an overview of who is a cancer survivor and basic concepts of cancer survivorship. Next, we will focus on areas of bioethics and how they relate to survivorship. Then we will discuss the impact of late and long-term effects on the survivorship trajectory. Throughout the chapter, case scenarios will be used demonstrate how ethical theories are linked to daily clinical practice within a survivorship paradigm.

WHO IS A SURVIVOR AND WHAT IS SURVIVORSHIP?

The terms cancer survivor and survivorship have evolved over time. According to Susan Leigh, a long-time cancer survivor and advocate, the insurance industry defined a cancer survivor as the spouse or family member of the patient who had died [13]. Interestingly, in the medical community, the term "cancer survivor" referred to a person who had lived up to 5 years with no evidence of disease. She also describes the work of the National Coalition for Cancer Survivorship (NCCS) to define a survivor as a person who had lived through the trajectory of the cancer experience: From the time of the initial cancer diagnosis and for the balance of life.

To date, inconsistency in the definition and use of these terms continues to exist. One contributing factor evolves from the worldview or context in which the person is using the term, i.e., a survivor, a clinician, or a caregiver. For example, the oncology world defines survivorship in both a qualitative and quantitative manner. Table 5.1 compares "survivorship" from both perspectives, and it must be emphasized that there is no absolutely right

TABLE 5.1 Quantitative and Qualitative Interpretations of the Survivorship Experience

Qualitative	Quantitative
Acute	Focuses on time of survival, i.e., 5 or 10 years
Extended	Refers to a phase or stage when curative treatment is completed
Permanent	Defines an outcome of the treatment, i.e., no evidence of disease

Adapted from Rodriguez MA, Zandstra F. Models of survivorship care. In: Foxhall L, Rodriguez MA, editors. Advances in cancer survivorship management. New York: Springer; 2014; Mullan F. Seasons of survival: reflections of a physician with cancer. New Engl J Med 1985;313(4):270-3; and Leigh, S. Cancer survivorship: a nursing perspective. In: Ganz P, editor. Cancer survivorship: today and tomorrow. New York: Springer; 2007. p. 8–13.

or wrong way to define these two terms or concepts. The first column stems from the "Seasons of Survival" published in 1985 by the *New England Journal of Medicine* [11]. The author, a cancer patient and physician, interpreted survival as a progressive journey from which a patient traveled through in a linear pattern. The second column uses a perspective from the oncology community and focuses on the outcome of the treatment rather than on the experience. In our case study, the patient with recurring cancer, used both terms to describe himself as a survivor. He used the quantitative perspective of 5 years with no evidence of disease and the qualitative to describe his current status as a "long-term survivor."

SURVIVORS AND PROVIDERS WORLDVIEWS TOWARD ETHICS AND SURVIVORSHIP

A person's worldview toward life and death are comprised of numerous layers of complex phenomena. These worldviews must be considered in any clinical encounter between a survivor, their family, and the providers. These worldviews increase the risk of cultural clash when ethical issues occur. In every clinical encounter, worldviews represent a patient, their family or caregiver, and their team of providers and interact to form "explanatory models" used to interpret illness, health, or wellness [14]. Published literature suggests that these perspectives could influence patient decision-making toward a diagnosis of cancer and its treatment [14,15]. Klucholm expanded on this model by suggesting worldviews are based on the following five elements: (1) Human nature—how people view humanity; (2) man and nature—how people view themselves in relation to nature; (3) time—how individuals view the past, present, and future; (4) activity—how people view being and doing; and (5) relational—how people view their interpersonal social relationships with family and other social support networks [15].

From an ethical perspective, a person's worldview shapes how people understand and respond to ethical conflicts. A person's worldviews or explanatory models can interact, overlap, contradict, or compete with one another. For example, truth telling about the recurrence of a cancer delivered by a health care provider can clash with the worldviews of family members who use a different value system to communicate bad news. The systems of explanatory models can have a significant role in establishing effective communication among patients, families, and providers.

Worldviews are critical aspects to consider when ethical situations arise, particularly because of the changing multicultural demographics of the United States [16]. When assessing and resolving ethical conflicts, the entire survivorship care team must take the survivors' multicultural worldviews into account. In addition, providers must also assess their own personal and professional cultural worldviews and their impact on their clinical care and decisions. Ethical conflicts occur in survivorship care when there is conflict among the worldviews, values, and principles of survivors, their families, and their providers.

ETHICAL CONSIDERATIONS IN SURVIVORSHIP CARE

Published literature is readily available on the ethical dilemmas that occur in the world of oncology, particularly during ongoing cancer care or at end of life [10,17]. In fact, Chapter 1, Ethics and Organizational Identity in a Cancer Care Setting, of this book provides an overview of ethical principles and other chapters address those specific various types of situations in which ethical conflict may occur. Because the survivorship experience encompasses the entire trajectory of a person's cancer journey, ethical issues will arise at different time points and for different reasons. Factors that contribute to ethical conflicts include social and economic inequalities, the rising cost of cancer care, the evolving field of oncology, i.e., human genomics and management of late effects due to the disease or treatment. Across these domains, core principles such as of truth-telling, justice, beneficence, and autonomy provide a roadmap for decision-making when ethical conflict occurs in clinical practice. The following section discusses the impact of specific factors when working with cancer survivors.

IMPACT OF SOCIOECONOMIC VULNERABILITY ON ETHICAL DECISION-MAKING

The high cost of cancer treatment financially burdens most cancer survivors. Not only is the cost of cancer treatment itself an issue, but data show that the after-treatment health expenses of cancer survivors are also disproportionate to those of age matched cohorts in the general population. The economic burden of the survivors' health care costs or coverage is further augmented by the burden of productivity losses, and in many cases by loss of employment, or by failure to integrate to the workforce if unemployed at the time of diagnosis. These factors combined make survivors and their families vulnerable to social and economic inequality. The ethical principle of justice is, thus, very much a concern, as there are no articulated meaningful strategies at present to address these issues of social inequality. In addition, survivors who are financially distressed are more likely to forego necessary medical and other health interventions, potentially resulting in patient harm.

Among the 1556 adult cancer survivors who participated in the 2010 National Health Interview Survey (United States) in 2010, 32% reported distress related to financial burdens. Statistical analyses of the data indicate that this burden was disproportionately high in those diagnosed at a younger age, members of a minority group, those with a history of more complex treatments, including chemotherapy and/or radiation, as well as those with higher tumor complexity (relapse or multiple cancers). Those with financial distress also had a significantly higher reported rate of either foregoing or delaying overall medical care, prescription medications, eyeglasses, dental care, and mental health care [18].

Two reports based on data from the medical expenditure panel survey (MEPS), which is sponsored by the Agency for Healthcare Research and Quality, supported the results of the NHIS. An initial report on data from 2008 to 2010 looked at the annual excess economic burden/person of those with a diagnosis of cancer. The cohort was stratified by age >64 years versus 18–64 years, and time from diagnosis >1 year versus 1 year or less. The results indicated that compared to age matched cohorts in the general population, the excess financial burden/person for recently diagnosed cancer survivors was $16,213 for those aged 18–64 years and $16,441 for those >64 years. For the group beyond one year of diagnosis (i.e., past acute treatment phase of survivorship), excess economic burden/person/year was $4427 for the younger subset and $4519 for those 65 years or older [19]. Guy et al. reported that the two main sources of payment were private insurance and Medicare (in those 65 years or older). However, the amount paid by private insurers varied across groups due to changes in employment or disability-related eligibility.

The Centers for Disease Control and Prevention (CDC) in 2014 issued a follow-up report, based on the 2008–11 MEPS, analyzing the yearly excess cost of health care/person as well as productivity loss for cancer survivors, this time stratifying male versus female survivors. The results indicated that both men and women who had a diagnosis of cancer had higher financial health care costs and productivity losses compared to age and gender matched cohorts in the general population. For men, the out of pocket health care costs and productivity losses/person respectively were $4187 and $1459, when compared to the matched cohort group. For women, these figures respectively were $3293 and $1330. Not surprisingly, persons with a diagnosis of cancer are the most likely among persons with critical illnesses to declare bankruptcy due to health care costs. Hence, in addition to the physical and emotional burdens of the disease itself, there are also anticipated economic losses for cancer patients and their families even beyond the cancer treatment costs, which can have major consequences for those who have less financial resources [20].

The CDC report recommended that health and employment intervention programs be developed with the aim of improving the overall outcomes of cancer survivors and their families, including socioeconomic wellness. There is little research, however, and relatively few interventions that have been studied or developed to address employment and economic outcomes of cancer treatment or cancer survivorship efforts. The NCI recently convened a meeting of experts to address a number of questions relating to cancer economics and health services research. One of the key recommendations from that meeting was that health economists should be included in the design and analyses of cancer care and survivorship intervention studies, to better evaluate cost burden and economic outcomes [21].

Another reason for economic distress among cancer survivors is the loss of employment after a cancer diagnosis, or difficulty in entering the workforce if unemployed prior to the cancer diagnosis. In women, aged 28–54 who were not working when first diagnosed, their rate of entry into the workforce 2–6 years out from diagnosis was reported as 12% less than age matched women without a cancer diagnosis. Even when employed, their work hours/week averaged 5 hours less than their counterparts [22]. Thus, it is likely that the cancer survivors were less likely to work full-time and, thus, had less income from employment than their peers.

Of those who are employed prior to the cancer diagnosis, most wish to return to work but often find it difficult to do so for a number of reasons. This was the general finding of a meta-analysis that included 39 qualitative studies published between 2010 and 2013 on cancer survivors' experiences of the process of returning to work. Synthesizing their results, there were nine key factors identified that related to the success of returning to work, within three major categories: Person-related [21]; environmental supports [20]; and occupation [19]. Examples of person-specific factors such as motivation, coping skills, ability to work, and post-treatment symptoms. Environmental factors included family support, workplace support, and a professional career. The occupation related factors were the demands on physical stamina made by the type of work and job flexibility, i.e., whether adjustment to duties was allowed [23].

The environment and flexibility of the workplace is very important in the success of those returning to work. A survey by the Center of Excellence for Public Health of Northern Ireland, issued to the country's employment advisors (50% response rate), found that of those who responded the majority (74%) had been consulted by a cancer survivor on the issue of returning to work. The most frequently reported barrier the survivors reported to returning to work, or staying at work, was fatigue, while the single most important factor in facilitating their return to work or continuing to work was a supportive employer. The counselors, however, reported feeling ill prepared on how to best advice the cancer survivors [24]. This highlights one of the issues discussed at the previously mentioned NCI sponsored meeting, and that is the lack of well-studied strategies or interventions to improve cancer survivors' socioeconomic well-being.

There are, however, new ongoing studies with hopefully meaningful results on the horizon. In Denmark, the Oncology Department of Aarhus University Hospital has partnered with job consultants in the municipalities of Silkeborg and Randers to conduct an intervention study. The strategy is to preemptively prepare cancer survivors who are deemed in remission to reintegrate to the work force after treatment. The intervention includes motivational, behavioral, and other therapeutic modalities, as well as job placement tools, to assist in reintegrating the patients to employment. The efficacy of the interventions will be measured by the rate of employment of the participants compared to patients from other municipalities not participating in this intervention [25]. A similar randomized study is currently underway in the Netherlands. In this study, cancer survivors who have been unemployed for 12–36 months will be randomized to either an intervention program of rehabilitation and psychosocial support/counseling or no intervention. The primary endpoint will be time it takes to achieve a sustainable return to work [26].

Hopefully, these studies will yield results to minimize or even ameliorate the vulnerability of survivors to socioeconomic inequality. However, knowledge alone is not sufficient to correct the problem. Dedicated resources, polices, and employment legislation will likely be needed to implement truly proactive programs that reintegrate and recover patients to productivity, and that help them avoid economic distress all are factors which will relate to principles of justice and equity in how survivors will access these programs and resources.

ETHICAL CONFLICTS IN LATE EFFECTS DISCUSSIONS

Most cancer treatments have acute effects that manifest during the process of active therapy, and these are usually well described in the literature and are disclosed as part of the consent process and during treatment planning. The late or latent effects of treatment, however, are less predictable and in many cases even unknown. Hence, these are often not discussed with patients at the time of treatment planning, with some exceptions, for example, risk of loss of fertility in patients of reproductive age. However, as the number of survivors has significantly increased over the past four decades, we have learned more about certain patterns of late effects. As a result, the disclosure of late effects is a current subject of discussion in the cancer survivorship literature, more specifically when and to whom are late effects disclosed, for which treatments, and for which genetic risks. The rapidly evolving field of human genomics, as well as the development of new cancer therapeutic agents, makes this topic one where maintaining an up-to-date base of new information is imperative. Nonetheless, there are well observed and noted risks for certain late effects, and the patients' right to know as well as the right to self-determination, are the ethical principles at the core of the issue of disclosure, communication, and education of patients and families about these potential complications.

The most robust body of information on late effects stems from the literature of childhood cancer survivors. As the possibility and hope of long-term survival after cancer therapy for children with certain malignancies, such as leukemia, became a reality in the 1970s and 1980s, the concept of survivorship care and close observation post therapy, emerged within the field of pediatric oncology. The best known late effects, therefore, have been observed and reported in survivors of childhood cancer who are now adults. The risks of radiation, depending on site and dose of radiation, as well as effects of certain chemotherapeutic agents have been learned from long-term observational studies of children growing into adulthood. Among these may be diminished skeletal growth, premature bone loss, hormonal deficiencies, premature cardiac disease manifestations, sterility, and second primary malignancies, among others. Predictive models that take into account treatment categories, age of the patient at treatment, sites treated, and familial history among other factors can be beneficial in planning strategies for early diagnosis and management, as well as life style changes that could potentially benefit the patient [27–29].

The degree to which information is disclosed and the type of information disclosed to children is a complex issue, as it could have implications for the child's psychological well-being. Thus, most adverse treatment information is disclosed to the parents, and not necessarily the child, depending on the child's age. However, as the child grows into adulthood, there will be a time when discussion and disclosure of information will be appropriate. In a recently published study by investigators from the University of Oslo, young adults who were survivors of childhood lymphoma were asked about their preferences of how and when to be informed of potential late effects of the treatment they'd received as children. Most did not favor disclosure as adolescents, unless the late effect was manifesting and it needed to be explained. They otherwise wanted to be educated in adulthood, around age 25, of potential late effects as well as life style strategies to prevent or help them manage the late effects. Furthermore, they wanted this information delivered both face-to-face and in writing. They saw their survivorship experience as a trajectory but preferred a "step by step" introduction to the topic of late effects [30].

For adult cancer patients, the question of what information about late effects must be disclosed is less complex psychologically than in children or adolescents. However, it is more difficult in that there are overlapping factors that could influence the risk of late effects, such as the patients' age, overall health, lifestyle, disease complexity, and treatment complexity. There are, nonetheless, certain patterns of late effects that are more prevalent and, thus, potentially predictable. A recently published systematic review of risk models for late effects found 14 studies that met the reviewers' criteria for clinically applicable information. They identified nine late effect syndromes that appeared consistently in the models and could occur after treatment of the following malignancies: Prostate cancer—the possible persistent and/or late effects were erectile dysfunction and urinary incontinence; breast cancer—cardiomyopathy and cardiac events, psychological distress, and arm lymphedema; head and neck cancer—swallowing dysfunction; Hodgkin's lymphoma—breast cancer; and childhood malignancies—thyroid cancer [9].

A commonly reported side effect by patients even long after treatment is completed, and that surprisingly was not noted in the review, is fatigue. While fatigue is a very general symptom and can be caused by many factors, it can be a debilitating symptom even in young patients [31]. It is very difficult to manage, and can have significant effects on the survivors' ability to return to work as well. Another general and frequently reported symptom by patients, sometimes questioned as whether it was "real," is now being studied more in depth thanks to functional MRI (fMRI) technology, and that is the so called "chemo-brain" or "chemo-fog" phenomenon. In a cohort of breast cancer survivors who had been treated with chemotherapy and radiation up to 11.5 years before the studies were done, 6%–10% had a pattern of task-related hypo-activation in prefrontal and parietal areas on fMRI compared to control cohorts [32]. Thus, we are still learning about the physiology of toxicities that may persist for many years and that until recently we have only known from the patients' subjective experience. We also have yet to learn about management of late effects that may be masked or exacerbated by comorbid conditions, particularly in the older adult survivors [33].

IMPROVING OUTCOMES THROUGH CARE PLANS AND COORDINATION

It is important to understand the more common and frequent late effect risks to the health of cancer survivors, so that discussion and education on potential strategies to prevent or preemptively treat these conditions can be part of their plan of care. The intent of care plans for survivors is aligned with the core medical ethical principle of beneficence, meaning we hope for positive or beneficial outcomes. Recent studies have reported that when cancer survivors receive a summary of the key recommendations for their care, i.e., the care plan, they are more likely to keep timely medical appointments and follow cancer screening recommendations [34]. Whether the care plans will translate to improvement in health and a lesser health cost burden for the patients, however, is yet to be confirmed.

Another important element in the process of developing the plan of care for survivors is coordination of care with their primary care and other providers who may be participating in their care. Poor care coordination was one of the findings of the IOM report, *Lost in Transition: From Cancer Patient to Cancer Survivor*. This finding led to the recommendation that a summary care plan would help communicate and coordinate care among all providers [4]. To date, there is limited published evidence that care plans improve patient care coordination. Nonetheless, care plans provide a systematic process to follow and communicate with survivors about strategies to self-manage their health and wellness.

EMERGING ETHICAL CONFLICTS IN SURVIVORSHIP CARE THROUGH VIGNETTE CASE STUDIES

In this section, ethical challenges in survivorship care will be examined through a series of case studies. The reader will note that a common theme across all case studies is the need to provide communication, arrive at decisions, and understand the decisions as tailored to the needs of each survivor.

Vignette 2: A Survivor's Freedom to Say No

In May 2008, a 66-year-old African-American female presented to the breast surgery clinic with a newly diagnosed, high-grade, 3 centimeter mass in her left breast. A segmental mastectomy was performed with recommendations for re-excision, lymph node sampling, and radiation chemotherapy. The patient declined follow-up recommendation and was referred to the integrative medicine service for nonpharmacologic treatment options. The patient declined to schedule a return appointment to the surgery clinic, but she did agree to schedule an appointment in one year.

Two years later, the patient returned for a follow-up visit in the breast survivorship clinic. Her clinical exam was negative for masses or other changes in the breast; yet, her mammogram revealed a 1 centimeter mass located at the scar site with associated calcification and suspicion of reoccurrence of her disease. The patient declined additional diagnosis evaluation. She informed the staff she did not believe further evaluation was warranted since she could not feel a lump or see any changes in her left breast. During this visit the survivorship, physician provided extensive counseling which included discussion about the high probability of reoccurrence of breast cancer. The physician also discussed the implications of delaying treatment and the possibility of death if no further medical action was taken. The physician asked the patient for permission to talk to a family member. The patient agreed that her daughter could be contacted and informed about her condition. Despite encouragement from her daughter and the medical team, the patient continued to decline follow-up work, treatment, and to schedule an appointment. She repeatedly informed the staff that when she was ready, she would schedule an appointment.

One year after these conversations, the patient returned to the clinic after finding a palpable mass in the left breast. A recent mammogram and ultrasound indicated a highly suspicious mass for malignancy and an indeterminate lymph node in the left axillary region for which biopsies were recommended. Further evaluation revealed a 3.4 cm mass, which was positive for recurrent high-grade, triple negative breast cancer. The patient refused a referral to the breast medical oncology department to discuss neoadjuvant chemotherapy. She was uncertain about her decision to consider adjuvant chemotherapy. Yet, she was adamant about her decision not to receive any neoadjuvant therapy.

The patient, however, agreed to have a total mastectomy but declined further surgical management and chemotherapy. After extensive discussion and encouragement, the patient agreed to return for radiation treatment. She expressed reservations regarding chemotherapy and radiation therapy due to logistical issues. The patient was referred to the social work department for assistance with transportation. During a follow-up visit, the patient informed the health care team that she decided to "do something else" about her cancer, but she refused to provide details.

Later that year, the patient returned for her oncology appointment. She informed the team she had received immune recovery treatment at a facility outside of Texas. She did not want to provide medical records related to the chemotherapy or immune therapy received. She was offered standard of care therapy for which she declines. The patient has not returned for further follow-up visits.

Autonomy

Patient autonomy for self-determination can be defined as a person's right to make medical decisions without the interference of others. The patient took control over her personal health decisions. Data from the medical records, observations of the patient's behavior, and conversations with medical team indicated she had the capacity to understand the implications of her decisions. The patient stood firm about her treatment plan decisions. Mental capacity is a concept of autonomy related to the ability to make independent treatment decisions. The ethical issue in this vignette was: Does a person have the right, despite mental capacity, to make a treatment decision that contradicts the standard of best care and could endanger the person's own life? Is this a form of suicide [35]? Mental capacity is a concept of autonomy related to the ability to make independent treatment decisions [36]. This patient mental capacity's allowed her to understand her diagnosis, proposed treatment plan, and potential risks. Based on her beliefs and without hesitation, she firmly and consistently communicated her treatment decisions. Although her decisions were not in alignment with the proposed standards of care, her wishes were recognized and honored by the medical team.

Beneficence

In this vignette, beneficence and autonomy were diametrically opposite concepts. The team exercised their professional duty to "do no harm" and to make treatment recommendations in the best interest for the patient. The health care team provided extensive counseling to the patient on treatment recommendations and its benefits. They also had frank discussions about the consequences of delaying treatment and how this decision could exacerbate medical her condition and ultimately cause death. The physician, with the patient's approval, worked with the family member to communicate the risks associated with her medical decisions. Once the patient decided not to pursue treatment, the medical team respected her decision.

In these types of situations, health care professionals often struggle over the fine line separating autonomy from beneficence? Should the team have dug deeper to identify other reasons that could be affecting her decision? Would a psychiatric evaluation have helped or hindered the situation? There was no medical evidence that the patient had any physical (i.e., pain or cognitive impairment) or psychological issues (i.e., denial or depression), which may have affected her decision not to have further treatment. On the other hand, the health care team demonstrated compassion and commitment to deliver high quality and safe care throughout her treatment and decision-making process. At what point is futility justified in clinical situations such as described in this vignette?

Vignette 3: Lung Cancer Survivor Sees No Harm in Smoking

A 62-year-old woman is diagnosed with a squamous cell cancer of an unknown primary site. When she was initially diagnosed in 2006, she reported that she had been smoking 1.5–2 packs of cigarettes a day for the past 30 years. She was successfully treated with induction chemotherapy, followed by external beam radiation therapy, and finally completed two separate courses of adjuvant chemotherapy. She completed her treatments in 2008 and continued to smoke. When she presented to the Survivorship clinic in 2013, routine imaging identified a small irregularity in the right upper lobe of her chest. Over the past two years, this lung nodule had continued to grow. She had continued to smoke and had refused treatment for the new chest mass for two years. Several ethical dilemmas are apparent in this scenario.

Autonomy

Medical evidence indicates 90% of all lung cancer can be attributed to cigarette smoking, and reports indicate that approximately 22% of long-term cancer survivors continued to smoke after treatment [37]. Lung cancer patients, who are/were former smokers, often face ridicule from the public related to a diagnosis of lung cancer. Society often blames the smoker for a "self-imposed" cancer diagnosis such as lung cancer [38–40]. This patient made decisions that could negatively influence her outcome such continuing to smoke and not seeking treatment for the new chest mass. Anecdotal reports have surfaced regarding health care professional who refuse to treat patients who refuse to cease habits or health styles detrimental to their health [41,42]. An ethical dilemma arises when the health care team makes a decision not to treat smokers. The question then becomes whether this a value judgment and not an ethical decision? Additionally, as health care providers, if we exercise our values/morals regarding not treating smokers, does that also include patients with hepatic disease due to alcohol, or the young women who refused the HPV injections and later develop cervical cancer? In the best interest of this patient, the team employed multiple efforts related to refer/encourage her to quit smoking and counseled the patient regarding follow-up and treatment related to the suspicious chest mass. The patient had also refused to adhere to breast and cervical screenings recommendations. Nonetheless, the patient continued to keep her survivorship appointments just as the team continued to make recommendations for screening and follow-up care.

Vignette 4: Hodgkin's Lymphoma Survivor Postpones Cardiac Referral

A 58-year-old woman with a history of Hodgkin's lymphoma stage IIB diagnosed in 1996. She received ABVD (Doxorubicin/Bleomycin/Vinblastine/Dacarbazine) for six cycles followed by mantle radiation with a total dose on 39.6 Gy. Her comorbidities are obesity, hypertension, and type 2 diabetes. She does not smoke. She was seen in the Lymphoma Survivorship Clinic for the first time and was counseled on the late effects of chemotherapy and radiation, which include premature atherosclerosis, calcification of the heart valves, coronary artery disease (CAD), and carotid stenosis. She was referred to a cardiologist for an evaluation whose office was located in the medical center. She declined and stated she would see a cardiologist closer to home, which was about 100 miles away. She did not follow-up with the recommendations and had the added burden of work-related issues regarding off time requests. Six months following her initial survivorship appointment, this woman presented to her local emergency room with acute onset of chest pain. She was diagnosed with ischemic coronary disease and underwent emergent angioplasty.

Self-determination

The patient made the decision to schedule an appointment with a cardiologist closer to her home—her right to do so. She failed to follow through with scheduling an appointment, according to her, because of her work requirements. To what extent should the team have explored this patient's understanding that chemotherapy and radiation had significant cardiac effect? Was she in denial or overwhelmed with the late effects of her cancer therapies? The medical record indicated the patient participated in a "teach-back" session regarding her specific late effects of heart health as results of radiation and chemotherapy. Patient education is an important element in assisting patients to reach informed decisions and, thus, an element in the process of ethical decisions [43]. Given our very best efforts "to do no harm," to be truthful regarding oncology treatment and its therapies, transitioning patients to survivorship care, arming patients with educational materials, and acknowledging their understanding—how do we reconcile a patients' responsibility for their care and follow-up? How should we respond when a patient makes a decision not to follow their provider's recommendations?

Sadly, this is an ethical dilemma often encountered by health care professionals. However, health care professionals must respect and recognize a patient's autonomy and their ability to self-determine health-related decisions. Health care providers must be cautious about being paternalistic in their communication with survivors and accept their right for self-determination.

CONCLUSIONS

The vignettes and examples of ethical conflict selected for this chapter remind us how easily ethical conflicts can occur in clinical settings. The complex nature in working with cancer survivors is also evident in these vignettes. Physical, psychological, socioeconomic, and other medical issues of survivors tend to be intertwined rather than manifest separately. Ethical conflict occurs when difficult decisions must be made about the care of survivors. These situations often occur when conflicts threaten one's ethical principles or when all decisions being considered are unacceptable. Such occurrences often involve family members and multiple providers who may not always agree on the decision made or the outcome(s) proposed. There are times when these differences cannot be resolved and an ethicist is needed to facilitate communication and understanding among all involved in the decision-making. Because of the complex physical and emotional needs faced by survivors, decision-making is an integral part of their care. Survivors must make decisions about their follow-up care, risk reduction/prevention recommendations, changes in their lifestyle, and at times, about treatment for a new cancer. Hence, patient-centered communication is essential when caring for survivors. Epner (2014) suggests the following steps are helpful when establishing patient-centered communicating with cancer survivors: (1) address feelings before facts, (2) acknowledge emotions, (3) use empathy to deliver bad news or difficult conversations, and (4) conduct mindful and attentive listening [44]. Although communication is a critical component of survivorship care, ethical reasoning and decision-making must be learned and practiced in order to effectively link theoretical models with clinical practice.

In general, there are no easy answers to ethical dilemmas; however, there are models and process to facilitate decision-making. Examples of models used to facilitate ethical decision-making often vary by profession. For example, the American Medical Association established a Code of Ethics in 1847, which continues to reflect current trends in medicine and day-to-day practice [45]. Examples of other models used by ethicists include the Six-Step framework [46] or the CoRE-Values framework taught to medical students [47]. Social workers use yet a different model known as the Congress's Ethics Model [48]. The majority of ethical models are based on the four medical principles, autonomy, nonmaleficence, beneficence, and justice, which Beauchamp and Childress suggest are being universal or transcultural standards [12]. Yet, few ethical models address the role of cultural pluralism, diversity, or religious norms, which limits the multiculturalism character of our American society. Irrespective of the profession, each model provides a systematic process based on decision analyses and can be used in any clinical setting.

This chapter has provided a brief overview of the impact of ethical dimensions on survivors, patients, and their oncology health care providers. The field of cancer survivorship is relatively new and will continue to evolve in the future. It is evident that cancer survivors have many unique characteristics that contribute to the risk of ethical dilemmas. Health care providers must be aware of ethical principles, reflect on their ethical values, and understand that a cancer survivor's journey is complex, and at times, unpredictable. The use of ethical principles and models for decision-making can promote patient-centered communication and understanding among survivors, families, and providers when decision dilemmas arise and threaten the delivery of safe and high quality care.

References

[1] Peppercorn J. Common theme among ethical issues in oncology: the need to individualize advance cancer care, <http://www.cancernetwork.com/ethics-oncology/common-theme-among-ethical-issues-oncology-need-individualize-advanced-cancer-care>; 2013 [accessed 07.09.15].

[2] de Moor JS, Mariotto AB, Parry C, Alfano CM, Padgett L, Kent EE, et al. Cancer survivors in the United States: prevalence across the survivorship trajectory and implications for care. Cancer Epidemiology, Biomarkers & Prevention 2013;22(4):561–70.

[3] Rodriguez MA, Zandstra F. Models of survivorship care. In: Foxhall L, Rodriguez MA, editors. Advances in cancer survivorship management. New York: Springer; 2014.
[4] Hewitt ME, Ganz P. From cancer patient to cancer survivor: lost in transition. Washington, DC: National Academies Press; 2006.
[5] Kissane DW. Communication training to achieve shared treatment decisions. In: Kissane DW, Bultz B, Butow P, Finlay I, editors. Handbook of communication in oncology and palliative care. New York: Oxford University Press; 2011.
[6] Ruger JP. Ethics in American health 1: ethical approaches to health policy. Am J Public Health 2008;98 (10):1751–6.
[7] Schickedanz A. Of value: a discussion of cost, communication, and evidence to improve cancer care. The Oncologist 2010;15(Suppl. 1)):73–9.
[8] Moy B, Abernethy AP, Peppercorn JM. Core elements of the patient protection and affordable care act and their relevance to the delivery of high-quality cancer care. Am Soc Clin Oncol Educ Book 2012;e4–8.
[9] Salz T, Baxi SS, Raghunathan N, Onstad EE, Freedman AN, Moskowitz CS, et al. Are we ready to predict late effects? A systematic review of clinically useful prediction models. Eur J Cancer 2015;51(6):758–66.
[10] Smith TJ, Bodurtha JN. Ethical considerations in oncology: balancing the interests of patients, oncologists, and society. J Clin Oncol 1995;13(9):2464–70.
[11] Mullan F. Seasons of survival: reflections of a physician with cancer. New Engl J Med 1985;313(4):270–3.
[12] Beachamp T. Childress J. Principles of biomedical ethics. New York: Oxford University Press; 2009.
[13] Leigh S. Cancer survivorship: a nursing perspective. In: Ganz PA, editor. Cancer survivorship: today and tomorrow. New York: Springer; 2007. p. 8–13.
[14] Kleinman A. Patients and healer in the context of culture. Berkeley: University of California Press; 1980.
[15] Kluckhohn FS. Variations in value orientation. Evanston, Il: Row Peterson; 1961.
[16] Turner L. Bioethics and religions: religious traditions and understandings of morality, health, and illness. Health Care Anal 2003;11(3):181–97.
[17] Peppercorn J. Ethics of ongoing cancer care for patients making risky decisions. J Oncol Pract 2012;8(5): e111–13.
[18] Kent EE, Forsythe LP, Yabroff KR, Weaver KE, de Moor JS, Rodriguez JL, et al. Are survivors who report cancer-related financial problems more likely to forgo or delay medical care? Cancer 2013;119(20):3710–17.
[19] Guy Jr. GP, Ekwueme DU, Yabroff KR, Dowling EC, Li C, Rodriguez JL, et al. Economic burden of cancer survivorship among adults in the United States. J Clin Oncol 2013;31(30):3749–57.
[20] Ekwueme DU, Yabroff KR, Guy Jr. GP, Banegas MP, de Moor JS, Li C, et al. Medical costs and productivity losses of cancer survivors—United States, 2008–11. MMWR Morbidity and Mortality Weekly Report 2014;63 (23):505–10.
[21] de Moor JS, Alfano CM, Breen N, Kent EE, Rowland J. Applying evidence from economic evaluations to translate cancer survivorship research into care. J Cancer Surviv 2015;9(3):560–6.
[22] Moran JR, Short PF. Does cancer reduce labor market entry? Evidence for prime-age females. Med Care Res Rev 2014;71(3):224–42.
[23] Stergiou-Kita M, Grigorovich A, Tseung V, Milosevic E, Hebert D, Phan S, et al. Qualitative meta-synthesis of survivors' work experiences and the development of strategies to facilitate return to work. J Cancer Surviv 2014;8(4):657–70.
[24] Lawlor ER, Donnelly M. Use of an employment advisory service by cancer survivors. J Psychosoc Oncol 2015;33(3):219–31.
[25] Stapelfeldt CM, Labriola M, Jensen AB, Andersen NT, Momsen AM, Nielsen CV. Municipal return to work management in cancer survivors undergoing cancer treatment: a protocol on a controlled intervention study. BMC Public Health 2015;15:720.
[26] van Egmond MP, Duijts SF, Vermeulen SJ, van der Beek AJ, Anema JR. Return to work in sick-listed cancer survivors with job loss: design of a randomised controlled trial. BMC Cancer 2015;15:63.
[27] Moskowitz CS, Oeffinger KC. Predicting adverse health outcomes in long-term survivors of a childhood cancer. Children 2014;1(2):63–73.
[28] Latoch E, Muszynska-Roslan K, Panas A, Panasiuk A, Rutkowska-Zelazowska B, Konstantynowicz J, et al. Bone mineral density, thyroid function, and gonadal status in young adult survivors of childhood cancer. Contemp Oncol (Pozn) 2015;19(2):142–7.

[29] Armenian SH, Kremer LC, Sklar C. Approaches to reduce the long-term burden of treatment-related complications in survivors of childhood cancer. Am Soc Clin Oncol Educ Book 2015;196–204.
[30] Lie HC, Loge JH, Fossa SD, Hamre HM, Hess SL, Mellblom AV, et al. Providing information about late effects after childhood cancer: lymphoma survivors' preferences for what, how and when. Patient Educ Couns 2015;98(5):604–11.
[31] Sprauten M, Haugnes HS, Brydoy M, Kiserud C, Tandstad T, Bjoro T, et al. Chronic fatigue in 812 testicular cancer survivors during long-term follow up: increasing prevalence and risk factors. Ann Oncol 2015;26 (10):2133–40.
[32] Stouten-Kemperman MM, de Ruiter MB, Boogerd W, Veltman DJ, Reneman L, Schagen SB. Very late treatment-related alterations in brain function of breast cancer survivors. J Int Neuropsych Soc 2014;1–12.
[33] Rowland JH, Bellizzi KM. Cancer survivorship issues: life after treatment and implications for an aging population. J Clin Oncol 2014;32(24):2662–8.
[34] Jabson JM. Treatment summaries, follow-up care instructions, and patient navigation: could they be combined to improve cancer survivor's receipt of follow-up care? J Cancer Surviv 2015;9(4):692–8.
[35] Sessums LL, Zembrzuska H, Jackson JL. Does this patient have medical decision-making capacity? JAMA 2011;306(4):420–7.
[36] Price A, McCormack R, Wiseman T, Hotopf M. Concepts of mental capacity for patients requesting assisted suicide: a qualitative analysis of expert evidence presented to the Commission on Assisted Dying. BMC Medical Ethics 2014;15:32.
[37] Karam-Hage M, Cinciripini PM, Gritz ER. Tobacco use and cessation for cancer survivors: an overview for clinicians. CA Cancer J Clin 2014;64(4):272–90.
[38] Wilson A. Justice and lung cancer. J Med Philos 2013;38(2):219–34.
[39] Snelling PC. Who can blame who for what and how in responsibility for health? Nurs Philos 2015;16 (1):3–18.
[40] Marlow LA, Waller J, Wardle J. Does lung cancer attract greater stigma than other cancer types? Lung Cancer 2015;88(1):104–7.
[41] Scott N, Crane M, Lafontaine M, Seale H, Currow D. Stigma as a barrier to diagnosis of lung cancer: patient and general practitioner perspectives. Prim Health Care Res Dev 2015;1–5.
[42] Shaw BA, McGeever K, Vasquez E, Agahi N, Fors S. Socioeconomic inequalities in health after age 50: are health risk behaviors to blame? Soc Sci Med 2014;101:52–60.
[43] Redman BK. Ethics of patient education and how do we make it everyone's ethics. Nurs Clin North Am 2011;46(3):283–9.
[44] Epner DE. Communication between patients and health care providers. In: Foxhall L, Rodriguez MA, editors. Advances in cancer survivorship management. New York: Springer; 2014.
[45] Affairs., A.M.A.C.o.E.a.J. Code of Medical Ethics of the American Medical Association, 2014–15. Atlanta, GA: American Medical Association; 2013.
[46] Back AL, Anderson WG, Bunch L, Marr LA, Wallace JA, Yang HB, et al. Communication about cancer near the end of life. Cancer 2008;113(7):1897–910.
[47] Manson HM. The development of the CoRE-Values framework as an aid to ethical decision-making. Med Teach 2012;34(4):e258–68.
[48] Congress EP. What social workers should know about ethics: understanding and resolving practice dilemmas. Advances in Social Work 2000;1(1):1–25.

C H A P T E R

6

Integrative Oncology and Ethics

Richard T. Lee[1,2] *and Jessica A. Moore*[3]

[1]Case Western Reserve University School of Medicine, Cleveland, OH, United States
[2]UH Seidman Cancer Center, Cleveland, OH, United States
[3]The University of Texas MD Anderson Cancer Center, Houston, TX, United States

INTRODUCTION

This field of integrative oncology is in the formative years and involves many approaches to health and wellness from a diversity of perspectives that aim to address the needs of the whole person. Many health care institutions that offer complementary and integrative therapies believe that focusing on the patient as a whole person (biopsychosocial) not only aids in the healing process, but also improves his/her overall wellness and quality of life. The commitment to respect the patient as a whole person and to inform him/her of all of his/her health options is an important part of helping the patient become an active participant in his/her own recovery. This respect for the whole person and the commitment to provide services that address the various biopsychosocial (including spiritual) needs of the patient allow for healing of the whole person even if she/he is not cured of the presenting disease, such as cancer in this case. Challenging ethical situations may arise due to misunderstanding of the common complementary, alternative, and integrative medicine (CAIM) therapies and philosophics that exist today. We will first provide a general background on the field, integrative oncology, to provide a common base of understanding and perspective when dealing with ethical challenges.

Whole Person Care

Scientists, philosophers, and theologians alike have identified the human person's internal structure of needs to consist of physical, psychological, social, and spiritual aspects. Others have described a similar need for fulfillment through outward focused relationships with man, society, God, and the world [1]; or those relationships which man holds with self, others, *the Other*, and nature. Many refer to this perspective as a

Ethical Challenges in Oncology.
DOI: http://dx.doi.org/10.1016/B978-0-12-803831-4.00006-3

whole person approach to healthcare similar to the bio-psycho-social model first introduced by George Engle in 1977 [2]. In 1948, the World Health Organization declared: "Health is a state of complete physical, mental, and social well-being and not merely the absence of disease or infirmity" [3]. This definition is often criticized, but it has advanced the notion that health is not limited to physical concerns [4]. Physical health is an important aspect of human health, but it is only one aspect of total human health. The basic goal of health care is fulfilling the needs of the human person. Human persons are social beings who can only be healthy and whole in community where they have the capacity for the interpersonal relations that are essential for physical and psychological growth [4].

The biomedical model is based on the premise that ill health is a physical phenomenon that can be explained, identified, and treated with physical means. The biopsychosocial model takes into account patients' physical conditions, their thoughts and beliefs—particularly those influenced by culture and faith tradition and their social expectations. Illness is not solely a physical phenomenon; it is also influenced by people's feelings, their ideas about health, and the events of their lives. The biopsychosocial model is appealing for its thoroughness and personal concern [5].

The field of palliative care has helped promote the concept of biopsychosocial care into the field of oncology for more than 30 years, to the point that palliative care is now considered a standard component of modern cancer care. The roots of palliative care began in Great Britain in the 1960s—spearheaded by Dame Cicely Saunders—and the movement was later championed in North America by a Canadian urologist, Balfour Mount, MD. Although the field of palliative care initially focused on end-of-life care, it has evolved to become an integral part of cancer care from diagnosis through survivorship as well as end-of-life care [6,7].

The World Health Organization has published reports on palliative care in cancer, first in 1990 and later updated in 2002 [8,9]. The importance of palliative care has also been recognized by several national and international organizations (e.g., ASCO, National Comprehensive Cancer Network, National Quality Forum, and National Institute for Health and Care Excellence) [10–13]. The Institute of Medicine also addressed the topic in its report *Cancer Care for the Whole Patient: Meeting Psychosocial Health Needs* [14–16].

The goal of these institutions is to promote wellness and to preserve the wholeness and dignity of the patients, their families, and others. They strive to create a plan of care in collaboration with the clients and their significant others consistent with cultural background, health beliefs, sexual orientation, values, and preferences that focuses on health promotion, recovery or restoration, or peaceful dying so that the person is as independent as possible and all of his/her needs are addressed in the most complete manner possible.

Soon after this movement by palliative medicine to incorporate a more whole person approach within oncology, the field of alternative medicine also began to emerge. By the 1980s and 1990s, American public had a growing interest in pursuing therapies commonly labeled as alternative or complementary medicine. These approaches also embodied the concept of whole person care that focused on a biopsychosocial framework of health and well-being.

COMPLEMENTARY, ALTERNATIVE, AND INTEGRATIVE MEDICINE

Patients diagnosed with cancer commonly seek out all possible treatments options, and since the 1990s, surveys have documented the common use of nonconventional treatment approaches, often termed complementary and alternative medicine (CAM). The National Center for Complementary and Integrative Health (NCCIH) (previously the National Center for Complementary and Alternative Medicine) has defined CAM as "...diverse medical and health care systems, practices, and products that are not presently considered to be part of conventional medicine" [17]. The level of clinical evidence for many CAM therapies is limited (even nonexistent in some cases) and, thus, insufficient to bring them into the realm of *conventional* medicine. The terms *alternative, complementary, and integrative medicine* are commonly used interchangeably, but in fact are distinctly different. *Alternative medicine* by definition is when a patient makes use of a nonconventional treatment modality in place of conventional medicine. *Complementary medicine* is the combination of CAM together with conventional medicine regardless of evidence or appropriateness without any coordination. The field of CAM has been criticized for using a nonscientific approach to medicine and incorporating what is often labeled as "snake oil" and "quackery." In order to move away from this criticism, the term *integrative medicine* has been used as a way to distinguish a more coordinated and evidence-based approach. *Integrative medicine* can be described as a more unified approach incorporating both conventional and nonconventional therapies in an evidence-based manner that is deliberate and coordinated. The Academic Consortium for Integrative Medicine and Health has defined integrative medicine as: "The practice of medicine that reaffirms the importance of the relationship between practitioner and patient, focuses on the whole person, is informed by evidence, and makes use of all appropriate therapeutic approaches, health care professionals and disciplines to achieve optimal health and healing" [18]. Because alternative, complementary, and integrative medicine are ways to describe these types of unconventional therapies, the term CAIM will be used to describe these therapies in this chapter.

When CAIM is applied to cancer care, the term integrative oncology is commonly used. Integrative oncology makes use of both conventional and nonconventional medicine using an interdisciplinary approach to cancer treatment. These approaches provide cancer patients with care that is comprehensive, personalized, evidence-based, and safe in order to achieve optimal health and healing.

CAIM Utilization

According to the 2012 National Health Interview Survey (NHIS), the prevalence of CAIM use in the United States was approximately 33% for adults and 12% for children [19,20]. Among patients and families touched by cancer the use of CAIM is even higher than in the general population. An estimated 40%–69% of U.S. patients with cancer use CAIM therapies [21–23] and the prevalence increases among specific populations of cancer patients, such as those involved in phase I clinical trials [23,24]. CAIM is used in 70% of all oncology departments engaged in palliative care in Britain [25,26]. A survey of five clinics within a U.S. comprehensive cancer center found that CAIM therapies were used by 66% of patients (excluding psychotherapy and spiritual practices) [23]. Breast cancer patients have been found to have an elevated use of CAIM when compared to other cancer types, and two reviews on the topics have found a prevalence range of 17%–87%, with one study reporting a mean prevalence of 45% [27–29]. Use is also common in the survivorship period with prevalence rates between 69% and 87%, with an increasing trend over the past decade reported by Boon et al., from 67% in 1998 to 82% in 2005 [30–32]. Use of CAIM is also prevalent among cancer patients outside the United States in all regions of the world [33,34]. The range of prevalence patterns reflect differences in the patient populations studied and definition used.

CAIM Therapies

Patients may seek a variety of different CAIM therapies ranging from a simple vitamin to more complex and less scientific approaches such as energy healing. The motivations for seeking these therapies are equally as variable and include a desire to improve quality of life, prolong life, boost the immune system, and improve the efficacy of conventional medical treatments [35,36]. These therapies can be separated into several major categories.

Dietary Supplements

Dietary supplements are the most common CAIM in general and include a variety of substances such as herbs, vitamins, minerals, probiotics, and extracts. Supplements can be categorized into two major groups: (1) Vitamins and minerals and (2) plant derived products (a.k.a. herbs and nutraceuticals). Contrary to common belief, dietary supplements have the potential to cause harm. Issues to consider when discussing dietary supplements include quality control, metabolic interactions, treatment interactions, organ toxicity, and cancer growth. A regular dose multivitamin might be reasonable for a malnourished patient, or similarly, vitamin D supplementation might be required for a patient with a deficient level and bone metastases. However, use of vitamins and minerals, especially at high doses, have been associated with increased cancer [37–39] and, thus, should be used cautiously, only at recommended doses, and when a clear medical indication exists. Research is ongoing to investigate how these substances can be used safely along with conventional treatment approaches to improve cancer treatment outcomes. The proliferation of nonscientific claims by companies through aggressive marketing as well as commonly held beliefs regarding natural remedies often leads to conflicting opinions on the role of herbs and supplements for cancer care.

Mind–body

Mind–body techniques include relaxation, hypnosis, visual imagery, meditation, biofeedback, cognitive–behavioral therapies, group support, and spirituality as well as expressive arts therapies such as art, music, or dance. They also include practices such as yoga, tai chi, and qi gong, where focused attention on movement, breath, or sound can increase self-awareness and relieve stress and anxiety. Ernst et al. examined changes in the state of the evidence for mind–body therapies for various medical conditions between 2000 and 2005 and found that, over that period, maximal evidence had appeared for the use of relaxation techniques for anxiety, hypertension, insomnia, and nausea due to chemotherapy [40]. Research examining yoga, tai chi/qigong, and meditation incorporated into cancer care suggests that these mind–body practices help improve quality of life through improved mood, sleep quality, physical functioning, and overall well-being [41–44]. Music therapy uses music (music making, songwriting, singing, and listening) in a prescriptive manner for nonmusical goals, including improving quality of life. Trained music therapists choose a music approach most appropriate to help patients achieve a desired result. Evidence suggests music therapy can help with management of mood disturbances, including anxiety [45–47]. Mind–body practices have an excellent safety profile, with some practices requiring more physical activity than others. Mind–body practitioners with experience in cancer patient populations can provide guidance to help patients engage safely in practices such as meditation, yoga, and tai chi. Many patients may have been practicing these activities for years and wish to continue them during treatment, and the lack of understanding by clinicians may lead to concern during treatment.

Manual Therapies

Manual therapies include massage, chiropractic care, reflexology, and other forms of physical treatments. As a manipulative touch-based therapy, massage can benefit cancer patients when it is performed by individuals who have special training in oncology massage to safely deliver the massage. Risk of bruising, bleeding, or injury can be minimized by careful application of pressure, avoiding massage into the deep tissue or bone in selected patients, such as patients that have recently had surgery or radiation. In patients with extremities subject to lymphedema, therapists will need to adjust their technique to maximize safety. Patients may benefit from formal lymphedema therapy as part of a physical therapy program [48]. Research to date suggests that massage is helpful at relieving pain, anxiety, fatigue, and distress and increasing relaxation [49–51]. Benefit on mood and pain relief is limited to the more immediate effect of massage, with no current studies demonstrating long-term relief. Massage provided by caregivers may offer a unique opportunity for interaction between patient and caregiver that can help enhance well-being of both [52].

Acupuncture has been practiced as part of traditional Chinese medicine for thousands of years, and most commonly it involves the practice of inserting needles into the skin at specific points throughout the body to achieve a desired effect. Depending on the condition being treated, heat may be applied to the needles after insertion. For points on the ears, small round balls or seeds may be taped into place and stimulated by pressing the seed left in place for several days. For some patients acupressure may be used, which involves applying pressure to acupuncture points instead of puncturing the skin. A recent systematic review evaluated 42 randomized-controlled trials involving the use of acupuncture to help manage eight symptoms (nausea, pain, hot flashes, fatigue, radiation-induced xerostomia, prolonged postoperative ileus, anxiety/mood disorders, and sleep disturbance) in cancer patients [53]. This review found the strongest evidence to date is in support of its use for pain, nausea, and vomiting. Although nausea and vomiting are among the top three most commonly reported side effects of cancer treatment, pain is the most common reason cancer patients use acupuncture. When performed correctly, acupuncture has been shown to be a safe, minimally invasive procedure with few side effects [54,55].

The most commonly reported complications are fainting, bruising, and mild pain. Infection is also a potential risk, although very uncommon. Acupuncture should only be performed by a health care professional with an appropriate license, credentials, and expertise, preferably one who has had experience in treating patients with cancer. The physical nature of these therapies is a common concern by clinicians, especially among patients undergoing active anticancer treatments. Both patients and clinicians need to be aware of the risks and benefits of these therapies and discuss a unified approach if these are to be incorporated into care.

Nutrition and Exercise

Growing evidence supports the important role of physical activity and nutrition in the health of cancer patients, and these factors have been correlated with improved clinical outcomes [56–61]. Unfortunately, a large amount of misinformation exists regarding what patients should and should not eat as part of a healthy diet for patients with cancer. Patients are flooded with a number of dietary approaches and need guidance. Several dietary approaches have been described as "anticancer" (i.e., a therapy to treat cancer) and often claiming that a strict diet of a variety of things, including raw foods, juiced fruits and vegetables, animal protein sources, or alkaline foods, will result in a cancer cure. These claims have no evidence to date to support their use and may be harmful during active treatment, since such diets may lack important nutrients necessary for proper healing and recovery. Reliable resources are published by the American Cancer Society (ACS) and the American Institute for Cancer Research (AICR) and are supported by clinical studies. These institutions support a diet rich in fruits, vegetables, whole grains, and lean meats, with an avoidance of processed foods as part of a healthy lifestyle. Referral to a dietician can help patients and survivors identify how to best meet their nutritional needs during their cancer journey.

Exercise can be important to patients receiving active treatment as well as survivors, helping to improve physical function and quality of life [62–64]. Regular exercise during cancer therapies such as chemotherapy or radiation has the potential to decrease treatment related fatigue [65]. ACS guidelines for cancer prevention recommend avoidance of sedentary behavior, encouraging 150 minutes of moderate physical activity spread throughout the week [66]. Patients dealing with short- or long-term side effects from cancer or its treatment will benefit from a consultation with a physical therapist.

Traditional Medical Systems, Energy Healing, Homeopathy, and Other Less Researched Therapies

Common traditional medical systems include traditional Chinese medicine and ayurvedic medicine although other medical systems exist around the world. These traditional medical systems are mostly based on theories developed prior to today's modern medicine and are sometimes used as an alternative or complementary approach to cancer care. Although clinical studies of specific therapies within these traditional medical systems exist (e.g., acupuncture or a specific herb like curcumin), as an entire medical approach, few clinical studies have documented their effectiveness compared to modern medical approaches. Energy healing therapies may be referred to as Reiki, medical Qi Gong, touch therapy, and other similar approaches. Homeopathy involves the use of very dilute solutions to treat diseases by stimulating the body's own mechanisms to promote healing and is more prevalent in Europe. Aromatherapy involves the use of essential oils that have purported specific properties that may aid in the symptom management. For example, the use of lavender oil to help relax or sleep. Additional new systems of care have also arisen such as functional medicine. This approach also attempts to integrate conventional allopathic Western medical practices with CAIM and incorporates areas of nutrition, diet, and exercise; use of the latest laboratory testing and other diagnostic techniques; and prescribed combinations of drugs and/or botanical medicines, supplements, therapeutic diets focused on individual nutritional needs, detoxification programs, or stress-management techniques—including mindfulness, movement practices, and exercise. These types of CAIM therapies and others approaches generally have limited research and remain controversial as an effective therapy for patients with cancer, especially when used to treat cancer rather than to help with side effects.

ETHICAL CHALLENGES

Practicing clinicians are commonly asked about the use of CAIM for a wide range of ailments. Among patients with advanced cancer with a poor prognosis, CAIM approaches seem very appealing. Patients near the end-of-life are especially attracted to unconventional therapies that often falsely claim dramatic results with no side effects. The allure of these treatments is too strong for patients to ignore despite the potentially high costs, including financial, emotional, and the utilization of time and energy. Thus, the clinician is often involved in discussions about alternative therapies and may even be asked to assist with their implementation. Although physicians may consider such discussions a waste of time, patients are interested in their physicians' opinion about the topic as a trusted source of information. When patients are told new information about the potential harms of herbs, the majority would either stop the herb or talk with their physician about it [67]. Because patients seek information from their physicians and value this discussion, the remainder of this chapter focuses on providing the practicing clinician a framework in which to have these often challenging discussions with patients wanting to pursue CAIM in an appropriate and ethical framework. Three distinct types of ethical challenges arise commonly: (1) Communication and discussion about the use of CAIM therapies; (2) the request to partner with the incorporation of CAIM therapies during active treatment; and (3) the pursuit of alternative therapies alone despite having a reasonable chance for cure with conventional therapies.

Patient–Clinician Communication Regarding CAIM

Clinical Scenario: *Your patient Jane Doe has just started on chemotherapy for her stage III colon cancer. Before beginning her second cycle of chemotherapy she states she has one more quick question before she leaves, "What are your thoughts about natural herbs to help treat my stress? My neighbor is really encouraging me to take something to help calm my nerves."*

Research indicates that neither adult nor pediatric patients receive sufficient information or discuss CAIM therapies with physicians, pharmacists, nurses, or CAIM practitioners [68,69]. The most common reason patients state for not bringing up the topic of CAIM (even if they have questions or are taking CAIM) is that it never came up in the discussion or no one asked them. Thus, patients may believe it is unimportant. Patients, such as Jane Doe, may also fear that the topic will be received with indifference or dismissed without discussion [70,71] and health care professionals may fear being unable to respond to questions or may shy away from initiating a time-consuming discussion. As a result, it is estimated that up to 70% of patients are taking CAIM without informing any member of their health care team [23,72]. This lack of discussion is of concern because supplements such as herbs may interact with medical treatments.

A discussion about CAIM therapies can be a challenge for health care professionals who typically have limited knowledge of this area. When clinicians are faced with unfamiliar information about CAIM therapies, they may feel "de-skilled" by being forced outside their medical specialty, and this discomfort can lead to defensiveness and a breakdown in communication with the patient. Additionally, others may feel somewhat challenged or upset by patients' desire to seek out other medical approaches. At the same time, patients like Ms. Doe can become frustrated if they cannot discuss CAIM with their physician. This bilateral frustration can result in a communication gap, which damages the clinician–patient (CP) relationship. Clinicians should be receptive to patient inquiries and aware of subtle, nonverbal messages to create an environment in which patients can openly discuss potential CAIM choices [70,73].

A clinician may feel that such a discussion is not a part of their ethical obligation to the patients since it involves an area outside their own expertise and training. Is it really necessary to engage in a discussion about CAIM? Will this discussion be in some way approving such therapies? Existing research suggests that the majority of cancer patients desire communication with their doctors about CAIM [74], and there is general agreement within the oncology community that in order to provide optimal patient care, oncologists must not only be aware of CAIM use but also be willing and able to discuss all therapeutic approaches with their patients [75,76]. It is the health care professional's ethical and professional responsibility to ask patients about their use of CAIM [77], and the discussion should ideally take place before the patient starts using a CAIM treatment—whether it is a dietary supplement, mind–body therapy, or another CAIM approach.

A number of strategies can be used to increase the chance of a worthwhile dialogue. One approach is to include the topic of CAIM as part of a new patient assessment. For example, when asking about medications, physicians should inquire about everything the patient ingests—including over-the-counter products, vitamins, minerals, herbs, and even the patient's diet. Physicians may consider having the patients bring in the actual bottles of herbs and supplements for evaluation. When asking about a patient's past medical history, physicians may ask about the use other health care professionals to learn if the patients have visited with naturopathic or chiropractic practitioners.

When the topic arises about CAIM, physicians may unintentionally express an immediate strong negative response or show apprehension about a topic of which they have limited knowledge. The following responses to a patient may not be fully appreciated as offensive or harmful: "Oh that nonsense" or "You're not really thinking about taking that are you?"; These types of responses have the potential for weakening the relationship with the patient by signaling a strong negative opinion without discussion. Instead, one practical approach when the topic arises is to re-assure the patient that you are:

1. Open to objectively discussing the topic: "Sounds interesting, let's talk about this."
2. Invested in the best outcome possible for the patient: "I want to help you feel better."
3. Willing to provide appropriate support no matter what the patients decides, even if it is in opposition to the medical advice you provide. In many instances, this involves emotional support: "No matter what you decide, I'm here to help you in the best way possible as your physician" and "I can see you are scared and anxious about this diagnosis; please tell me more about how you are feeling."

This sends a clear message to the patient that the involved clinician is committed to their health and well-being and will not abandon them even if there is a difference of opinion. Additionally, by providing these statements at the beginning of the conversation about CAIM therapies, the patient understands that a disagreement about the benefits of an alternative therapy does not equate to abandonment. The idea of abandonment is poisonous to the CP relationship. Showing broad commitment to the clinician's role to help the patient will strengthen the bond between clinician and patient [78].

Clinicians need to develop an empathic communication strategy that addresses the patient's needs while maintaining an understanding of the current state of the science. In other words, this strategy needs to be balanced between clinical objectivity and bonding with the patient so that it can benefit both the patient and the health care provider. These patients need reliable information on CAIM from reliable resources, as well as adequate time to process this information. Underlying these specific strategies should be an open attitude combined with a willingness to review evidence-based references and consult with other health care professionals knowledgeable or trained in CAIM [79]. Studies indicate that these discussions about CAIM often lead to an improvement in the CP relationship [80].

Another key issue to consider is the patient's expectations from the CP relationship, especially if they come from a different cultural backgrounds. Western physicians often diagnose "disease" as a biological process, while patients experience an "illness," a culturally based and personal interpretation of symptoms resulting from the disease process. CAIM therapies may actually be more effective at addressing the patient's needs and improving their experience by decreasing their symptom burden. The danger lies in the realm where "feeling better" subjectively does not necessarily correlate with "getting better" objectively. Cultural competence is a requirement of all health care providers. A basic familiarity with the cultures and traditions of the patients for whom you most often treat is an important starting point for fruitful discussion aimed at shared decision-making. Medical humility, or openly expressing a desire to learn about a patient's religious and cultural beliefs, will acknowledge your knowledge gap and willingness to include these important issues into the decision-making process [81]. Asking questions about how the patient's culture and or faith affect medical decisions and the types of care they pursue rather than making assumptions based on a limited knowledge of a particular culture or faith tradition will begin a productive dialogue that will assist the clinician in providing effective therapies in a respectful manner consistent with the patient's values. Making assumptions will often lead to misunderstandings, a strained relationship, and closing the door to productive communication. When clinicians communicate empathetically and effectively, patients will often leave the encounter feeling better, regardless of the medical decisions made.

Discussion about the goals of care and expectations with the use of these CAIM therapies may provide helpful information regarding decisions to pursue such therapies. Consider, e.g., the use of an unknown herb alongside chemotherapy being used for curative intent. Explaining to the patient that the use of herbs may lead to an interaction that will decrease the efficacy of chemotherapy will often be enough to delay the use of such herbs. However, in a patient with refractory metastatic breast cancer that has advanced despite multiple treatment regimens and where treatment is administered for purely palliative reasons, the use of a herb to help with fatigue may seem more reasonable. Research shows that when patients are informed of potential risks of combining herbs and supplements during cancer treatment, the majority would be willing to stop taking the herb or supplement or would want to discuss this issue with their oncologist [67]. An open discussion regarding needs and goals of care will help both the physician and patient place treatment options in an appropriate and ethical context.

Today, most patients expect to be part of the decision-making process. Shared decision-making between the clinician and patient helps create a sense of control for patients, which has been cited as a reason for pursuing CAIM [82]. Whereas some patients may demand to be the final decision maker in their care, other patients may prefer a more passive role, allowing the physician to direct therapy. In either case, the personalized approach provides patients with a sense of ease about their goals of care and medical decisions. By utilizing a shared decision-making process and assessing goals of care, physicians have an opportunity to address common reasons why patients seek CAIM therapies: Sense of control and unmet needs. Having patients and family play an active role in decision-making and treatment planning can provide patients with a sense of empowerment and responsibility, leading to a more powerful therapeutic relationship. Although the clinical endpoint may remain the same, all parties experience an enhanced sense of success and satisfaction.

Reviewing the Risks and Benefits

The common belief by patients that "natural" means a product or therapy is safe needs to be addressed with education. Some herbs and supplements have been associated with multiple drug interactions [83], as well as increased health risks such as organ toxicity. Many CAIM therapies can potentially cause adverse outcomes and are, thus, a major concern among health care professionals. CAIM therapies, such as massage or acupuncture, often have minimal risks when performed by trained health professionals. In contrast, herbs and supplements should be considered more similar to prescription medications in that they have the potential to have powerful effects on the natural biological processes of the body. This is especially true when natural plants are processed into concentrated powders, liquids, or pills. The pathways by which CAIM therapies may lead to negative clinical outcomes include metabolic interactions, treatment interactions, direct organ toxicity, direct biological effects on the disease process, and unregulated manufacturing of pharmaceutical products. Even the quality of herbs and supplements can be variable.

Studies show as much as 50% of products do not contain the amount indicated on the product label [84].

While most CAIM therapies may become a financial burden, as most are not covered by insurance, there are other hidden costs. These include the delay of potentially curative treatments, which if used too late, will have diminished benefit. Perhaps the most important costs for those patients with a limited or poor prognosis is time and energy. Consider, for example, a terminally ill patient who desires to go to another country to pursue an untested alternative therapy for two months. In addition to the cost of the actual therapy, the patient will need to spend time, energy, and money to travel to a foreign country. Would these two months be best spent with family and friends or in a hospital setting away from the patient's strongest psychosocial support? Would the patient feel comfortable dying in a foreign hospital away from family and friends if an unexpected medical emergency occurred? These types of questions must be asked in order to have a full discussion about the potential outcomes of pursuing CAIM therapies.

Unfortunately, for many CAIM therapies, no reliable clinical data exists to inform the discussion about risks and benefits. With this in mind, it is important to acknowledge that three potential outcomes are possible: Beneficial, harmful, or no effect on the goals of care. The estimate of promising treatments that look helpful in preclinical studies that actually are found to be clinically helpful is in the range of 10% or less. Patients are commonly unaware of the differences between United States Food and Drug Administration (FDA) approved medications (which require evidence of efficacy, safety, and quality control manufacturing) and supplements, which are governed not by the FDA but by the Dietary Supplement Health and Education Act (DSHEA) of 1994. Supplements under this legislation are exempt from the same scrutiny the FDA imposes on medications; furthermore, these supplements are not intended to treat, prevent, or cure diseases. Patients will often argue that because these therapies are nonpatentable or have no profit driven company to fund the research, none of the CAIM therapies will have good clinical research. Reminding patients that many of our commonly used medications are derived from plants or micro-organisms, including aspirin, morphine, antibiotics, and even chemotherapies such as taxanes and anthracyclines can help counter these claims perpetuated by the alternative community. In regard to research, dozens of clinical trials have been funded by the government to research areas such as fish oil, acupuncture, and vitamins with some results being positive and others negative. This type of myth needs to be addressed by facts. For instance, the supplement industry is estimated to be a 30 billion dollar industry. These same companies are almost all entirely profit driven and do not support clinical research of their own products. These facts may help patients view their possible beliefs in a different light.

A key issue often ignored by patients is the quality of the available evidence. The lay public is often overwhelmed and confused by the amount of information available on cancer and possible treatments. This is further worsened by the enormous amount of erroneous information available on the Internet. Taking time to explain that a treatment proven for breast cancer from a phase III trial with several thousand patients should not be considered in the same way as data from a phase II trial with less than a hundred patients or as pre-clinical studies in mice can be a worthwhile discussion.

Discussions about the use of CAIM therapies can be time consuming and draining for the physician. Unless an urgent deadline exists, providing patients enough time to process the information is always helpful and should include a plan for follow-up in 1–2 weeks. Another important strategy is to offer other options which help them achieve their goals. Patients commonly ignore the most natural approaches to their needs such as fatigue or insomnia. Good evidence exists that physical activity can improve fatigue levels and growing research shows that sleep hygiene and yoga or meditation can improve sleep quality. Many would consider these approaches more natural than taking a processed pill. If questions still exist, offer to investigate the topic or refer them to a specialist.

If a decision is reached to add a complementary therapy to the treatment of cancer, the physician's role has not ended. The physician still has the responsibility of verifying, with some degree of certainty, the reliability of the specific therapy in question. Physicians may want to verify a product's reliability and safety by referencing independent websites (www.consumerlab.com) or conferring with a knowledgeable colleague, particularly if the clinician has access to an integrative medicine service. Once a plan is determined, one should be sure to communicate the integrative treatment plan to the other health care professionals involved with the patient's care. Additionally, regular follow-up is needed to monitor adverse effects and effectiveness and make adjustments, as with any other conventional treatment [85].

Patients Asking for Assistance With CAIM Therapies

Clinical Scenario: *John Smith is about to begin on chemotherapy for his diffuse large B-cell lymphoma. He tells you, "I've been seeing an acupuncturist for years and when he heard I had cancer, he referred me to his naturopathic colleague. The naturopath has recommended I have a variety of blood work done to figure out what supplements I need. He gave me a list of blood tests I need and suggested I ask you to order them. Would that be okay?"*

Beyond a discussion, some patients may request your assistance with receiving CAIM therapies. Many CAIM practitioners are not trained clinicians and/or lack any formal licensed training that limits the medical services they are able to provide. For example, a patient receiving an evaluation and treatment from a CAIM practitioner may request certain laboratory tests or even imaging to be completed. These requests often include tests not considered standard of care and sometimes are more accurately described as experimental. In other cases, these tests are commonly utilized but for different conditions, such as laboratory tests related to diabetes care rather than cancer care. Thus, the clinician is challenged with whether or not to assist in the treatment provided by an outside practitioner that you disagree with being provided. Ordering such tests may result in failure of insurance companies to pay and thus the costs must be absorbed by the facility or charged to the patient.

The initial reaction to such a request is to immediately disagree, but as discussed previously, this may leave your patient feeling disappointed and abandoned. At a minimum, this should lead to an open dialogue to inquire about why the patient is seeking such therapies. Assuming after a full discussion, the patient insists on an alternative or complementary approach, you may be required to consult with other health care professionals and even contact the outside CAIM practitioner to learn more about their recommendations [79]. A willingness to talk with involved CAIM practitioners shows your commitment to work collaboratively and provide the best care, which the patient will be undoubtedly thankful no matter the outcome of this communication. Involvement of a CAIM practitioner can complicate communication because of a triangular relationship: patient–physician, patient–CAIM practitioner, and the physician–CAIM practitioner. A productive and fruitful communication process requires all three relationships to be intact [86].

The discussion with a CAIM practitioner can be one of the most challenging, in part because one may encounter a spectrum of attitudes toward modern medicine. Because the CAIM practitioner may have a negative attitude or even antagonistic attitude toward any conventional approaches, it is worthwhile to focus on the mutual patient with a goal for the most effective and safe care plan. Additionally, emphasizing an evidence-based approach will also help keep the discussion framed in science rather than in opinion or personal experience. Disagreements will arise and, thus, the conversation may conclude with agreeing to disagree, but sometimes can lead to changing the timing of care so as to minimize any risks. Regardless of the outcome of this conversation with the outside CAIM practitioner, it is important to update the patient of your conversation.

In many situations, you may still be left with the question of whether or not to aid in the treatment requested by the outside CAIM practitioner and your patient. Unfortunately, there is no easy answer to this situation. Agreeing to order extra blood work or other tests, may signal your approval of such a CAIM approach (where none exists) or may lead to more requests in the future. Who is responsible if the test results are abnormal? In the worst case, if your patient experiences a detrimental clinical event due to the CAIM treatment, will the clinician bear any responsibility for this outcome? Will he or she be practicing within acceptable standards of care? On the other hand, by not agreeing to these tests, will your patient hide from you the other CAIM therapies he or she is receiving or even decide to seek out another medical clinician who is willing to accommodate the request? Much of this will depend on your level of comfort with what is being asked and the strength of your relationship with your patient. Each clinician will have a different threshold at which point you will decide not to agree to your patient's request. In some cases, the principle of beneficence may be at the forefront in the form of safety and doing what is best, especially in cases of incurable cancer. In other cases, the principle of nonmaleficence will be the driving principle to prevent pursuit of a potentially harmful treatment. These are challenging cases and consultation with an integrative oncologist you trust or other colleagues may be beneficial to help clarify the issues relevant to the patient's case. Questions regarding liability may be best addressed to a risk management professional or health care attorney. Many health care institutions employ these professionals to provide expert guidance in difficult situations. These questions are addressed by Michael Cohen in a number of his publications on this topic [87].

Patients Who Insist On Using Alternative Therapies

Clinical Scenario: *Jane Smith is a patient with chronic myelogenous leukemia currently on imatinib therapy with a good response after several months of therapy. Today, she arrives to your* *clinic with a binder of information. She tells you, "I've booked an airplane ticket to visit this clinic in Tijuana, Mexico. The clinic tells me that they have cured a lot of people with my cancer. What do you think about this?"*

Occasionally, a patient may insist on pursuing an alternative therapy, a treatment in place of what you would recommend. This is even more challenging when the patient has an early stage cancer with a reasonable chance for cure. Generally, arguing with the patient that they should not try an unproven therapy, which they are already convinced will be helpful, is not very productive. In fact, it is likely to damage the therapeutic relationship and drive the communication process underground. Rather, frank, nonjudgmental discussion with the patient is necessary to inform the patient effectively about the known risks and benefits of the therapies. A key point to emphasize during this time is nonabandonment which is difficult when a patient is refusing your best advice. This discussion should acknowledge that more definitive information will not be available in the near future regarding most alternative therapies and that decisions will need to be made that balance risks and benefits, many times without adequate quality research. Often, reasonable evidence exists for the use of an herb alone, but minimal information exists regarding an herbs use in combination with chemotherapy or radiation. Thus, the risk–benefit ratio may change depending on how the herb is used, specifically the timing of its use relative to other treatments. These discussions are usually aided by reviewing all the therapeutic options in the context of the patient's goals. For patients with curative intent, reminding patients that alternative therapies may significantly impact their goal of cure is enough to have them reconsider.

It is worthwhile to review the goals of the therapy in these situations. For instance, if a patient with cancer brings up the topic of acupuncture, the natural reaction might be that acupuncture has no role in treating cancer—this would be correct if acupuncture was used to cure the cancer. However, if the patient's goal was to treat pain or nausea, strong evidence exists that acupuncture is a reasonable treatment option. It may be equally important for patients to discuss ongoing symptoms or other issues for which the patient might be seeking alternative therapies. An open discussion regarding needs and goals of care will help both the physician and patient place treatment options in the appropriate context.

Accepting Patient Choices: Respecting Autonomy

Despite a clinician's efforts to inform patients about the risks of alternative therapies and encouragement to pursue the safe evidence-based options, some patients will still wish to pursue alternative therapies. One should not feel any sense of failure or shame. Often times, patients may have strongly held emotions or beliefs about certain therapies. Just as Jehovah's witnesses generally refuse blood transfusions, we should not feel our job is to change their beliefs. Instead, we should still aim to provide them with the most accurate information based on the most current science to allow them to make an informed decision. Afterwards, regardless of their decision, we should provide them the best care possible within the limits provided by the patient and the limits of medically appropriate, ethical, and safe guidelines. The next decision will be what type of follow-up plan is needed to evaluate for safety and efficacy—again emphasizing continued care and nonabandonment. Unlike Western trained physicians, CAIM practitioners may not incorporate appropriate follow-up assessments for safety and efficacy. Discuss with the patient what potential risks exist and how it would be best to monitor for these. In addition, it is useful to decide prior to the treatment the goals and criteria for success or failure of the alternative therapy within a certain time frame. Without well-defined criteria for success, these types of CAIM therapies could be continued indefinitely without any clear benefit. As an example, if a young woman with stage II breast cancer decides to pursue an alternative treatment, consider follow-up imaging within 2–3 months. Creating a follow-up plan serves to help keep the patient safe and minimizes the duration of ineffective treatment. This strategy provides an avenue for the patient to return to the clinician to seek further help if ultimately these alternative treatments prove to be ineffective. Otherwise, patients without a follow-up plan may feel abandoned and when these CAIM treatments fail, are too embarrassed to return to see the physician for further medical care.

Because of the difficult nature of these discussions, a referral to an integrative medicine specialist will provide patients with an avenue to explore their interest in CAIM and provide safe and effective treatment choices. Another strategy we utilize for patients in this situation is to convey to them that you are most interested in helping them with their goals of care and providing them your best medical advice. Acknowledging to the patient that he or she as the patient has the right to make their own decisions, and if they choose alternative therapies, we do not abandon them and remain available to help monitor them for potential side effects and treatment response. This allows an open opportunity for the patient to return to your care if and when such alternative treatments fail to provide an effective cancer treatment.

SUMMARY

Discussing the use of CAIM therapies with patients can be a daunting task, especially for physicians with limited knowledge about the field. Begin with creating a strong CP relationship built upon excellent communication. At the onset, clearly state your interest in discussing the topic of CAIM therapies, interest in treating the patient in the best way possible and commitment to helping them even if they choose different options from what is recommended. From this foundation, clinicians should proceed with an open-minded, empathetic, and ethical approach to the patient along with a comprehensive assessment of the CAIM therapy and clinical situation. Inquiring why patients are seeking these therapies may uncover unmet needs of the patient, including the need for emotional support. Establish realistic goals of care through a shared decision-making process. As the discussion proceeds to evaluating different CAIM treatments, the clinician should ascertain if patients understand what is meant by reliable information. Review the topics of risks, benefits, and costs—especially time and energy to the patient. Ultimately, if the patient chooses to pursue unproven alternative therapies despite your best advice, protect the patient from further harm by formulating a follow-up plan that will assess safety and efficacy in the context of the goals established. Because many of the unproven alternative therapies will eventually fail, the time and energy spent constructing a strong CP relationship will foster a clear path for the patient to return to receive further cancer care from a trusted and reliable health care professional.

References

[1] Voegelin E. Order and history. Baton Rouge: Louisiana State University Press; 1956.
[2] Engel GL. The need for a new medical model: a challenge for biomedicine. Science 1977;196(4286):129–36.
[3] "Constitution" The First Ten Years of the World Health Organization, 1958, Geneva, WHO, p459.
[4] Ashley BM, O'Rourke KD. Health care ethics: a theological analysis. 4th ed. Washington DC: Georgetown University Press; 1997 p3.
[5] DuPre A. Communicating about health: current issues and perspectives. Mountain View, CA: Mayfield Pub. Co; 2000. p. 376
[6] Sanft TB, Von Roenn JH. Palliative care across the continuum of cancer care. J Natl Compr Canc Netw 2009;7 (4):481–7.
[7] Clark D. From margins to centre: a review of the history of palliative care in cancer. Lancet Oncol 2007;8 (5):430–8.
[8] World Health Organization. National cancer control programmes: policies and managerial guidelines. 2nd ed. Geneva: World Health Organization; 2002. p. 180
[9] World Health Organization. Cancer pain relief and palliative care: report of a WHO expert committee. World Health Organ Tech Rep Ser 1990;804:1–75.
[10] Ferris FD, et al. Palliative cancer care a decade later: accomplishments, the need, next steps—from the American Society of Clinical Oncology. J Clin Oncol 2009;27(18):3052–8.
[11] Levy MH, et al. NCCN clinical practice guidelines in oncology: palliative care. J Natl Compr Canc Netw 2009;7(4):436–73.
[12] Ferrell B, et al. The national agenda for quality palliative care: the National Consensus Project and the National Quality Forum. J Pain Symptom Manage 2007;33(6):737–44.
[13] National Institue of Clinical Excellence. Guidance on cancer services: improving supportive and palliative care for adults with cancer; 2004.
[14] Institute of Medicine. Ensuring quality cancer care; 1999.
[15] Institute of Medicine. Improving palliative care for cancer; 2001.
[16] Institute of Medicine. Cancer care for the whole patient: meeting psychosocial health needs; 2007.
[17] NCCIH. <https://nccih.nih.gov/health/integrative-health> [accessed 28.5.2015].
[18] Academic Consortium for Integrative Medicine and Health. Introduction <https://www.imconsortium.org/about/about-us.cfm>; 2015 [accessed 28.12.2015].
[19] Black LI, et al. Use of complementary health approaches among children aged 4–17 years in the United States: National Health Interview Survey, 2007–2012. Natl Health Stat Report 2015;78:1–19.
[20] Clarke TC, et al. Trends in the use of complementary health approaches among adults: United States, 2002–2012. Natl Health Stat Report 2015;79:1–16.
[21] Navo MA, et al. An assessment of the utilization of complementary and alternative medication in women with gynecologic or breast malignancies. J Clin Oncol 2004;22(4):671–7.
[22] Mao JJ, et al. Use of complementary and alternative medicine and prayer among a national sample of cancer survivors compared to other populations without cancer. Complement Ther Med 2007;15(1):21–9.
[23] Richardson MA, et al. Complementary/alternative medicine use in a comprehensive cancer center and the implications for oncology. J Clin Oncol 2000;18(13):2505–14.
[24] Dy GK, et al. Complementary and alternative medicine use by patients enrolled onto phase I clinical trials. J Clin Oncol 2004;22(23):4810–15.
[25] Ernst E. Complementary therapies in palliative cancer care. Cancer 2001;91(11):2181–5.
[26] White P. Complementary medicine treatment of cancer: a survey of provision. Complement Ther Med 1998;6:10–13.
[27] Astin JA, et al. Breast cancer patients' perspectives on and use of complementary and alternative medicine: a study by the Susan G. Komen Breast Cancer Foundation. J Soc Integr Oncol 2006;4(4):157–69.
[28] DiGianni LM, Garber JE, Winer EP. Complementary and alternative medicine use among women with breast cancer. J Clin Oncol 2002;20(18 Suppl.):34S–8S.
[29] Morris KT, et al. A comparison of complementary therapy use between breast cancer patients and patients with other primary tumor sites. Am J Surg 2000;179(5):407–11.
[30] Ganz PA, et al. Quality of life in long-term, disease-free survivors of breast cancer: a follow-up study. J Natl Cancer Inst 2002;94(1):39–49.

[31] Boon HS, Olatunde F, Zick SM. Trends in complementary/alternative medicine use by breast cancer survivors: comparing survey data from 1998 and 2005. BMC Womens Health 2007;7:4.
[32] Matthews AK, et al. Complementary and alternative medicine use among breast cancer survivors. J Altern Complement Med 2007;13(5):555–62.
[33] World Health Organization: Fifty-Sixth World Health Assembly. "Traditional Medicine: Report by the Secretariat", Doc. A56/18. March 31, 2003. Available at: <http://apps.who.int/gb/archive/pdf_files/WHA56/ea5618.pdf>. [accessed 11.30.2015]
[34] Molassiotis A, et al. Use of complementary and alternative medicine in cancer patients: a European survey. Ann Oncol 2005;16(4):655–63.
[35] Richardson MA, et al. Discrepant views of oncologists and cancer patients on complementary/alternative medicine. Support Care Cancer 2004;12(11):797–804.
[36] Verhoef MJ, et al. Reasons for and characteristics associated with complementary and alternative medicine use among adult cancer patients: a systematic review. Integr Cancer Ther 2005;4(4):274–86.
[37] Larsson SC, et al. Multivitamin use and breast cancer incidence in a prospective cohort of Swedish women. Am J Clin Nutr 2010;91(5):1268–72.
[38] Lawson KA, et al. Multivitamin use and risk of prostate cancer in the National Institutes of Health—AARP Diet and Health Study. J Natl Cancer Inst 2007;99(10):754–64.
[39] The Alpha-Tocopherol, Beta Carotene Cancer Prevention Study Group. The effect of vitamin E and beta carotene on the incidence of lung cancer and other cancers in male smokers. N Engl J Med 1994;330(15):1029–35.
[40] Ernst E, et al. Mind-body therapies: are the trial data getting stronger? Altern Ther Health Med 2007;13(5):62–4.
[41] Mustian KM, et al. Multicenter, randomized controlled trial of yoga for sleep quality among cancer survivors. J Clin Oncol 2013;31(26):3233–41.
[42] Chen Z, et al. Qigong improves quality of life in women undergoing radiotherapy for breast cancer: results of a randomized controlled trial. Cancer 2013;119(9):1690–8.
[43] Chandwani KD, et al. Randomized, controlled trial of yoga in women with breast cancer undergoing radiotherapy. J Clin Oncol 2014;32(10):1058–65.
[44] Bower JE, et al. Yoga for persistent fatigue in breast cancer survivors: a randomized controlled trial. Cancer 2012;118(15):3766–75.
[45] Palmer JB, et al. Effects of music therapy on anesthesia requirements and anxiety in women undergoing ambulatory breast surgery for cancer diagnosis and treatment: a Randomized controlled trial. J Clin Oncol 2015;33(28):3162–8.
[46] Hilliard RE. The effects of music therapy on the quality and length of life of people diagnosed with terminal cancer. J Music Ther 2003;40(2):113–37.
[47] Archie P, Bruera E, Cohen L. Music-based interventions in palliative cancer care: a review of quantitative studies and neurobiological literature. Support Care Cancer 2013;21(9):2609–24.
[48] Ezzo J, et al. Manual lymphatic drainage for lymphedema following breast cancer treatment. Cochrane Database Syst Rev 2015; 5:CD003475.
[49] Cassileth BR, Vickers AJ. Massage therapy for symptom control: outcome study at a major cancer center. J Pain Symptom Manage 2004;28(3):244–9.
[50] Collinge W, MacDonald G, Walton T. Massage in supportive cancer care. Semin Oncol Nurs 2012;28(1):45–54.
[51] Kutner JS, et al. Massage therapy versus simple touch to improve pain and mood in patients with advanced cancer: a randomized trial. Ann Intern Med 2008;149(6):369–79.
[52] Collinge W, et al. Touch, caring, and cancer: randomized controlled trial of a multimedia caregiver education program. Support Care Cancer 2013;21(5):1405–14.
[53] Garcia MK, et al. Systematic review of acupuncture in cancer care: a synthesis of the evidence. J Clin Oncol 2013;31(7):952–60.
[54] MacPherson H, et al. The York acupuncture safety study: prospective survey of 34 000 treatments by traditional acupuncturists. Bmj 2001;323(7311):486–7.
[55] Ernst E, White AR. Prospective studies of the safety of acupuncture: a systematic review. Am J Med 2001;110(6):481–5.

[56] Pierce JP, et al. Greater survival after breast cancer in physically active women with high vegetable-fruit intake regardless of obesity. J Clin Oncol 2007;25(17):2345–51.
[57] Pierce JP, et al. Influence of a diet very high in vegetables, fruit, and fiber and low in fat on prognosis following treatment for breast cancer: the women's healthy eating and living (WHEL) randomized trial. JAMA 2007;298(3):289–98.
[58] Meyerhardt JA, et al. Impact of body mass index and weight change after treatment on cancer recurrence and survival in patients with stage III colon cancer: findings from Cancer and Leukemia Group B 89803. J Clin Oncol 2008;26(25):4109–15.
[59] Meyerhardt JA, et al. Association of dietary patterns with cancer recurrence and survival in patients with stage III colon cancer. JAMA 2007;298(7):754–64.
[60] Jeon J, et al. Impact of physical activity after cancer diagnosis on survival in patients with recurrent colon cancer: findings from CALGB 89803/Alliance. Clin Colorectal Cancer 2013;12(4):233–8.
[61] Meyerhardt JA, et al. Impact of physical activity on cancer recurrence and survival in patients with stage III colon cancer: findings from CALGB 89803. J Clin Oncol 2006;24(22):3535–41.
[62] Courneya KS, et al. Effects of aerobic and resistance exercise in breast cancer patients receiving adjuvant chemotherapy: a multicenter randomized controlled trial. J Clin Oncol 2007;25(28):4396–404.
[63] Courneya KS, et al. Effects of exercise dose and type on sleep quality in breast cancer patients receiving chemotherapy: a multicenter randomized trial. Breast Cancer Res Treat 2014;144(2):361–9.
[64] Courneya KS, et al. A multicenter randomized trial of the effects of exercise dose and type on psychosocial distress in breast cancer patients undergoing chemotherapy. Cancer Epidemiol Biomarkers Prev 2014;23(5):857–64.
[65] Cramp F, Byron-Daniel J. Exercise for the management of cancer-related fatigue in adults. Cochrane Database Syst Rev 2012;11:CD006145.
[66] Kushi LH, et al. American Cancer Society Guidelines on nutrition and physical activity for cancer prevention: reducing the risk of cancer with healthy food choices and physical activity. CA Cancer J Clin 2012;62(1):30–67.
[67] McCune JS, et al. Potential of chemotherapy-herb interactions in adult cancer patients. Support Care Cancer 2004;12(6):454–62.
[68] Friedman T, et al. Use of alternative therapies for children with cancer. Pediatrics 1997;100(6):E1.
[69] Swisher EM, et al. Use of complementary and alternative medicine among women with gynecologic cancers. Gynecol Oncol 2002;84(3):363–7.
[70] Tasaki K, et al. Communication between physicians and cancer patients about complementary and alternative medicine: exploring patients' perspectives. Psycho-Oncology 2002;11(3):212–20.
[71] Wyatt GK, et al. Complementary therapy use among older cancer patients. Cancer Pract 1999;7(3):136–44.
[72] Robinson A, McGrail MR. Disclosure of CAM use to medical practitioners: a review of qualitative and quantitative studies. Complement Ther Med 2004;12(2–3):90–8.
[73] Kao GD, Devine P. Use of complementary health practices by prostate carcinoma patients undergoing radiation therapy. Cancer 2000;88(3):615–19.
[74] Verhoef MJ, White MA, Doll R. Cancer patients' expectations of the role of family physicians in communication about complementary therapies. Cancer Prev Control 1999;3(3):181–7.
[75] Robotin MC, Penman AG. Integrating complementary therapies into mainstream cancer care: which way forward?. Med J Aust 2006;185(7):377–9.
[76] Berk LB. Primer on integrative oncology. Hematol Oncol Clin North Am 2006;20(1):213–31.
[77] Sugarman J, Burk L. Physicians' ethical obligations regarding alternative medicine. J Am Med Assoc 1998;280(18):1623–5.
[78] Epner DE, Ravi V, Baile WF. When patients and families feel abandoned. Support Care Cancer 2011;19(11):1713–17.
[79] Cohen L, et al. Discussing complementary therapies in an oncology setting. J Soc Integr Oncol 2007;5(1):18–24.
[80] Lee RT, et al. National survey of US oncologists' knowledge, attitudes, and practice patterns regarding herb and supplement use by patients with cancer. J Clin Oncol 2014;32(36):4095–101.
[81] Epner DE, Baile WF. Patient-centered care: the key to cultural competence. Ann Oncol 2012;23(Suppl. 3):33–42.

[82] Richardson MA, et al. Discrepant views of oncologists and cancer patients on complementary/alternative medicine. Supportive Care in Cancer 2004;12(11):797–804.
[83] Ulbricht C, et al. Clinical evidence of herb-drug interactions: a systematic review by the natural standard research collaboration. Curr Drug Metab 2008;9(10):1063–120.
[84] Newmaster SG, et al. DNA barcoding detects contamination and substitution in North American herbal products. BMC Med 2013;11:222.
[85] Frenkel M, et al. Approach to communicating with patients about the use of nutritional supplements in cancer care. South Med J 2005;98(3):289–94.
[86] Frenkel M, Ben-Arye E. Communicating with patients about the use of complementary and integrative medicine in cancer care. In: Cohen L, Markman M, editors. Incorporating complementary medicine into conventional cancer care. Totowa: Humana Press; 2008. p. 33–46.
[87] Cohen MH. Using legal and ethical principles to guide clinical decision making in complementary/integrative cancer medicine. In: Cohen L, Markman M, editors. Integrative oncology: incorporating complementary medicine into conventional cancer care. Totowa, NJ: Humana Press; 2008 [Chapter 3].

CHAPTER

7

Clinician Experience: Some Current Ethical Considerations

Robert S. Benjamin[1] *and Nancy B. Benjamin*[2]

[1]The University of Texas MD Anderson Cancer Center, Houston, TX, United States,
[2]US District Courts, Houston, TX, United States

INTRODUCTION

Money, rather than quality of care, has become the key determinant of medical care in the United States. All the participants in the current health care system in the United States, including the federal government, private insurance companies, pharmaceutical companies, entities providing medical treatment, physicians and their advance practice providers, and patients bear some responsibility for the ethical quandaries with which health care providers are struggling. Medicine is much more complex and expensive than it used to be. Diagnostic approaches have advanced from the stethoscope and the chest X-ray to the MRI and the PET/CT. An expensive drug used to cost $100–300 per month; it is now $10,000–20,000. Physicians used to make house calls and provide care for patients regardless of their ability to pay. The only consideration was their own time and effort. Today how much essential information can be obtained in a house call with no laboratory or imaging data available? The present challenge is to find balance and compromise between quality care and financial reality to create and sustain a responsible, ethical, and viable approach to medical care for the United States.

Physicians in countries with socialized medical systems have long been accustomed to considering cost when addressing therapeutic options. In the United States, physicians have long considered predominantly quality. As costs have risen and there is no obvious end in sight, it has become clear that there are limits to funding medical care, and physicians will have to start making decisions based on cost-effectiveness as well as clinical effectiveness, and patients will need to address the same issue.

Ethical Challenges in Oncology.
DOI: http://dx.doi.org/10.1016/B978-0-12-803831-4.00007-5

MEDICAL PRACTICE OR HEALTH CARE INDUSTRY

In most of the world, health care is supported by government and is paid for by taxation. There are rules and limits on what is or is not supported. Some branch of the state decides what treatment and tests can be performed, and it is limited by its health care budget. Drugs are less expensive in countries other than the United States because the state can negotiate price with the manufacturer and even decide that an effective drug is too expensive to be used. Physicians in such systems are more restricted in what they can or cannot do for their patients, and the cost of care has always been a consideration since more money spent in one area means that less is available for another. It is increasingly difficult, if not impossible, for physicians to prescribe off-label medication. There is no problem for patients who fit the label, but often there are patients with rare diseases, who have been shown to benefit from the same drug, but who were not included in the studies leading to the drug's approval. Those patients cannot receive treatment since they do not fit the labeled indications.

The objective of providing health care to sick patients used to be focused on making the patients better or at least minimizing patients' suffering. While that is still the objective of many or even most physicians, both physicians and administrators have learned that medicine is a very profitable business, and many aspects of patient care have changed focus from simply helping patients to how to get paid as much as possible for doing so. What used to be called the practice of medicine is now referred to as "the health care industry." The sophistication and complexity of medical care and research have transformed private practice into large groups, or even corporations, designed to make profit. Even nonprofit institutions have to support huge administrative and, sometimes, research infrastructure from the income from patient care, so an overriding objective of patient care is focused on financial gain.

COVERING THE COST

In the United States, private health insurance pays for most patients, and physicians have been able to decide what tests and treatment were in the patient's interest. Patients with good insurance would get whatever treatment or testing the doctor prescribed. Once patients turn 65, the government-run Medicare system takes over, but most people with private insurance also continue with Medicare supplemental insurance policies to cover what Medicare would not. As health care costs have increased, the insurance companies, whose business seeks profit over expanding medical care, have become more restrictive as to what they will pay for and what they will not. A physician can still order whatever test he or she feels is appropriate and whatever medications, whether on- or off-label, but when the tests and medications are expensive and the insurance companies deny coverage, >99% of patients cannot afford to pay or choose not to do so. Most patients are happy to get treatment that costs $20,000 per month, as long as someone else is paying the bill, but if they have to pay themselves, the situation is altogether different, even if the patients need the treatment to stay alive. The same situation exists with Medicare, where budget restrictions limit coverage. To make things worse, the supplemental private policies that used to cover costs not covered by Medicare now limit that coverage to charges approved by but not paid in full by Medicare. Off-label prescriptions are no longer covered because they are expensive, and the insurance companies' profits go up when they do not have to pay.

When The University of Texas MD Anderson Hospital and Tumor Institute (it only later became The University of Texas MD Anderson Cancer Center) was formed, its charter stated that it would care for any patient from the state of Texas without regard to his or her ability to pay. For that indigent care, the institution received a small level of funding from the state legislature. The expenses of modern medical care today so far exceed those imagined by the founders of that institution that the institution has found it cost-effective to hire an extensive administrative staff to be sure that any patient afforded free care truly qualifies and that all efforts are pursued to collect as much as possible from other potentially responsible third-party payers. Unfortunately, during the time it takes to assure that the patient is financially appropriate for care, a potentially curable patient's cancer may have progressed to the point of incurability. At what point does the general solution to providing financial stability of the institution fail the patient, by missing the chance for cure, and hurt the institution financially by making it responsible for the much more expensive chronic treatment of an incurable patient?

Since insurers may not pay for the tests or treatment physicians prescribe and patients cannot afford to pay without insurance, institutions and practices hire additional staff solely to assure that insurance authorization has been obtained prior to performing a service. That further increases the cost of health care. If the treatment cannot be pre-approved, it does not get done. If the physician thinks that the test or treatment is critical, he or she can appeal to the insurance company to authorize payment by what is referred to as a peer-to-peer appeal. In such an appeal, the primary physician is required to talk to a physician at the insurance company, taking the treating physician away from his or her patient care activities, thus creating both an inconvenience and a financial incentive for the treating physician simply to acquiesce to the insurance company's demands. Furthermore, if the treating physician goes to the trouble of pursuing the peer-to-peer appeal, the physician at the insurance company may approve the diagnostic or therapeutic plan once the rationale has been explained, but the insurance company has rules for its physician to follow that often require the physician caring for the patient to perform unnecessary and inferior testing on the patient in order to justify the test he or she knew was needed in the first place. In that situation, there is still further increase in the cost of health care. As costs of health care increase, the cost of insurance increases, and as those costs increase, employers, who pay for the insurance of their workers, and the federal government, whose health care costs increase, complain and try to focus on cost containment rather than quality of care.

Insurance companies are no greedier than other large businesses, but somehow that greed raises greater ethical concern because insurance is supposed to provide benefits for its customers who are unfortunate enough to be severely ill. Although many previously uninsured patients have access to health care due to the Affordable Care Act, it is not necessarily true that medical care is better for most people. Many patients, whose private insurance used to cover their care at major cancer centers like MD Anderson, Memorial Sloan Kettering, or Dana Farber, are no longer covered at those centers of excellence because those centers are too expensive. Instead, they are directed to institutions that charge less. While the insurance carriers cannot require patients to use cheaper alternatives, their failure to cover the costs at the better but more expensive centers effectively does just that. Patients get inferior care and have inferior outcomes, and then are directed to the centers of excellence after it is too late for effective care.

PHARMACEUTICAL PRICES

Drugs are the most expensive item in the health care budget. Prices of drugs have increased at a higher rate than other health care expenses because pharmaceutical companies in the United States can set any price they want for their new drugs. Congress has prohibited the Centers for Medicare and Medicaid Services (CMS) from negotiating price with the pharmaceutical companies. Prices in the United States are further inflated by the expenses that pharmaceutical companies incur on the overly expensive trials required by the FDA to get approval in the first place, and they have to recoup these costs, both for drugs that get approved and those that fail to obtain approval, in addition to their profits. So if the entire process could be streamlined, cutting down the cost of drug development, and if the price at the back end could be more effectively negotiated, the cost of drugs might come down.

Pharmaceutical companies, like insurance companies and health care providers, should have a responsibility to the patients they serve and not just their stockholders; yet, drug prices continue to rise even after drug approval (as do CEO salaries). In some cases, the very patients in the studies that led to drug approval are asked to pay for their medications while continuing on the study. Some institutions are now beginning to require that statements indicating the extent potential financial risk or guaranteed provision of free drug is included in the informed consent form for the studies. There is also a huge advertising budget for new drugs, not only to physicians, but also to the public, who can pressure their physicians to order the newest, most expensive drug for them, even if a standard alternative would work just fine. Free speech is wonderful, but the older system of advertising or promoting drugs to those who can actually understand what the new drugs do or do not offer and can authorize their purchase is a more ethical approach, resulting in a more effective, less costly treatment strategy, albeit at the cost of some pharmaceutical profits.

THE BUSINESS OF HEALTH CARE

Another factor influencing the cost of health care is the current business model. It is hard to quantify a physician's value, yet the physician is central to all medical care. An expert cannot charge patients more because of his or her knowledge and experience, so the AMA agreed to charges based solely on the extent of documentation, whether or not it is needed to provide adequate or even superb patient care. Physicians (and their institutions) can charge higher prices just by padding the documentation, and if they can get paid more for providing the same level of care, why charge less? Physicians rationalize this practice by thinking that their patients have not suffered because someone else is paying the bill. Ultimately, however, health care costs will continue to spiral out of control, and although physicians are not the primary culprits, they contribute to the problem. In order to maximize payments, physicians (or their employers) hire additional midlevel providers, one of whose major roles is to be sure that documentation is performed to maximize charges, and they hire a billing and coding staff, whose job is to make sure that they get every dollar back that they can. The physicians get paid enough more to cover the expense of the additional personnel. As fully computerized and complicated medical records require more time during each patient visit for completion of the documentation and because physicians spend more of the time during each patient visit with the computer, there will be less time to spend with patients. Notes may be completed more efficiently, improving communication with others, and bills may go out much more quickly, improving the profit margin, but patient care will suffer unless more time is allotted to each patient visit.

The rules are about to change, somewhat with payment to institutions determined by overall patient outcome after correction for the problems being managed. Medical oncologists, for example, will not be faulted if their patients with metastatic cancer die, but presumably they will be rewarded for how long and how well they live. The cost of care may also play a role in the reward system. Current CMS proposals will decrease the rate of reimbursement for hospital- or office-administered drugs to incentivize the use of cheaper alternatives, whether those are really better choices for the patient or not. It is hard to anticipate how the new reimbursement process will sort itself out, but two trends are clear: Institutions and physicians will figure out how to make the most money they can under the established rules and the most profitable approach to patient care will not necessarily be the best for the patient. Most physicians do and will continue to take good care of their patients, but optimal patient care cannot be driven by a goal of profit.

In the United States, cost did not become a concern until recently. As costs have spiraled, many are finally realizing that the money tree is drying up. Ignoring the costs of health care (not the profits) can no longer continue. The recent rate of increase in health care spending cannot continue without bankrupting the nation, so some cost containment policies must be established, and better they be established actively by physicians and other health care professionals than by those with purely financial motivation.

HEALTH CARE COSTS AT END OF LIFE

The majority of health care costs occur during the final year of life. While some of this expense is totally justified, much probably is not. The medical community needs to make some decisions as to when to initiate further costly medical interventions with little or no chance of meaningful benefit. While that statement is very hard for me as a physician to make as it goes against my general treatment approach, I am increasingly aware of its necessity. It does not mean that we should not treat patients who are likely to die anyway, but it does mean that we have to put limits on expensive, last-ditch efforts to buy our patients a few more days. Perhaps the most obvious example is resuscitation and artificial life support of patients at the end of a chronic illness. Patients, and more often family members who do not want to lose their loved ones, frequently insist on resuscitation and intubation when there is absolutely no chance of a patient's recovery. Physicians need to be gatekeepers.

When a patient asks a surgeon to perform an impossible or useless operation, the surgeon can and should say no. When a patient asks a physician to prescribe a drug that would be useless, harmful, or both, the physician can and should say no. In many states, however, physicians are obliged to follow patient and or family wishes for resuscitation, regardless of their logic. Luckily in Texas, two physicians can attest to the medical futility of such wishes and choose to disregard them. But many physicians even in Texas feel compelled to follow family directives in situations where they represent only misguided hope. Patients cannot demand specific medications or surgical interventions—that is practicing medicine without a license—but they can, in many states, demand futile resuscitation and ICU management at an exorbitant cost and then ask someone else to pay for it. Maybe there should be a role for insurance pre-approval in such circumstances.

CONCLUSION

Modern health care providers have access to the best diagnostic and therapeutic tools in the history of medicine. Those tools come at a high price, the price is inflated due to a multitude of additional financial factors, and the price is rapidly becoming unaffordable. All participants in the system, the pharmaceutical industry, the insurance industry, federal and state governments, health care providers—hospitals, large medical groups, and physicians—and even patients, bear some responsibility for the problem, and each group must address the issue, with or without governmental pressure. The challenge for the clinician working within the system is to provide the best possible care for his or her patients while addressing, for the first time, issues of cost-effectiveness and looking at the big picture of what the nation can afford. Physicians have to be their patients' advocates, but they have to be able to tell their patients that what the patients want may not be an option. Until the escalating cost of health care is controlled, it may be impossible to take advantage of the medical advances that are continuing at a rate never seen before.

CHAPTER

8

The Ethical Imperative of Providers to Engage Patients and Families in Improvement

John Bingham[1] *and Doris Quinn*[2]

[1]The University of Texas MD Anderson Cancer Center, Houston, TX, United States,
[2]Consultant in Process Improvement, Emmitsburg, MD, United States

THREE WORDS THAT CHANGE EVERYTHING

It was one of those *kyros* moments—"*moments in time that give meaning to the rest of time*." A call came in from someone very close to both of us. "Hi," she said rather softly. Something was different from the way her voice sounded. She said, "I need to talk to you—about me." There was a long pause and it sounded as if she was crying. Her words were barely audible. She finally said, "I have cancer."

I said, "Karen, I am so sorry" for I knew only too well what was in store for her. "What is the diagnosis?" I asked. I could hear her softly sobbing as she said "I have colon cancer." Once again my heart sank as I knew that among the many cancer diagnoses, this could be one of the worst. She said she had received the diagnosis on Friday and had just told her two boys, ages 26 and 22. Their Father had died four years earlier from cardiomyopathy. The deep emotion behind the new meaning of each word was evident as her voice quivered and the heart wrenching sobs came forth. Everything around me receded into the distant background as I processed the meaning of what she had just shared and what an emotional blow this was to her. She went on to describe the type and stage of cancer—"It is colon cancer, stage IV that has metastasized to the lung, liver, and lymph nodes." This diagnosis could not have been more devastating for me to hear because I knew colon cancer was a difficult cancer to treat at an advanced stage. Stage IV with four organ metastasis, and the aggressive onset of the disease, was not a pretty picture. There was more bad news. She said, "My cancer is a rare type of colon cancer called 'signet ring cell colon cancer'" I knew that a rare colon cancer diagnosis did not bode well. My friend was a nurse practitioner and I know she took fantastic care of herself. How could this be

Ethical Challenges in Oncology.
DOI: http://dx.doi.org/10.1016/B978-0-12-803831-4.00008-7

happening to her? She was very familiar with what a diagnosis of cancer meant and her few words spoke volumes.

As the Chief Quality Officer of the world's leading cancer center at the time, I had often started my talks with "three words that change everything." I would describe the types of questions that I thought patients would want answers to when they heard the words, "You have cancer." Questions like—"Will I see my son or daughter get married?"; "Would I be able to go on our family summer vacation?"; and "What will be the quality of my life and will I even be able to work?" I could not have imagined that I would be hearing these three words from someone so dear to me and she was only 58 years old. Those three words did indeed change everything!

INTRODUCTION

The Beryl Institute states there are two types of suffering for patients: The suffering from their diagnosis and the suffering from the systems in which they receive their care. Do we understand the suffering of our patients? Do we understand the suffering of caregivers? How can we truly relieve the suffering from both the diagnosis and the systems of care [1]?

Karen's journey provides an example of why the authors feel there is an ethical obligation for providers to engage patients and their families not only in improving their health but also to engage them in improving the health care system.

This ethical imperative is grounded on the core ethical principles of *autonomy, beneficence, nonmaleficence, and justice*, and balancing benefits and burdens discussed elsewhere in this book. In particular, this imperative embraces the core ethical principle of *patient autonomy*, which refers to the capability and right of patients to control the course of their own medical treatment and participate in the treatment decision-making process.

Improvement, as defined in this ethical imperative, means not only the important role providers play in partnering with patients to improve their health but also the important role patients have in helping health care professionals improve the system of care. Use of the term "imperative" emphasizes the sense of urgency that engagement of patients and their families in their care and in improving the system of care must have. The current condition of the health care system in the United States, in terms of financial viability, efficacy, affordability, and many other dimensions, speaks to the importance of urgency.

This chapter is based on a construct entitled, *"Framework for Improving as a System"* developed by the authors (Fig. 8.1). It incorporates the six Institute of Medicine (IOM) aims for improvement (i.e., *care that is safe, timely, efficient, effective, equitable, and patient-centered*) [2] and Michael Porter's value-based health care delivery into a cohesive framework [3].

It is an adaptation of the production as a system, introduced by W. Edwards Deming to Japan in 1951 [4]. A version of this system of production was subsequently adapted and introduced by Paul Bataldan [5]. Our diagram is a cancer version of this framework. It illustrates the patient's journey from access to end of life. The patient's experience of care can be quantified by obtaining data on the *measures that matter* using the six IOM aims for improvement domains. Equally important are measures that get to the hearts and minds of caregivers and their experience within the system. These too can be quantified by three

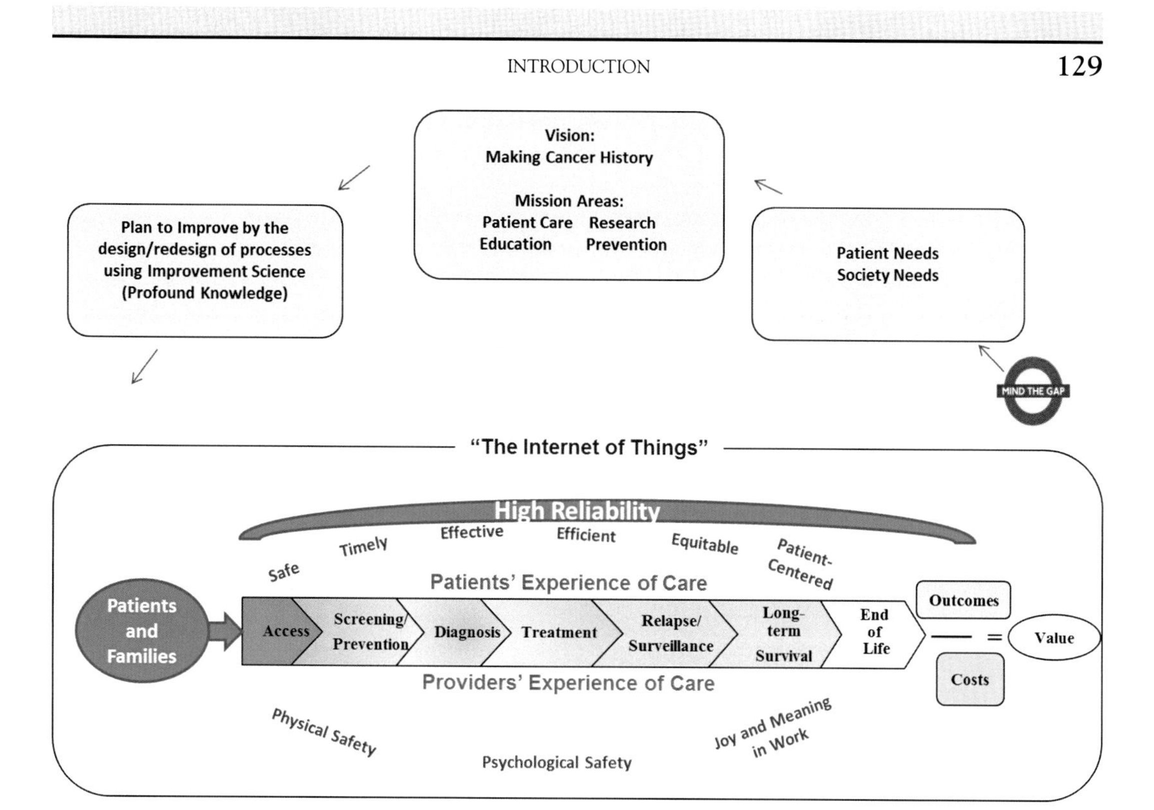

FIGURE 8.1 A framework for improving as a system. *Source: Based on W. Edwards Deming's "System of Profound Knowledge."*

measures: Would staff say they are physically safe working here? Would they say they are psychologically safe here? Would they say that they experience meaningful work?

The system of production generates both costs and outcomes that are cumulative across the continuum of care as patients go through the system. It is here that Michael Porter's value proposition model is applied, where the outcomes of care and the costs of care are measured to determine the value produced by the system [6].

Data for the measures that matter are accumulated along the care journey from patients and their families as well as caregivers within the system. These process measures are evaluated to help determine the reliability of the system of care. The desired state is one where the processes that lead to optimal outcomes are performed with high reliability and consistency over time. They can be aggregated and analyzed to determine the gaps and opportunities for improvement. These gaps between what patients want and need and the performance of the system inform leaders about how well the organization is meeting the identified need. All of this input is used to create a new shared vision of what the organization needs to improve in order to meet patients' needs. The last step is to identify the opportunities for improvement that can impact the patients' and caregivers' experience of the system. We must focus on what should be done to optimize the achievement of health for patients and the achievement of meaningful work for caregivers. Each of these is mutually dependent and equally important, for one cannot be achieved without the other.

Although implementation of the framework for improving as a system provides significant opportunities for any organization to improve, it is not sufficient without patients and families engaged as full partners. The unique perspectives that patients and families have—about what works and doesn't; what adds value and what doesn't; what needs improvement; and what is good enough for now—are essential components of the framework.

Using this framework, we will elaborate on the ethical imperative of engaging patients and families in improvement and how it can be actualized within health care. Snippets from my friend's journey with cancer will be used to illustrate what could and should be part of the health care system. We will argue that there is no alternative more compelling and more impactful than to fully engage patients and families in partnership to improve their health and improve the health care system.

ACCESS/DIAGNOSIS

The journey through the health care system always begins with access: Access to a physician, to facilities, to information, and to a diagnosis. These first two phases of the continuum of care are often very difficult for patients and families. It seems that when they are suffering from an illness, there is a giant castle with a mote of barriers that must first be crossed before treatment can even begin.

Karen had suffered a sudden onset of symptoms that led to a 6 week fragmented journey—and that was just getting to a diagnosis. She had easy access to her primary care physician, probably because she worked every day with primary care physicians in her role as a school nurse, but that was not sufficient. There was no definitive diagnosis yet, but the early opinions were that it was probably "diverticulitis" and conservative treatment was implemented. She was referred to her gynecologist for follow-up but that too was inconclusive. From here, she was referred to a gastroenterologist who provided more definitive guidance but was still inconclusive. The final answer came when she was referred to a medical oncologist along with the appropriate confirmatory pathologist report.

She had bounced from one clinical setting to another with her paper trail of diagnostic efforts following her in a system fraught with delays, difficult scheduling, and missed signals of her impending crisis. Her difficulties in gaining access to the right provider at the right time were evident in the disparate and disjointed referral and appointment processes and lack of timely communications. Although she was an informed consumer and an Advanced Practice Professional with a career spanning multiple health care settings including hospitals, clinics, public health, and school nursing, her health care experience was not that of a system. She was exhausted from just trying to get the initial diagnosis. Although highly activated to participate in her care, she was thwarted by a dysfunctional system that could not easily get to a timely diagnosis.

After 6 weeks, Karen finally had a definitive diagnosis and a plan of care—but she wanted to be sure. She called me to arrange for a second opinion, since I worked at one of the top cancer centers in the world. She wanted our physicians' take on what her providers did, what they found and what they recommended. The faxed reports and paper documents from her multiple providers poured in over the next 12 hours. She had to physically obtain many of them due to "office policies" and "legal requirements" for release of information. The disparate puzzle pieces she was assembling illustrated the challenge she faced in the very first steps of her journey in what we cynically call "the health care system." She searched for anything resembling a "system."

Her records were passed on to three world class cancer physicians for diagnostic and recommended treatment review. Their verdict was returned within hours: Yes, it was a rare form of aggressive cancer called "signet ring cell stage four colon cancer." The tests revealed that the colon was the primary site and that the cancer had metastasized to her lungs, lymph nodes, and liver. No, she would not be a candidate for surgery at our institution given the disease status, and yes, the chemotherapy regime her oncologist had prescribed was identical to what we would have recommended. It was further recommended that she should come to our institution if the first regime of chemotherapy was unsuccessful to see if we had anything additional in the form of clinical trials to offer her. She was also encouraged to get her personnel affairs in order as she began her treatment. We understood the meaning of those words! Her diagnostic phase was now definitive.

The ethical question for us here is whether she was engaged in her care and in improving the system and to what degree? What would the evidence show if the following questions could be answered and supported with data:

- Did her providers know the important information about her medical history?
- Was it readily available to them?
- Did they spend enough time with her to really listen?
- Was she able to get answers to her medical questions in a timely manner?
- Would a more timely diagnosis have improved her treatment options?
- Was she a partner with her physicians during this phase of care?
- Did they have any idea of the journey it took to get to the diagnosis?
- Did they have any data to show that her care was safe, timely, efficient, effective, and patient-centered?
- Did they know if her wishes and choices were honored?
- Did the nonphysician providers know how well their processes worked?
- What percent of the total 6-week time period was face-to-face with providers?
- Did the providers talk to each other to learn what they could improve?
- What did the caregivers learn from her that they could improve?
- Were they clear about what went well and what needed improvement?

TREATMENT

Thus began her *treatment* phase—a summer from hell with chemotherapy. The first 2 weeks of chemotherapy brought numerous trips to the Emergency Department for nausea and pain control. The loss of strength and weight took its toll and required a 2 week delay in treatment just to be able to continue the chemo treatment. Eventually, the medications were adjusted enough for her to tolerate completion of her first course of chemotherapy. Her two college sons and her best friends escorted her as best they could in and out of the system of care—adjusting their work, college, and family schedules to accommodate a single mother's coping with chemo brain and all that goes with it. She finally made it to the halfway point of her first 12-week regime of chemotherapy.

She wanted to take a break and spend some time away from all the medicine and side effects just before her 6-week checkup. She had decided to go to her brother's place in Frisco, Colorado for a week. Although she experienced a few really tough 12-hour visits to the emergency department with neutropenia from the chemotherapy and some episodes of nausea, she was really quite upbeat and optimistic that the drugs were working. Her physician had been thorough in his explanation of what the course of treatment and potential complication would be like and, so far, she was progressing down that path as prescribed. Somehow, this Frisco visit was not just a break from it all at the halfway point.

Upon seeing her, I was stopped in my steps at how much thinner she looked. The stark contrast of her weight loss since I had last seen her and her appearance washed over me like a wave of cold water. I knew there was more to this "time away" plan. I had the sense that this was a "saying good-bye trip" for Karen and that she knew much more about her condition than she had so far let on. The hugs communicated volumes about why this weekend in Frisco was so important in so many ways. The chemotherapy was not impacting the tumors in her lung as expected, and, thus, her breathing was becoming more difficult.

Karen and her family continued to enjoy a few special days together in the high mountains of Colorado. Karen remained upbeat yet somehow quietly stoical about the future—as if the near term future was very clear to her. All returned to their lives back home and Karen to her second half of chemotherapy treatment.

Another large question we need to be asking is "What does it mean to be patient-centered?":

- Was her care "delivered" to her or was she invited to "coproduce" her care?
- Were her two sons and her friends invited to be partners in her journey?
- How much did her caregivers know about her situation?
- Did they know whether their treatment was patient-centered?
- Did they know the roles her two sons were playing?
- Were her wishes and choices being heard?
- Was she intimidated by traditional patient/physician authority gradients?
- Did she feel vulnerable?
- What was happening in the minds and hearts of her caregivers?
- Were their leaders aware of what patient-centered care means?
- Did they know how if they were delivering patient-centered care?
- Was their work meaningful to them as caregivers?
- Were the caregivers able to share with leadership the shortcomings of their systems?
- What did the leaders of the organizations know about how it went for her?
- Did they know about the IOM aims for improvement?
- Did they have measures of their performance in place?
- Did they know about the Colorado visit and its meaning to Karen and her family?

END OF LIFE

Although my regular calls to Karen for her progress continued, a special Labor Day call from her changed everything again. The call came on my cell phone and Karen's name in pixels yanked me back into her reality. I knew she had been in to see her physician to determine the effectiveness of the first 12 weeks of chemotherapy and this would be an important call.

Her words shattered my hopes for some good news. She said, "I am going into Hospice this week. The tumor has spread into both of my lungs and my left lung has collapsed. The cancer is now taking over the right lung and the tumor in my colon has started growing again. Hospice is the best alternative for me now. Can you come to be with me?" I knocked on the front door and her son greeted me with a big hug—a hug filled with the heavy burden of a 22-year-old son on a hospice journey with his young mother.

Karen was asleep on the worn leather couch in her family room. Joey, their aging dog got up from where he was at her feet to say hello and check me out. Karen began to awaken and I quickly went over to give her a hug. She tried to get up but I could tell that between the lack of strength and what I presumed was the effect of the medication, it was difficult for her to even sit up. She gave me what I had come to understand as her "knowing smile" that seemed to communicate everything about her situation.

I sat with her to catch up on everything that was happening. She slowly became more alert, and I could tell she wanted to talk. She thanked me for coming and said how much it meant to her and to the boys. I asked if she was in pain and if she needed anything right now. She said she was fine but that since she had just started hospice care—just 2 days into it—there were a lot of adjustments that would take some time to get just right. She was on an oxygen machine with a 50 foot cord that allowed her to move from her bedroom into the family room and kitchen. The pain was under control as she was on oral morphine every 4 hours. I looked closely at her youngest son and could see the dark rings under his eyes from lack of sleep; the blood shot eyes from tears shed in private; and the strain on his face from trying to deal with the reality of his mothers' condition. My heart broke for this young man who was trying to cope with such an unfair disruption in the normal sequence of a typical family aging pattern.

He filled me in on all that was going on with Hospice and what the drill was on the medications, the oxygen, and the scheduled visits from hospice caregivers. Karen seemed so proud that her son was such an able caregiver. After a while, he went in to take a shower and Karen sat up and became very focused and intent. She said she had some important information to share with me. I had come to dread these moments when Karen prefaced her little talks with words like those.

She looked directly into my eyes, and I could see from the intensity of her blue eyes that this was a really important communication. Her next words were, "I want you to understand that I have completely accepted the fact that my cancer is not stopping and is spreading through my body and that I am in hospice care. I fully understand what hospice care means and I want you to understand my wishes. I want to stay at home and when it is time for me to go, I want to die peacefully at home, preferably asleep, with my boys at my bedside." She went on in that very direct manner that I had become so accustomed to, "I don't want any heroic medical interventions, except to keep me as free from pain and as comfortable as possible. I fully understand what is happening with the spread of my tumors in my remaining lung and I want you to accept this situation as quickly as you can and to honor my wishes. I have shared this exact message with my boys and I want to make sure you understand as well?" She went on, "I have one important request of you. I want you to help my boys get through this and to be there for them after I am gone.

They will need other adults in their lives and you have been the greatest friend our family could have hoped to have. Thank you!"

It was difficult to say anything else, as the clarity and finality of her statements left little room for interpretation or clarification. Given her medical background and mine as well, we both knew exactly what she meant. She talked about her life with her husband and her boys; her career and she said she felt she had lived a full life and that it was her time to go. She said she was "Good with God" and had been able to have wonderful conversations with her church's associate pastor about what was happening and how she felt and where she was with her spiritual life. Her eyes closed as she seemed exhausted from the conversation but relieved. Her breathing became more regular and she drifted off to sleep.

The next few days were a whirlwind of activities and emotions. The day after I arrived, Karen's primary caregiver from Hospice made a visit. This was my first up close and personal experience between a hospice caregiver and their patient—my first glimpse at what coproduction of health care could be. Karen asked me to be present along with her son, so we could see and learn more about her course of care and how she planned for it to be. I was so impressed with the gentle compassion and calm discussion of Karen's condition; what the various medications were for; what the side effects might be; and how we all could help Karen. The hospice caregiver was thorough, professional, and comforting. He engaged all of us in her care and made sure that everyone was aware of and comfortable with the plan of care that Karen had developed for herself. We all felt we had perfect clarity on what was happening; what our roles were; how we could get help if needed; what the course of the next few days would be; and what Karen might be experiencing. Of course, the fact that Karen was a nurse practitioner helped all of the communications in a uniquely meaningful and clear way. I began to understand more fully what it meant to fully engage patients in their care. Karen was coproducing her end-of-life care along with her health care team from hospice and her family.

I spent the last 10 days of Karen's life caring for her along with her boys. There is no question that she coproduced her end-of-life care. Under her direction, I adjusted medications and oxygen levels; took care of her biological and physical needs; empathically listened and cared for her emotional and spiritual needs and helped her execute her plan. We spent many moments together in the sunlight of her front porch—the sunset of her life evident in her gaunt face and frame and protruding collar bones—her drifting in and out of consciousness and escaping into what she called her own "private movie" from the medications and organ failure. We did everything she asked for during this phase of her life. Looking back at her last 30 days of her life and with more knowledge and understanding of what coproduction of health (and dying) means, I know that I witnessed an incredible example of what health care can be. While I doubt that her access, diagnosis, and treatment phases of her journey with cancer were coproduced, I am sure that she coproduced her end-of-life care.

I learned about how her life-sustaining systems would begin to shut down and the changes that would occur in her physical, mental, emotional, and spiritual health. But the instant of her last breath and her passing was the most poignant moment I have ever experienced! She died precisely at 9:00 am on Tuesday, October 8, 2013—as if that is when she planned it. Her two boys were at her bedside, along with her dog Joey, and me. She passed quietly and peacefully, and I still remember the honking sounds of geese coming through her open window on that memorable fall day—as if they came to escort her soul away.

Why Do We Connect the Patient Experience and Quality Improvement?

We believe that the "awareness" of the importance of engaging patients as partners in their care is rapidly growing. While not yet considered an "imperative" by most organizational leaders and providers, there has been clear national guidance for some time on how important it is. In its 2001 *Crossing the Quality Chasm Report*, The IOM made this point abundantly clear in the forward of the report:

> "Fundamental reform of health care is needed
> to ensure that all Americans receive care that is
> safe, effective, patient-centered, timely, efficient, and equitable."
>
> Forward in *Crossing the Quality Chasm*
> by Ken I. Shine, M.D.
> President, Institute of Medicine
> March 2001

The report included "new rules to redesign and improve care" [7].

Those "Ten rules of what patients should expect from their health care" describe how care should extend beyond the traditional patient visit to care in the form that patients want. These rules describe a world in which:

- The patient as the source of control and that the mantra of "nothing about me without me" guides all transactions.
- That patients should expect that their time and resources will not be wasted but will deliver good value that is constantly being improved over time.
- Patients needs will be anticipated and those providing care will engage patients as full partners.
- Experiences in the health care system will be seamless and patients will never feel lost or forsaken.

In the 2012 IOM report, "***Best Care at Lower Cost: The Path to Continuously** Learning Health Care in America*," the IOM became even more explicit about engaging patients in its fourth recommendation which stated [8]:

- "Involve patients and families in decisions regarding health and health care, tailored to fit their preferences.
- Patients and families should be given the opportunity to be fully engaged participants at all levels, including:
  - Individual care decisions;
  - health system learning; and
  - interventions to promote health."

And even more recently, the 2013 IOM *Report Delivering High Quality Cancer Care: Chartering a New Course for a System in Crisis* [9] which had the following recommendation:

> The cancer care team should collaborate with their patients to develop a care plan that reflects their patients' needs, values, and preferences and considers palliative care needs and psychosocial support across the cancer care continuum.

Many professional codes of ethics/policy documents for professionals include language that specifically addresses patient engagement. For example, The American College of Healthcare Executives Professional Policy Statement on health care executive's role in quality and patient safety states that:

> ... health care executives should lead a comprehensive approach to ensuring patient safety and quality, including developing processes to hear the voices of patients and families and applying their input in the design and improvement of care processes [10].

Defining what patient engagement means is evolving over time as more evidence and understanding is accumulated. A more recent definition of patient engagement endorsed by the authors here is "...patients, families, and their representatives and health professionals, working in active partnership at various levels across the health care system—direct care, organizational design and governance, and policy making—to improve health and health care" [11]. This definition emphasizes the patients as the source of control working in "active partnership" at various levels and across the continuum. The authors of this definition propose a robust framework that conceptualizes both the continuum of potential engagement of patients families and their representatives (consultation, involvement, partnership, and shared leadership) to the potential levels of engagement (direct care, organizational design and governance, and policy making) (Fig. 8.2). This framework provides an excellent heuristic for illustrating the dimensions of engagement in any effort, initiative, and intervention or policy designed to improve health.

In addition to the above concepts of patient engagement, there are four principles of patient and family-centered care identified by the Institute for Family Centered Care and used by the American Hospital Association in its *Strategies for Leadership in Patient and Family-Centered Care* [12].

These four principles are:

1. *Dignity and respect*—the extent and depth to which health care providers listen to, honor and incorporate patients and families perspectives and choices into all aspects of the organization.
2. *Information sharing*—the extent to which providers share complete, timely, unbiased, and useful information with patients and families that optimizes the partnership.
3. *Participation*—patients and families are encouraged and supported to participate in care and decision-making at the level they choose.
4. *Collaboration*—the extent to which patients families and providers collaborate in policy and program development implementation and assessment of facility design, professional education, and delivery of care.

These four principles can then be embedded in 10 domains for organizational assessment that further provide a heuristic for assessing the degree to which patient and family centered care exists in an organization.

These 10 domains are:

1. Leadership
2. Mission and definition of quality

3. Chartering and documentation
4. Patients and families as advisors
5. Patient and family support
6. Patterns of care
7. Quality improvement
8. Information and education for patients and families
9. Personnel
10. Environment and design

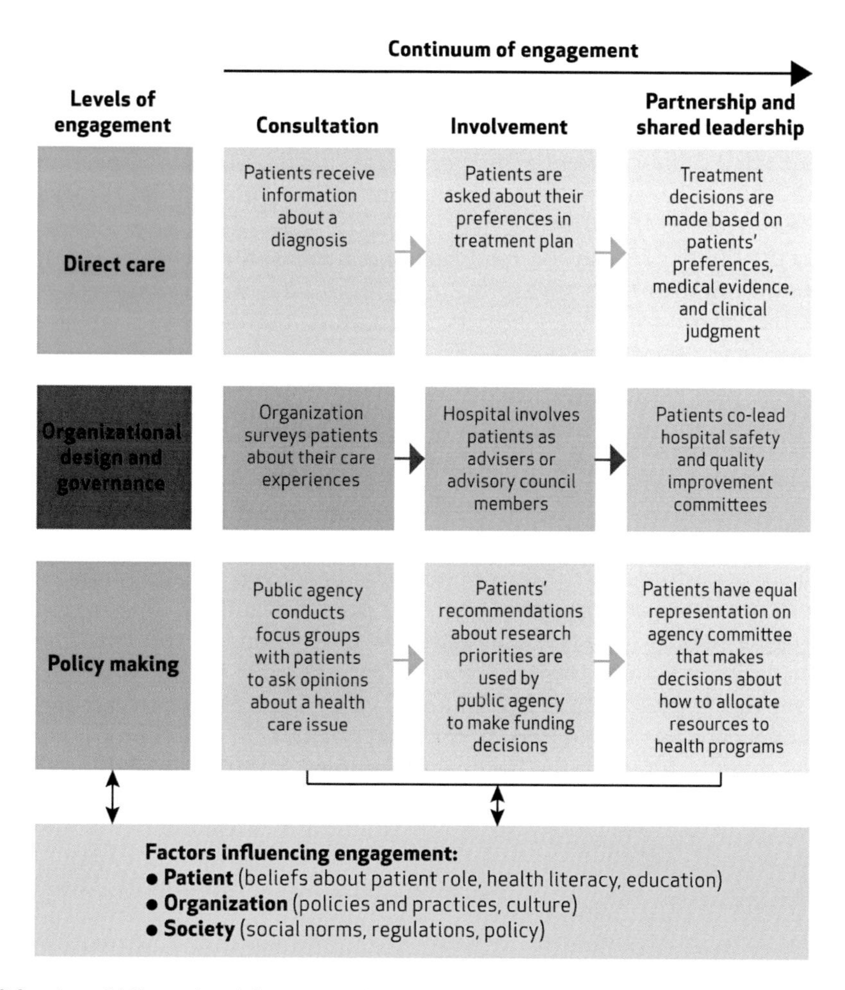

FIGURE 8.2 **A multidimensional framework for patient and family engagement in health and health care.** *Source: From Carmen et al. Patient and family engagement: a framework for understanding the elements and developing interventions and policies. Health Affairs 2013;32(2):223–231.*

The assessment tool provides a set of questions within each domain that provides members of an organization to self-assess their current state on these dimensions of patient and family-centered care.

In his book *Service Fanatics*, James Merlino, MD, former Chief Experience Officer of the Cleveland Clinic, provides great insights into the evolving understanding of what it means to make patients our partners. He describes the evolution of patient involvement concepts from *involve, educate, empower, engage*, to the emerging term of *activate* [13]. These terms add emerging clarity of thought about the potential roles of patients and families with health care professionals and the system of care. They also illustrate the historical paternalistic thread of systems that have been far too provider/organization centric.

Patient activation refers to "understanding one's role in the care process and having the knowledge, skills, and confidence to manage one's health and health care." A tool has been created called "Patient Activation Measure," which gauges a person's self-concept as a manager of his or her health and health care. The measure is both reliable and valid across different cultures and demographic groups, languages and health statuses and has shown evidence that higher patient activation scores have a positive impact on health outcomes, health care experiences, and health care costs (Hibberd JH, Stockard "Patient Activation Measure (PAM) health services research"). Because the evidence also shows that patient activation scores can be modified and increased over time, this tool makes a significant contribution to both measuring and improving patient engagement [14]. The point here is that there are valid methods to quantify the degree to which patients are engaged as partners in their own health care.

Coproduction of Health and Improvement

In the article *Coproduction of healthcare service* [15], the authors propose a conceptual model for coproducing health care service that captures the essence of what we are trying to articulate in this chapter. The central tenet of their model is that "health care services are always coproduced by patients and professionals in systems that support and constrain effective partnership." The opportunity is in seeing and understanding and minimizing or optimizing as appropriate, what supports or hinders this coproduction.

Their conceptual model (Fig. 8.3) illustrates the interactions of supports and constraints impinging on individuals and organizations in coproducing health. Core components of their model are the "coexecution," "coplanning," and "civil discourse" among health care professionals and patients. The understanding that health care is a coproduced service and not a "product manufactured by the health care system" is critical in shifting the paradigm to the coproduction of improved health and health systems. The boundaries of this model become endless when the synergies of the components to do good is optimized.

Path Forward: Where Do We Begin?

Understanding the system in which we work is the first step toward being able to address the IOM aims of safety, timeliness, efficiency, and, especially, patient-centeredness. The real issue starts with the fact that we do NOT understand the details of the many processes in which we work. Tempers flare and people are blamed for all that is wasteful and frustrating in health care. However, every time there is needless waste, rework or delay, the question we should always ask is "what is the process and what part of it is broken?" We

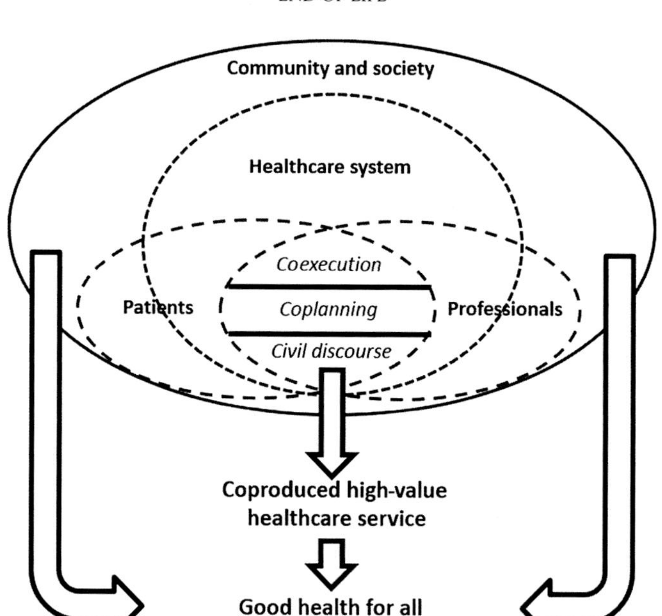

FIGURE 8.3 **Conceptual model of health care service coproduction.** *Source: From Batalden et al. Coproduction of healthcare service. BMJ Qual Saf 2015. doi:10.1136/bmjqs-2015-004315.*

constantly work in systems that are faulty and yet we ignore them. What is even more frustrating is the fact that busy clinicians will tell us they do not have time for process improvement. The interpretation is "I am so busy dealing with waste, rework, and delays that I do not have time to eliminate them from my daily work." Patients tell us what is wrong and we make excuses because we do not KNOW what is wrong. With poor processes everyone loses: patients, staff, clinicians, managers, and the organization.

Don Berwick, MD, in his keynote address to the 2011 Institute for Healthcare Improvement Conference in Orland, Florida stated: "We must mobilize clinicians to make necessary changes needed for health care, or those in Washington, DC will make the health care cuts in order to eliminate waste."

The first step to improve health care is to understand the processes that all link together to create a system of care for patients. Flowcharts are not new, they were used by engineers as early as 1947 and then became popular when computer programming propelled us from the typewriter to the computer. It is a logical method of identifying the steps needed to make something happen. To improve systems, the first step is to capture current work processes from those actually doing the work. Documenting the realities of these processes in a flowchart validates how difficult it can be to meet the patients' needs. These flowcharts teach system thinking as well as promote discussions among the multidisciplinary team about how each of their own processes help or hinder the rest of the team. Only when everyone can see these interdependencies, will they be able to work together more efficiently. An important point about creating flowcharts is that the staff must feel very comfortable providing the current reality of the processes, NOT what should be happening. They must feel "psychologically safe" to tell it like it is! Because the flowcharts represent the reality of work, these conversations should always be nonthreatening and in fact, they can actually be very empowering! Everyone involved in the exercise learns about the environment in which they work: Physicians do not know what the front desk processes are and can be curt with the staff when patients are not checked in quickly; business managers get annoyed when all the information is not available for billing; nurses often function as assistants in order to get the daily tasks done; and mid-level providers can also be utilized below their skill level because of the poor processes in place. This list goes on and on creating a costly system. Therefore, it is imperative that as the flowcharts are developed, the "opportunities for improvement (OFI)" are clearly identified by all those involved and documented on the chart itself where they occur.

Flowcharts are also very important when practices must report outcome metrics. Outcomes can only improve when all the processes are capable and there are process measures to show this. For example, if outpatient centers must document medicine reconciliation, it is helpful to see where that function occurs on the flowchart, especially if that center is scoring poorly with this metric. For the sake of our patients, we must decrease wait times. This can only be accomplished if the current process is understood and can, therefore, be "leaned" [16], which means eliminating the waste and rework causing delays. Otherwise, working harder will not produce any improvement, only more frustration. Another example is administration of antibiotics before surgery. A medical center was shocked to see their own publicly reported time for this mandate. With an understanding of the process, the staff was able to determine who the best person was to administer the antibiotic to achieve the 1-hour window before surgery. Then a process was put in place to assure that this step was met reliably and reported on an ongoing basis.

We have commented on the suffering caused by the health care processes/systems. Let us briefly discuss the suffering to the staff and clinicians. People go into health care because of a desire to help others. They want to do the best job possible, but when the system fights back and they find themselves powerless to help patients, they often become cynical and lose the joy in their work. We have seen this in medical schools where the first year medical students are excited and full of promise. By the fourth year they have completed their clerkships and have seen how difficult it is for them to provide the best care when tests are delayed or drugs are missed and everyone seems to lack the ability to create a seamless system of care. Nursing students also learn how difficult their practice can be and are often the ones accused of not being able to care for patients properly. Support staff are often asked to do multiple tasks like check in patients, answer the phone, and fill requests for clinicians. They often cannot do all the tasks and become angry when blamed for not doing their jobs. A brief understanding of the many tasks and the time needed to accomplish these tasks would have alerted the manager of this impossible demand. Thus, the blame cascades down the authority gradient until staff begin to forget the patient and the suffering they must bear with their diagnosis in addition to our less then optimal systems. Patients tell us stories of grumpy clinicians who do not want to take the time to answer their questions and allay their anxieties. When patients hear confusing care orders, they feel anxious because no one seems to be on the same page. Patients and families should not have to worry that clinicians do not know what is happening to patient at all times.

Another benefit of doing flowcharts is the ability to identify variation in practice. When we have completed flowcharts of a clinic area such as Breast Center or primary care clinic, we often find that several clinicians will want different tests, different medications, etc., for the same group of patients (regardless of individual patient needs). Often this variation is based on traditions of the individual clinicians: Where they trained, who trained them, etc. [17]. Clinicians do not see the problem with needless variation, the staff must remember who wants which test, in which sequence, etc. This variation adds complexity to their often chaotic lives. When variation is surfaced in a process flowchart, it becomes a focus point to identify the best evidence-based practice and who is adhering to the practice. If we are to measure outcomes, there cannot be needless variation as a confounding factor in the process. This addresses the IOM aim of "effective" care. How do we know what we are doing is making a difference to the care of the patient?

Karen's story makes a poignant cry for us to include the voice of the patient in how we design our processes. The ideal way of doing flowcharts is to include patients in the exercise or at least have them review what the staff has created! They will identify areas of improvements that only those who carry the burden of suffering of a serious diagnosis and dealing with our systems will know. Patients/families can tell us what we need to change but also what we are doing well and should continue. A wonderful example of this was a flowchart done in a Breast Center with the staff and two patients. The center had three processes: Routine screening, second view because of suspicious findings, and confirmation of malignancy. Routine screenings were efficient, relatively stress free, and patients were complementary of this process. However, both patients had had cancer, so they were able to comment about the two other processes of second view and diagnosis of cancer. They both told us how stressful it was to be left alone in a room with no knowledge of what was going on. At times, the exam rooms were needed for other patients, so the staff would give the women bathrobes and ask them to wait in the staff kitchen, where they could get coffee or tea and sometimes interact with other women who were anxiously waiting for their results. The staff apologized to the women about having to put them in the kitchen but our two patients said it was the best thing they could have done! This is a great example of how little we understand our patients' needs. Another significant finding on the flowchart was what happened when the woman was told she had cancer. The next question the physician would ask the woman was to decide on a treatment plan. Both women said they could not think about treatment since they were still trying to come to grips with their diagnosis! They needed a survivor to come and be compassionate, provide a shoulder to cry on, and tell them how many women survive breast cancer. Deciding on a treatment plan was the last thing these women could think about, yet the physicians thought that the women would want to get started as soon as possible. On this particular flowchart we created a "stress meter" to identify where the patients felt the most stress. A very clever visual to alert the staff to what was happening to their patients.

Process flowcharts proved a very important tool to disarm a serious situation where care has been suboptimal. One example involved a very prominent individual in the community who was very unhappy about what had happened during his care. He wrote to the medical center leaders and wanted to know how we were going to remedy the situation. The leaders knew that one of the authors (DQ) was involved in process improvement and asked her to intervene. She was able to bring a detailed flowchart currently being used for improvement. The patient walked through the flowchart identifying more issues than the staff could have imagined. The encounter proved very successful. Rather than having to defend ourselves in a court of law, we had an ally for improvement. A similar scenario occurred at another institution, where a patient was not able to get into a particular clinic until his lawyer-wife figured out that he could sign up for a clinical trial and subsequently get to the right clinic. Once more, the patient's family requested that this poor situation be addressed. Luckily, a flowchart of this particular clinic was in progress and the wife was very happy to review all the steps and tell us where we had failed her husband as he battled a diagnosis of cancer. Hospitals that have taken to heart the need to improve, like Cincinnati Children's Hospital, include patients and families on all their improvement teams. There is no doubt that the future of patient-centered care will demand that we use these customers of our systems to help us fix them.

A tool for identifying potential safety issues in a process/system is called a "Failure Mode Effect Analysis" (FMEA). The same flowcharts used to identify opportunities for improvements can also be used to identify possible failures. Such an analysis should also include an in-depth examination of where we fail our patients and where care is not safe, timely, efficient, effective, equitable or patient-centered! With this analysis, the list of improvements will be long, but we will have a much deeper understanding of the suffering our current systems cause our patients and their families and all those who work in our systems. As we stated above, we have an ethical responsibility to decrease the suffering caused by the disease but an equally important responsibility to eliminate the suffering caused by our cumbersome and confusing systems.

Limitations and Future Directions

The authors want to acknowledge that there are significant reasons for the lack of progress in achieving patient engagement and that there are major barriers going forward that must be addressed to achieve the goal.

Some of these are:

- A fragmented delivery system of multiple sites/ownership/governance structures.
- A health care industry centric culture that will require years to change.
- Insufficient time with patients and families.
- Lack of adequate communication systems in a fragmented delivery system.
- Payment system variation that leads to uncoordinated care.
- Variability in accountability structures that confuse patients.
- Insufficient definition of outcomes that matter to patients.
- The lack of data systems to support measures that matter.
- Lack of efficacy in defining process measures that impact outcomes that matter.
- Lack of patient understanding of medical terminology and information overload.
- Lack of training in communication skills.
- Variation in cultural and emotional needs of patients.
- Differing information needs of patients.
- Variation in patient's and families' desires to be engaged.

Final Reflections (JB)

Having served as both a hospital Chief Executive Officer and as a Chief Quality Officer for many years, I have come to appreciate the value of reflection as a great source of learning. I have spent a great deal of time reflecting on Karen's interactions with the health care system and what it truly means to engage patients in their health care journey and in improving the health care system. I know that I witnessed a powerful personal example of the importance of engaging patients in their health, in improving the health care system, and in their end-of-life care. During her last 30 days of life, Karen was fully engaged with her hospice team and her family in coproducing her health care. She provided great insights to her hospice team on where they could improve and what was working and what wasn't. She did it with grace, humor, and kindness. I only hope that I can leverage my lessons learned to benefit others on their personal journeys of improving their health and the health care systems.

There is no question that those acute or chronic moments, where human suffering meets healing and restoration of health, are powerful life-changing moments. In no other place does the nexus of knowledge, experience, and desire to help others become more needed, essential, and poignant for patients. However, if one calculates the moments a person spends in a physicians' office, in ambulatory settings, in a hospital—it quantitatively shows that health care professionals are currently just "bit players" in the lifetime health journey of a person. But imagine what could be when patients and their families have the star role as both actors (actresses) and coproducers of their health. Imagine when our society's forces are resources are optimized to produce better health and health care systems. It is this model of care to which we need to aspire.

Final Reflections (DQ)

I have been doing flowcharts for almost 30 years. Each time a team has come together, I have seen incredible moments of awareness where frustrated clinicians and staff come to the realization that no one knew the whole picture. Each person does the best he/she can without knowledge what others must do. We often admonish our staff to work better as a team, to be more respectful of each other, we send them to training, but in my experience, nothing brings a team together and disarms the animosity like seeing everyone's role in the care of patients and the frustrations often caused by their own colleagues. Waste, rework, delays, and workarounds all come to light, and so much more so, if the patients can tell us their journey through our systems.

References

[1] Lee TH. The word that shall not be spoken. N Engl J Med 2013;369:1777–9. Available from: http://dx.doi.org/10.1056/NEJMp1309660.

[2] Institute of Medicine. Crossing the quality chasm: a new health system for the 21st Century; 2001.

[3] Porter ME, Olmstead TE. Redefining health care: creating value-based competition on results. Harvard Business School Press; 2006.

[4] Demings WE. Out of the crisis. Massachusetts Institute of Technology; 1982.

[5] Batalden PB, Mohr JJ. Building Knowledge of Health Care as a System. Quality Management in Health Care 1997;5(3):1–12.

[6] Porter ME, Olmstead TE. Redefining Health Care: Creating Value-Based Competition on Results. Harvard Business School Press; 2006.
[7] Institute of Medicine. Crossing the quality chasm: a new health system for the 21st Century; 2001. 63.
[8] Institute of Medicine. Best care at lower cost: the path to continuously learning health care in America; 2012.
[9] Institute of Medicine. Delivering high-quality cancer care: charting a new course for a system in crisis; 2013.
[10] American College of Healthcare Executives. The healthcare executive's role in ensuring quality and patient safety. Healthcare Executive 2009;24(2):88.
[11] Carman KL, Dardess P, Maurer M, Sofaer S, Adams K, Bechtel C, et al. Patient and family engagement: a framework for understanding the elements and developing interventions and policies. Health Affairs 2013;32 (2):223–31. Available from: http://dx.doi.org/10.1377/hlthaff.2012.1133.
[12] Institute for Family-Centered Care. Strategies for leadership, patient- and family-centered care; 2004.
[13] Merlino J. Service fanatics: how to build superior patient experience the Cleveland Clinic way. McGraw Hill Professional; 2014. ISBN:978-0-07-183325-7.
[14] Hibbard JH, Greene J. What the evidence shows about patient activation: better outcomes and care experiences; Fewer data on costs. Health Affairs 2013;32(2):207–14. Available from: http://dx.doi.org/10.1377/hlthaff.2012.1061.
[15] Batalden M, Batalden P, Margolis P, et al. Coproduction of healthcare service. GMJ Qual Saf Published Online First, <http://dx.doi.org/10.1136/bmjqs-2015-004315>; 2015 [accessed 16.09.15]
[16] Graban M. Lean Hospitals: Improving Quality, Patient Safety, and Employee Satisfaction. New York, NY: Productivity Press; 2009.
[17] The Dartmouth Institute for Health Policy and Clinical Practice. What kind of physician will you be? Variation in health care and it's importance for residency training; 2012.

CHAPTER

9

Ethical Considerations in Human Subjects Research: Emerging Issues in Cancer Research

Tyron C. Hoover and Aman Buzdar
The University of Texas MD Anderson Cancer Center, Houston, TX, United States

INTRODUCTION

Recent advancements in medical, technological, and scientific fields bring unprecedented potential for the prevention of disease and alleviation of human suffering as well as new, and newly important, ethical considerations. We live in a time of remarkable access to information and data; this underlies some emerging ethical scenarios in the field of human cancer research.

Today, data are generated at an astounding rate. It is consumed by a much larger and broader audience than ever before in human history. Depending on one's perspective, this may be a good thing or a bad thing, but it is unquestionably a thing which manifests itself in some of the specific ethical quagmires that we present in this chapter. Where data exist, ownership, access, privacy, and security issues naturally follow.

In cancer research, data may be generated via traditional methods such as clinical measurements and laboratory analysis (some of which may yield new types of data such as genetic test results). New means of data generation are on the horizon as well thanks to online social media and personal biometric monitoring devices that are becoming increasingly common.

The relative positions of persons interested in research data, regardless of how it is generated, may lead to differing views of how to best utilize and manage it. Some of these differences may be easily resolved, while others may have no obvious best choice.

Here, we present first the basics of what, in the United States, constitutes human subjects research followed by a summary of the legal and ethical framework underpinning that research. With that backdrop, we then move on to discuss some current and

Ethical Challenges in Oncology.
DOI: http://dx.doi.org/10.1016/B978-0-12-803831-4.00009-9

emerging ethical scenarios encountered with increasing regularity in the oversight of human cancer research.

HUMAN SUBJECTS RESEARCH DEFINED AND HISTORY OF REGULATIONS

Research lies at the heart of modern medical science. The Merriam-Webster dictionary defines research as a "(1) careful or diligent search; (2) studious inquiry or examination, *especially*, investigation experimentation aimed at the discovery and interpretation of facts, revision of accepted theories or laws in the light of new facts, or practical application of such new or revised laws; and (3) the collecting of information about a particular subject" [1]. Stedman's Medical Dictionary defines research as "(1) the organized quest for new knowledge and better understanding, e.g., of the natural world, determinants of health and disease. Several types of research are recognized: observational (empiric), analytic, experimental, theoretic, applied. (2) To conduct such scientific inquiry" [2].

Both dictionaries make clear that the key features of research are actions coupled with specific intent. Without searching, inquiring, examining, collecting, or questing [1,2], a mere intent to discover facts, revise theories, or apply them for better understanding would not be sufficient. It's more than simple serendipity, more than one-off experimentation, and more than simply intending to learn something new. It is a more regimented and standardized approach to fact-driven knowledge development whose intent is to actually effect change.

With few exceptions, advancements in health care develop as a result of research. The field of cancer medicine is no exception. The great majority of clinical advancements continue to arise as the result of research conducted under the framework of clinical trials and related studies. For example, ClinicalTrials.gov, a service of the US National Institutes of Health (NIH) which provides open access to information about a large percentage of clinical trials in the United States, reports that 198,340 research studies have registered with ClinicalTrials.gov as of September 14, 2015 (of which 46% are conducted at non-US only locations, while 38% are conducted only in the United States) [3].

These research studies generate vast quantities of data, much of it genetic. For example, Illumina, a prominent biotechnology company focused on genomics, believes a particular type of genetic data alone, next-generation sequencing data, has doubled in volume every year since 2007 (over a 1000-fold increase); many other sources of data exist and are also rapidly proliferating (proteomics, metabolomics, social media-derived data, personal biometric data, medical records) [4]. This data must be properly managed. Management begins with adherence to applicable laws and regulations.

Perhaps more than most other bodies of law, the US laws governing human subjects research are based solidly on ethical principles. The laws and rules themselves contain explicit references to ethics (see, e.g., the charge to the authors of the Belmont Report codified in the National Research Act of 1974). As such, any discussion of current ethical issues involving medical research, at least in the United States, should start and be bounded by the four corners of these laws.

The US legal system has codified a definition of research in The Federal Policy for the Protection of Human Subjects, or, as it is more commonly known, the Common Rule, a set of very similar laws governing human subjects research and codified in various places of the Code of Federal Regulations by at least 16 different federal departments and agencies [5]. The Common Rule embodies the US federal government's system for the protection of human research subjects and "is heavily influenced" by the Belmont Report, which, written in 1979 by the National Commission for the Protection of Human Subjects of Biomedical and Behavioral Research, states the oft-cited basic ethical tenets which should underpin human research, respect for persons, beneficence, and justice [5]. The Belmont Report in turn represented an advisory report to the federal government in response to and following abuses and atrocities committed under the guise of research by the Nazis, increased public awareness of the need for protection of human patients as a result of injuries resulting from uninformed exposure to the drug thalidomide (which while approved in Europe had not been approved in the United States by the FDA and led to the so-called Kefauver Amendments to the Food, Drug and Cosmetic Act and required drug manufacturers to prove effectiveness of drugs for the first time), the Declaration of Helsinki (a 1964 statement by the World Medical Association to guide medical research), and the unethical conduct of the United States Public Health Service's Tuskegee Syphilis Study (which resulted not only in a class-action lawsuit that settled for $9 million and an eventual apology in 1997 by President Clinton but to the passage of the National Research Act in 1974). The National Research Act established the National Commission for the Protection of Human Subjects of Biomedical and Behavioral Research. The Commission (charged with identifying the basic ethical principles which should underlay human research, developing guidelines to be followed based on those ethical principles, and recommending administrative action to the federal government to apply those principles to research supported by the federal government) issued the Belmont Report, which serves, along with subsequent federal regulations from the 1970s through the last decade, as the ethical and legal basis for the Common Rule [6,7]. It is the Common Rule that provides the basic legal and ethical framework for conducting medical research, including clinical trials and biobanking studies, in the United States.

Clinical trials are a form of human subjects research designed to evaluate the effects of interventions on the health of human beings [8]. The National Cancer Institute broadly categorizes clinical trials into treatment trials, prevention trials, screening trials, and quality-of-life/supportive care/palliative care trials [9]. Treatment trials evaluating drugs and their impacts on the human condition are the most common type of clinical trial seen in oncology research today.

The evaluation of new drug interventions typically occurs via evaluating the drug's effects during each of three standard phases of trials in humans: Phase I–III trials. The findings from these studies provide evidence of the drug's safety and effectiveness in applications to the FDA seeking approval to market the drug.

Phase I trials typically involve 15–30 subjects and are designed to determine a safe dose, how best to give the drug (orally, intravenous, etc.), and how the drug affects the body (typically in otherwise healthy individuals). Phase II trials involve more subjects ($\sim$100 typically) and are designed to evaluate the effects of the drug on a certain cancer (as well as continue to assess how it affects the body). Phase III trials can involve thousands of patients and are designed to determine how well the drug compares with existing treatments. [9]. Progression through these trials typically takes about a decade and can cost hundreds of millions of dollars. However important, it should be remembered that clinical trials are just one of multiple steps in the development and sale of a new drug ([10]; for a concise, infographic summary of the FDA drug approval process, see Ref. [11]). It is within this conceptual framework of human subjects research that the ethical issues considered below arise. Before we move into those topics, however, it is worth considering a brief review of the current legal and ethical framework that guide the conduct of these trials, and, thus, impact certain aspects of the associated ethical issues.

CURRENT LEGAL AND ETHICAL FRAMEWORK

Under the Common Rule, human subjects research must be undertaken only with appropriate informed consent of the research participant and must, with limited exceptions, be approved by an authorized research oversight body, the Institutional Review Board (IRB) [12].

Another increasingly important Federal law impacting the conduct of human subjects research is the Health Insurance Portability and Accountability Act of 1996 (HIPAA). HIPAA privacy and security rules often impact cancer research studies relying on the identifiability of data associated with research participants. At the very least, HIPAA has changed the way informed consent documents address consenting to data use, requiring, separate from the consent to participate in the study, a specific research authorization to permit the transfer of protected health information to certain entities [13]. Not infrequently, complying with HIPAA can create administrative and ethical questions for some studies, as HIPAA compliance does not necessarily equate to Common Rule compliance and vice versa.

The Healthcare Information Technology for Economic and Clinical Health Act (HITECH), passed in 2009 as part of the American Recovery and Reinvestment Act, essentially broadens the scope of the privacy and security protections under HIPAA and increases potential liability for noncompliance, adding among other things, strict liability for violators maximum penalties of $1.5 million for all violations of identical provisions of the Act [14].

Related specifically to the use of genetic information by employers, the Genetic Information Nondiscrimination Act of 2008 (GINA) prohibits employers and insurers from using genetic information as a basis for discrimination. There are some holes in the protections GINA affords; GINA does not cover life insurance, disability insurance, or long-term care insurance [15].

CLIA, the Clinical Laboratory Improvement Amendments of 1988, and laboratory accreditation requirements relating to the retention of certain types of biomaterials and data also regularly must be considered. CLIA mandates minimum federal standards for all US laboratories that test human specimens for health assessment or the diagnosis, prevention, or treatment of disease [16]. Essentially under Federal law, test results not performed in a CLIA-certified lab may not be disclosed to anyone for those purposes. As part of these rules, CLIA mandates that test results and specimens be retained for minimum time periods, ranging from 2 to 10 years (see, e.g., a summary of CLIA retention requirements—Ref. [17]). However, the College of American Pathologists (CAP), the primary body authorized to inspect and accredit medical labs in the United States, often requires more than the minimum CLIA retention times, typically 10 years (see, e.g., a summary of CAP requirements—Ref. [18]).

One should also consider potential common law actions that could impact one's ability to conduct human subjects research without incurring civil liability. Common law actions are those legal causes of action whose bases are not in statutes or legislative edicts but which have arisen over the years as a result of accumulations of specific court decisions. These are important, and in some ways more threatening than regulatory or statutory violations, in that individual research subjects may have grounds to sue a researcher or institution directly. Typical common law causes of actions that could be available include civil theft (conversion), lack of informed consent (battery), and breach of fiduciary duty [19].

Finally, there are, at least in the arena of banking human research tissue, emerging standards to take into account. Although these all currently remain no more than mere best practices, they are increasingly referenced and do cover ethical aspects of donating and using tissue and data for research. The most commonly considered voluntary standards are the National Cancer Institute's Best Practices for Biospecimen Resources (specifically Section C "Ethical, Legal, and Policy Best Practices"), the International Society of Biological and Environmental Repositories' Best Practices (specifically Section L "Legal and Ethical Issues for Biospecimens"), and the CAP biorepository accreditation program's checklist (though the CAP accreditation is primarily concerned with operations and sample quality rather than ethics) [20–22].

EMERGING ETHICAL ISSUES IN HUMAN CANCER RESEARCH

The Big Deal About Big Data

More data is generated now than ever before in human history. It's estimated that the world's total data are doubling every 2 years. Every 15 minute, the amount of data generated by the human race increases by about 20 petabyte (the same amount of information as would be contained in three Libraries of Congress or half of all written works in human history in all languages) [23]. Eric Schmidt, the former CEO of Google, has said that "[f]rom the dawn of civilization until 2003, humankind generated five exabytes of data. Now we produce five exabytes every two days ... and the pace is accelerating" [24]. Health-care and medical research has not been immune to this explosion of big data, which can be thought of as the accumulation and analysis of large amounts of information, attempting to extract meaning from it [25].

The cost and time to sequence the human genome has fallen from around $1 billion and 13 years to somewhere between $1000 and $1500 and just a few days [26,27]. Much of these genetic data are finding its way online, potentially accessible to many more people than just the original research team. For example, the NIH's Genomic Data Sharing Policy strongly encourages all NIH-funded large-scale human data be transferred to the NIH so that it can be used in subsequent research questions not part of the original participant's study [28].

There are pros and cons to this kind of accessibility. The Federal government's position is that sharing data more openly enhances statistical power by allowing the combining of multiple studies, facilitates reproducibility and validation of data, and facilitates the development of research tools used to analyze such data [28]. The majority of clinical researchers agree with these views, provided there are some measures taken to properly control access and use of the data. It is probably unrealistic to expect that risk of inappropriate access or use of research data will ever be zero. A recent New York Times opinion/editorial piece penned by a University of Maryland law professor reports that since 2009, there have been over 1100 large-scale data breaches at health-care organizations and that "[h]ealth information firms may suffer temporary reputation damage when data escapes, but the people identified by the data could be consigned to a lifetime of worry. Altruistically donating information, they could end up "rewarded" with little more than vulnerability to discrimination" [29]. Increasingly, legislation, such as HIPAA and GINA, aims to protect individuals from misuse of health (and research) data. Laws, however, do not eliminate all risks, cover all possible fact situations, or change the behavior of wrongdoers.

Some data security experts foresee the misuse of biologic and genetic data. At least one has stated that because of advances in technology and data access "[t]he ability of a single person to affect many for good or evil, is now scaling exponentially," and "[e]ven a technology as welcome as personalized cancer treatments may turn out to be the flip side of personalized bioweapons—which means terrorists may soon be able to bioattack anyone *individually*" [30]. Where data are concerned, there is clearly a need to balance the societal benefit of access to large and detailed databases with the rights of individuals whose data constitutes those data sets. Exactly where the balance will or should be struck is open for debate. But it seems that the trend and momentum clearly is in the direction of increasingly open access and availability of research data to more and more potential users.

Tissue and Biobanking

An area of increasing importance in cancer research is biobanking. One can define biobanking as the collection and storage of human-derived biospecimens under a research protocol for future secondary research purposes not initially stated or known at the time of procurement. Research biobanking may occur under its own research protocol or as part of another research protocol (e.g., where a clinical treatment study also collects and stores extra blood or tissue samples).

Biobanking human samples involves ethical issues related to ownership and control of donated samples and data, an unresolved area of the law. Because of this there are increasingly disputes over ownership and control of banked samples and data, and, in general, the risk of potential disputes increases the longer a sample has been banked. The stakeholders most often at odds in such allocation disputes are researchers and institutions. There is an increasing trend for patients and their families to attempt to assert control over previously donated samples and data (e.g., by attempting to direct that samples donated for research be sent to labs of their choosing unrelated to the research study). Resolving such disputes can be time consuming as well as costly so when one can prevent them from happening in the first place it is preferable.

The most often cited legal decision relating to ownership and control of donated tissue for research is the *Moore v. Regents of the Univ. of California* case [31]. In that case, the California Supreme Court held that the patient donor did not retain any property rights in tissue provided for research even where the patient was not aware that the tissue was being taken for research and not for his clinical care. It was essentially a public policy decision by the court to favor the interests of research, and the resultant potential greater good to the public, over individual property rights of the patient. While the matter has not been decided with finality by the federal court system, the handful of cases that have subsequently addressed similar facts have followed the *Moore* court's logic [32]. Yet, the question remains: could there ever be a scenario where a patient retains control and ownership over research samples and data?

From a legal standpoint, it is possible; it would only require the right facts in the right jurisdiction. The few reported legal cases are fact-dependent and could be read to leave the door open to a fact situation where, following the same logic of *Moore*, the patient could own their research samples. It could also happen in state or federal jurisdictions that haven't yet considered the matter as they are not bound by court decisions in other jurisdictions (just as the United States Supreme Court would not be if it ever considers the matter).

From an ethical standpoint, it could also be a possibility. Where a patient provides more than just the samples and data, it is possible that a court could find they still own at least some rights to it (e.g., where the patient also provides funding, assists in recruiting other patients or their family, actively steers the study, etc.). It might be the just view to take in such a scenario.

In either case though we believe that it is best to ensure, as much as possible, that the patient's expectations and understanding of what will happen to research samples they provide match the expectations of the researchers and the hospital. This is best accomplished via a robust informed consent process that includes explicit mention of ownership and control of samples and data. It is our experience that not all research study documents, particularly those of studies opened many years ago, adequately address ownership and control for donated samples.

Clinical Tissue vs Research Tissue

Another issue that arises with some regularity almost seems so basic that it's surprising that some well-educated and well-intentioned researchers, clinicians, and regulators get derailed by it. Yet it does happen, but, in their defense, there may be an increasingly valid reason muddying the line between clinical care and research activities—personalized medicine.

From a purely legal and regulatory standpoint, it's rather simple to delineate patient care from research activities. Under the Common Rule, research is defined as "a systematic investigation, including research development, testing and evaluation, designed to develop or contribute to generalizable knowledge" [33]. A human subject is defined as "a living individual about whom an investigator ... conducting research obtains data through intervention or interaction with the individual, or identifiable private information" [34].

Any activity meeting those two definitions qualifies as human subjects research, and not clinical care (aka "standard of care"), even if there also happens to be a treatment aspect to the research. For example, a clinical trial often includes the administration of a drug, an act commonly performed under the heading of clinical care (albeit under differing circumstances or a different drug) and entails monitoring of subjects using many of the same modalities used for standard clinical care of patients not on research protocols. Yet, because it meets the Common Rule definition of human subjects research, it would still be considered research.

Analogously, there can be confusion about whether human tissue samples from a patient are "clinical" tissue or "research" tissue. The distinction is important because if they are clinical tissue whose primary purpose is to aid in patient health-care management, they are subject to clinical standards of handling and retention and are under the authority of the pathologist, whose duty is first to the patient's clinical care. Under clinical standards the pathologist must retain clinical tissue samples for 10 years [35]. This means that, with few exceptions, clinical samples are not available for research absent specific IRB approval and agreement by the pathologist to provide a portion of clinical tissue.

If the samples are procured only for the purposes of research, then the samples are immediately, and most likely irrevocably, reserved only for research.

Federal law requires that samples tested for clinical care be maintained and tested in a CLIA-certified laboratory [36]. Many research labs lack CLIA-certification and therefore test results obtained in these labs cannot be conveyed to anyone for patient care purposes.

The growing ethical issue is that the line between clinical care and research, even given the apparent bright line regulations above, continues to blur. Personalized medicine aims to tailor a patient's treatment based on their unique characteristics often determined by testing tissue or blood from the person and "matching" it with a proposed treatment. We are still in the early days of this shift away from one-size-fits-all approaches to treatment, so we expect an increasing blurring of the line between clinical and research tissue samples. For example, where a patient agrees to donate a blood sample for genetic testing as part of a research study whose primary purpose or plan, and so explained to subjects as part of the informed consent process, does not include returning research results to anyone, but where it is later learned that a test result is likely to be clinically actionable and meaningful to the patient, it will be critical to know if the genetic testing, even though performed solely for research purposes, was performed in a CLIA-certified laboratory; if it was then, the results could be disclosed under current federal law (although it is not a given that they must be disclosed under current regulations); if it wasn't then, the results could not be disclosed (although it might be possible to "re-run" the test in a CLIA-certified laboratory to allow for compliant disclosure). Conversely, consider the situation where a new laboratory-testing platform, properly validated and operating in a CLIA-certified patient care laboratory, performs as part of its routine analyses of samples a number of tests in addition to those required and ordered for patient management. What happens when some of these "extra" tests have no current clinical utility or are not widely accepted or requested, yet, they are routinely performed and results are by necessity generated (whether or not they are ever reported to the ordering physician or placed into the patient's medical record) because it is more economical to perform them as part of the batch assay? Are the "'extra" results deemed to be clinical because they are performed as part of a clinical assay performed as a whole for patient care? Should the "extra" results be maintained or simply deleted? If maintained, should they be accessible in the future? By whom and for what purpose? By researchers? With or without patient knowledge? By patients? Would this create a new duty (if so for who? researchers, for labs, or for health-care providers?) to continually revisit prior test results to determine whether or not an "extra" result is now clinically actionable when it was not at the time of testing? It is likely that the answers to these kinds of questions will continue to take shape as these kinds of scenarios increasingly play out in practice. However, currently these remain difficult and mostly case-by-case determinations at many institutions.

In the field of cancer medicine, many attempts at personalized care are by necessity performed under the umbrella of research protocols with sample testing occurring in research labs lacking CLIA-certification. To arbitrarily say that a terminal cancer patient's treatment intervention under such a research study, when the patient has exhausted all nonresearch standard of care options, is only research may not suffice. Yet research labs may be bound by their lack of CLIA-certification and unable to notify anyone of potentially actionable and meaningful test results that could impact patient care; the best they can do is perhaps, with IRB approval, suggest that the testing be repeated in a CLIA-certified laboratory.

We believe the trend will be for research labs to obtain CLIA-certification as a means to provide a laboratory environment amenable to the return of research results to patients, physicians, or others as a part of personalized cancer research and treatments. But as before, it is imperative that the plans to do this are made clear to potential research subjects during the informed consent process.

Informed Consent

Using biobanking as an illustrative model once again, informed consent is an emerging ethical area. Biobanking studies, unlike the typical clinical trial, may (1) last indefinitely and (2) contain no inherent treatment aspects. This brings up an ethical concern with properly informing potential research participants about the benefits, if any, vs the risks of donating tissue for research. Specifically, how does one inform a person about risks and benefits if no one knows what those may be at the time of consent? And, how does one adequately inform a person what it means to donate tissue to research when it may be an entirely new concept to them?

Even though the risks involved with participating in a typical biobanking study are minimal and primarily of a nonphysical nature (such as privacy risks related to associated data), because the samples may exist forever, the risks grow over time. As more data become associated with a given sample in a longitudinal fashion, the samples become more useful for researchers, increasing the risk further. Specific future uses of a donated and banked sample may not be known at the time of consent.

Some would argue that it's impossible to inform someone of risks and benefits of unknown future uses of samples and data under a so-called broad consent, where a patient agrees simply to donate tissue to be used in the future for some vaguely defined category of research. They would require researchers to obtain from the original donor specific consent for each future use, a so-called dynamic consent model, to keep donors involved in each use of samples; however, the majority view is still that a broad consent is preferred. Even though one can make an argument for increased autonomy with dynamic consenting, the flip-side could be that a system of self-policing is created whereby participants become ethical and scientific gatekeepers [37]. This kind of gatekeeping by each individual donor of research tissue could ultimately chill research. Indeed, this is the reasoning behind the California Supreme Court's unwillingness to find that patients have any property interest in donated research tissue—the needs of the many outweigh the needs of the few or the one essentially.

In our experience, patients prefer a broad consent and do not desire to be involved in subsequent decisions related to the use of samples. Others have formally studied cancer patients' perceptions and views and have come to the same conclusion [38].

The second difference that can increase the difficulty of obtaining informed consent for banking studies can be overcome during the consent process itself. In our experience, most patients are not already familiar with the concept of research biobanking. When they have heard the term biobank, they usually envision a clinical blood bank or a tissue bank for transplantation. While patients may have a basic grasp of what a treatment study entails [39], general beliefs and knowledge about research biobanking is likely not comparable. A recent study found that less than half of cancer patients opting-in to donate samples to a biobank understood that they were also consenting to allow access to their medical data [40].

Also, attitudes toward biobanks vary with the patient population, something to be strongly considered when developing an informed consent process. A study in Nigeria found that lay persons had limited knowledge about biobanks but did accept the concept when educated about it. [41]. A study in China that found that public trust in medical institutions was low enough that only a little more than 60% of those surveyed wound even consent to donate residual clinical samples, a figure much lower than in the typical North American or European biobank [42,43].

This has led some to begin to consider what it means to consent to provide samples and data to a biobank and how best to go about the informed consent process. Perhaps there could be an argument made that under existing law, the Common Rule, for example, biobanking could be considered merely preparatory to research and not research. There is federal guidance that supports the view that once released from a bank to a recipient researcher that can't easily determine the donor's identity, then that research should not be considered human subjects research [44]. But there is no such guidance regarding the accumulation of the banked materials and data in the first place.

If one subscribes to the view that banking samples for research is not in and of itself research, then obtaining consent, at least the kind we think of under the Common Rule, would not be necessary. Indeed, there have been biobanks that did not use a Common Rule opt-in prospective consent. For example, the BioVU bank at Duke University at one time employed an opt-out or presumed consent model for residual blood having been deemed not to be human subjects research by the IRB and the Federal Office of Human Research Protections [45]. This however is not the majority view, even among proponents of opt-out consents for banking, who still advocate some form of information exchange prior to banking samples [46].

Still, the perceived benefits to medical science (via the availability of more banked samples and data) and high consent rates to biobanks in general relative to the high cost of consenting can make a seemingly compelling argument for an opt-out consent model. [47].

Attractive though it may be, it would not be advisable to institute anything except a prospective informed consent process for biobanks. It's clear, at least for the foreseeable future, that US law will continue to require prospective consent in most cases prior to banking samples and data for future use [48].

CONCLUSION

As long as the ethical tenets underpinning our current regulatory framework remain carefully considered and reflected in applicable laws, and as long as any changes in any of the above ethically ripe areas of opportunity operate in a legally compliant fashion, then such changes should be as welcome as are the advancements in our knowledge and resultant improvements in our patients' care.

References

[1] Merriam-Webster Online Dictionary. <http://www.merriam-webster.com/dictionary/research>, emphasis in original [accessed 15.09.2015].

[2] Stedman's Medical Dictionary Online. <http://stedmansonline.com/content.aspx?id= mlrR0500013718&termtype=t> [accessed 15.09.2015].

[3] ClinicalTrials.gov. Trends, charts, and maps. <https://clinicaltrials.gov/ct2/resources/trends> [accessed 15.09.2015].

[4] Labant M. Big sequencing beclouds big data. Genet Eng Biotechnol News 2015;35(11):1.

[5] U.S. Department of Health & Human Services. <http://www.hhs.gov/ohrp/humansubjects/commonrule> [accessed 15.09.2015].

[6] National Research Act of 1974, Public Law 93-948, as found at <https://history.nih.gov/research/downloads/PL93-948.pdf> [accessed 15.09.2015].

[7] U.S. Dept. of Health & Human Services. <http://www.hhs.gov/ohrp/humansubjects> [accessed 15.09.2015].

[8] Hodgson DC, Tannock IF. Guide to studies of diagnostic tests. In: Tannock IF, et al., editors. Prognostic factors, and treaments, in the basic science of oncology. 4th ed. New York: McGraw Hill; 2005. p. 494.

[9] <http://www.cancer.gov/about-cancer/treatment/clinical-trials/what-are-trials/types> [accessed 15.09.2015].

[10] <http://www.fda.gov/ForPatients/Approvals/Drugs/ucm405622.htm>, where clinical trials are only the 3rd of a 5-step process from discovery to marketing of a drug [accessed 15.09.2015].

[11] <http://www.fda.gov/downloads/Drugs/ResourcesForYou/Consumers/UCM284393.pdf>.

[12] Common Rule, as found at 45 C.F.R. 46 (2009). <http://www.ecfr.gov/cgi-bin/text-idx?tpl=/ecfrbrowse/Title45/45cfr46_main_02.tpl> [accessed 15.09.2015].

[13] HIPAA Privacy Rule Authorization, 45 C.F.R. 164.508(c).

[14] HITECH Act, Public Law 111-5, as found at <http://www.hhs.gov/ocr/privacy/hipaa/understanding/coveredentities/hitechact.pdf> [accessed 15.09.2015].

[15] GINA, 42 U.S.C. 2000ff; 2008.

[16] CLIA, 42 U.S.C. 493; 1988.

[17] <http://www.dph.illinois.gov/sites/default/files/publications/clia-laboratoryrecordretentionrequirements.pdf> [accessed 15.09.2015].

[18] <http://www.ncleg.net/documentsites/committees/PMC-LRC2011/December%205,%202012/College%20of%20American%20Pathologist%20Retention%20Policy.pdf> [accessed 15.09.2015].

[19] The seminal *Moore v. Regents of the Univ. of California* case, found in its entirety at <http://www.eejlaw.com/materials/Moore_v_Regents_T08.pdf> [accessed 15.09.2015].

[20] NCI best practices for biospecimen resources, as found at <http://biospecimens.cancer.gov/bestpractices/2011-NCIBestPractices.pdf> [accessed 15.09.2015].

[21] ISBER best practices for repositories: collection, storage, retrieval and distribution of biological materials for research, as found at <http://c.ymcdn.com/sites/www.isber.org/resource/resmgr/Files/ISBER_Best_Practices_3rd_Edi.pdf> [accessed 15.09.2015].

[22] CAP accreditation for biorepositories; <http://www.cap.org/apps/docs/laboratory_accreditation/lap_info/bio_brochure_042011.pdf> [accessed 15.09.2015].

[23] Enriquez J. Reflections in a digital mirror, the human face of big data, against all odds productions; 2012. p. 18.

[24] Smolan R, Erwitt J. The human face of big data, against all odds productions; 2012. p. 1.

[25] Smolan R, Erwitt J. The human face of big data, against all odds productions; 2012. p. 14.

[26] <https://www.genome.gov/11006943>, <http://www.livescience.com/28708-human-genome-project-anniversary.html> [accessed 15.09.2015].

[27] <http://www.bloomberg.com/bw/articles/2014-12-04/dna-sequencing-craig-venter-says-genomic-era-is-just-starting> [accessed 15.09.2015].

[28] The NIH's Genomic Data Sharing Policy summary at <https://gds.nih.gov/pdf/NIH_GDS_Policy_Overview.pdf>, [accessed 15.09.2015].

[29] <http://www.nytimes.com/roomfordebate/2015/03/02/23andme-and-the-promise-of-anonymous-genetic-testing-10/insure-people-against-genetic-data-breaches> [accessed 15.09.2015].

[30] Smolan R, Erwitt J. The human face of big data, against all odds productions; 2012. pp. 74–7.

[31] *Moore v. Regents of the Univ. of Cal.* 793 P.2d 479; 1990.

[32] *Washington Univ. v. Catalona.* 437 F.Supp.2d; 2006 (where Washington Univ. and not the physician researcher was held to own biospecimens that had been provided, as determined by the court, as *inter vivos* gifts to the university), and *Greenberg v. Miami Children's Hosp. Research Inst.* 264 F.Supp.2d; 2003 (where donor's retained no property rights in research samples).

[33] 45 C.F.R. 46.102(d), as found at <http://www.hhs.gov/ohrp/humansubjects/guidance/45cfr46.html#46.102> [accessed 15.09.2015].

[34] 45 C.F.R. 46.102(f), as found at <http://www.hhs.gov/ohrp/humansubjects/guidance/45cfr46.html#46.102> [accessed 15.09.2015].

[35] Nat'l Cancer Institute's summary of CAP and CLIA retention requirements in their General Policy Manual at <http://home.ccr.cancer.gov/LOP/intranet/PolicyManual/GeneralPolicy/CAPCLIA.asp> [accessed 15.09.2015].

[36] A summary of CLIA certification requirements for research laboratories at <https://www.cms.gov/Regulations-and-Guidance/Legislation/CLIA/Downloads/Research-Testing-and-CLIA.pdf> [accessed 15.09.2015].

[37] Steinbekk KS, et al. Broad consent versus dynamic consent in biobank research: is passive participation an ethical problem? Eur J Hum Genet 2013;21:897–902.

[38] Bryant J, et al. Oncology patients overwhelmingly support tissue banking. BMC Cancer 2015;15:413.

[39] Leroy T, et al. Factual understanding of randomized-clinical trials: a multicenter case control study in cancer patients. Invest New Drugs 2011;29(4):700–5.

[40] Macini J, et al. Consent for biobanking: assessing the understanding and views of cancer patients. J Natl Cancer Inst 2011;103(2):154–7.

[41] Igbe MA, Ademamowo CA. Qualitative study of knowledge and attitudes to biobanking among lay persons in Nigeria. BMC Med Ethics 2012;13:27.

[42] Ma Y, et al. Consent for use of clinical leftover biosample: a survey among Chinese patients and the general public. PLoS ONE 2012;7(4):e36050.

[43] Johnsson L, et al. Patients' refusal to consent to storage and use of samples in Swedish biobanks: cross sectional study. BMJ 2008;337:a345.

[44] FAQs. Terms and recommendations on informed consent and research use of biospecimens the Secretary's Advisory Committee on Human Research Protections (SACHRP), July 20, 2011; <http://www.hhs.gov/ohrp/sachrp/commsec/attachmentdfaq'stermsandrecommendations.pdf.pdf> [accessed 15.09.2015].

[45] Pulley J, et al. Principles of human subjects protections applied in an opt-out, de-identified biobank. Clin Transl Sci 2010;3(1):42–8.

[46] Giesbertz NA, et al. Inclusion of residual tissue in biobanks: opt-in or opt-out?. PLoS Biol 2012;10(8): e1001373.

[47] Forsberg JS, et al. Changing defaults in biobank research could save lives too. Eur J Epidemiol 2010;25(2): 65–8 [Note: where the authors suggest that research oversight bodies should presume that individuals "wish to contribute to the advancement of healthcare through biobank research on previously taken samples" rather than require consent].
[48] Notice of Proposed Rulemaking to the Common Rule; <http://www.gpo.gov/fdsys/pkg/FR-2015-09-08/pdf/2015-21756.pdf> [accessed 15.09.2015] [Note: where the clear intent is to continue to require prospective consent for federally funded research].

ACRONYMS AND ABBREVIATIONS

CAP College of American Pathologists
CLIA Clinical Laboratory Improvement Amendments of 1988
FDA Food and Drug Administration
GINA Genetic Information Nondiscrimination Act
HHS US Dept. of Health and Human Services
HIPAA Health Insurance Portability and Accountability Act of 1996
HITECH Health Information Technology for Economic and Clinical Health Act
IRB Institutional Review Board
ISBER International Society for Biological and Environmental Repositories
NCI National Cancer Institute
NPRM Notice of Proposed Rulemaking

CHAPTER

10

Ethical Concerns When Cancer Patients Become Human Research Participants and Are Treated on Clinical Trials

Richard L. Theriault
The University of Texas MD Anderson Cancer Center, Houston, TX, United States

INTRODUCTION

Discussing the ethical concerns of the patient or participant, or the commonly used term "subject" in clinical research, is a challenge for many reasons. The most is, perhaps, clarifying what we mean by the use of these designations—"the patient," "the participant," and "the subject." Each has connotative value and each term differentiates roles and responsibilities between the person who is suffering and the person who is sought for care. The word patient, derived from Latin "patiens," has the common meaning of a person who is under medical care or treatment. An older meaning is that of one who suffers and is the bearer of misfortune. When persons become unwell and seek medical care, we refer to them as "patients." When an individual person is enrolled in a clinical trial, they become a "subject" that is "under the control of another as an object of investigative medical care or experiment."

Some professionals prefer the terminology "participant" rather than "subject' since this implies that the previously noted patient is now "sharing" with others in a research endeavor. When the patient becomes a participant a change in the relationship between physician/researcher and patient occurs and the ethics of care are modified.

Patient care, biomedical education, and clinical research are foundations of biomedical practice, the development of new knowledge and "evidence based medicine." When a patient is considered for becoming a "participant" in a clinical trial the patient may be

Ethical Challenges in Oncology.
DOI: http://dx.doi.org/10.1016/B978-0-12-803831-4.00010-5

viewed differently by the treating physician. The assumptions of the new participant and those of the research clinician regarding the roles, purposes, and expectations of the trial may differ. A resolution of the understanding of the participant's expectations is part of the larger realm of concerns when cancer patients become willing participants in human subjects' research.

The physician/researcher assumes new responsibilities related to the clinical trial, its safe conduct and reporting of results and may no longer be perceived as the patient's physician.

Among the ethical concerns to be considered are access to clinical trials, understanding of the purposes of clinical research, the role of Institutional Review Boards (IRB), informed consent (document and processes), (perceptions of) the value of clinical research, knowledge of the structure, and implementation of the research and conflicts of interest. Relatively new in this environment is the use of genetic information (tumor and patient) in research, privacy concerns with genetic information, specimen acquisition, storage and use, and implications of genetic information for family members.

Of necessity, this chapter will be limited in scope and observations will be based on the ethical principles espoused in the Belmont report and other internationally recognized research ethics documents [1,2].

ACCESS, AVAILABILITY, AND PATIENT/PUBLIC PERCEPTIONS OF CLINICAL TRIALS (PRINCIPLE—JUSTICE)

Cancer clinical trials form the basis for developing the evidence and knowledge needed to advance care for patients. A patient cannot become a participant without access to and availability of clinical research trials. Availability of and access to clinical trials is a substantial issue for cancer patients as well as research clinicians and is crucial for establishing the validity and potential generalizability of research results. Barriers impeding patients becoming participates may include language, geographic, financial, educational (literacy), racial/ethnic minority status, study designs, and objectives [3–9]. In a meta-analysis and systematic review of patient-related factors impeding clinical trial participation, Mills et al. identified three categories of barriers to participation in clinical trials: protocol-related, patient-related, and physician-related. Included in the patient-related issues were lack of family support, discomfort with experimentation, loss of control of decision-making, feelings of uncertainty, fear or mistrust of the researchers, and concerns regarding research costs/transportation [10]. Identified physician barriers included negative effect on the doctor patient relationship, belief that the physician should make the decisions, feeling coerced to join, and physicians' attitude toward the trial. Protocol-related barriers were the largest group and included dislike of randomization, unease with research processes, complexity and stringency of protocol requirements, inconvenience to life style, potential side effects, and the sense that trials were not appropriate for serious disease. Of interest was the reported perception that the trial has no benefits or did not offer the best option.

Geographically, many centers that conduct clinical research are in major metropolitan areas of the United States. Transportation and time commitments may be difficult for patients to overcome and family or other social support may not be easily available. Financial coverage for clinical trial participation may be limited or prohibited by insurance provider contract (third-party payers). Expenses related to trial participation may be dependent on trials' design requirements as well as participant requirements for attendance and trial procedures. For example the need for interventional biopsies and special or additional studies done for research purposes may impair participant enrollment or continued participation once enrolled. Once enrolled in a clinical research program, the care and decision-making is largely determined by the protocol; participant and physician decisions regarding therapeutic interventions are prescribed and may not be seen by participants as consistent with "their best interests" as a patient.

An additional concern regarding research access is availability of research at public hospitals. Research infrastructure at publically funded hospitals may not be available or adequate for supporting clinical trial initiation, enrollment, oversight, evaluation, and reporting. The requirements for IRB availability for review and approval of studies and the infrastructure required for success of research may be additional concerns. Research nurses, personnel, and offices for documentation and monitoring of research may be too expensive for some public facilities. Those patients cared for in public hospitals may receive medical care but may not have access to research investigations.

Understanding of the purposes and goals of clinical research and its potential benefit for individuals and communities may not be well understood by the public and potential participants. A study of public attitudes toward cancer clinical research by Comis et al. examined a national sample of independently living adults and noted that 32% of those interviewed by telephone were willing to participate in clinical research [11]. The authors concluded that the low accrual rate to clinical trials is not patients' attitudes but unavailability of appropriate clinical trials and disqualifications of a large number of potential participants. Ninety-two percent of those interviewed felt that trails would be of benefit for themselves and be of future benefit, while 64% felt that a clinical trial might provide "better initial treatment" and was associated with "high-quality medical care." Developing new knowledge may not be an immediate thought for the patient with cancer. In a systematic review of attitudes of patients with advanced cancer, Todd et al. noted both positive and negative attitudes toward research participation [12]. Common among the positives were a sense of altruism, hope, and self-benefit. Negatives included concerns regarding symptom, cost (either direct or related to insurance denials) control and increased hospital admissions. This review was limited by the lack of "real" clinical trial data since most of the studies examined patient attitudes about participating in hypothetical trials.

Strategies to improve the effectiveness of recruitment of underrepresented populations have been assessed by Lai et al. [13]. The authors conducted a systematic literature review and concluded that there was limited evidence of "efficacious" or "effective" strategies for recruiting underrepresented populations to cancer clinical trials. Barriers enumerated by Heller and colleagues included fear and lack of trust in research as well as lack of awareness, referral and access to clinical research [8]. These authors recommended a multistrategy approach for enhancing recruitment to trials including addressing study designs, provider enrollment site review and both individual and community evaluations. The goal of clinical research as noted in the 1964 Helsinki document is to "improve diagnostic, therapeutic, and prophylactic procedures and the understanding of the etiology and pathogenesis of disease. Because it is essential that the results of laboratory experiments be applied to human beings to further scientific knowledge and **to help suffering humanity** [2].

There are varied perspectives on this understanding and the concept of being a guinea pig has not been eliminated from public perception.

There is an overarching, long-standing suspicion of research based on prior poorly conducted and ethically unsupportable research, especially among minority populations. Previous reporting of egregious research misconduct and recent publicized research misconduct continues to impact the public trust needed for clinical research to flourish [14–17].

Strategies for overcoming barriers to clinical trial enrollment have been subject to systematic review. Barriers to recruiting "underrepresented populations" to cancer clinical trials specifically have also been evaluated in a systematic review [10]. A recent National Cancer Institute—American Society of Clinical Oncology Cancer Trial Accrual Symposium addressed the challenges to clinical trial accrual [18]. The report identified patient/community, physician/provider, and site/community factors challenging clinical trial enrollment. The symposium participants identified 11 "best practices" focused on patient/community-centered factors. "Consider the patient point of view when reviewing and implementing trials" was the first recommendation. The third recommendation was "simplify the informed consent documents and enhance personal communication during the informed consent process." Evaluation and implementation of the results of these systematic reviews and thoughtfully addressing the NCI-ASCO symposium recommendations may have the potential to increase access to cancer clinical trials.

UNDERSTANDING OF THE INTENT AND PURPOSES OF RESEARCH (PRINCIPLES—BENEFICENCE, NONMALEFICENCE, AUTONOMY)

There may be varied perspectives of the intentions, purposes, and expectations of research among patients/participants, research physicians, and the public. Each of those who have a role in the development, implementation, and conduct of research may approach human subjects' research with self-referential objectives. The ethical foundation of clinical care is based on individual patient well-being and "best interests." This presumes that medial practice, as noted in the Belmont Report, "refers to interventions that are designed solely to enhance the well-being of an individual patient or client and that have a reasonable expectation of success." This is consistent with the long history of medicine and has been codified in the recent American College of Physicians/American Society of Internal Medicine physician charter and position paper. "By contrast, the term research designates an activity designed to test a hypothesis, permit conclusions to be drawn, and thereby to develop or contribute to generalizable knowledge" [19]. Although not specifically noted as one of the vulnerable populations enumerated in the Code of Federal Regulations, patients with a recent diagnosis of cancer or recurrence/progression of disease may be especially vulnerable.

The knowledge disparity between patient and physician, the fear engendered by the existential threat of the "new" disease and the threat of loss of life and personhood may all add to the suffering of the patient [20]. In this environment, the research physician has a challenging opportunity for the patient if a clinical trial is available and the patient meets eligibility criteria for enrollment. Careful, deliberative communication is necessary regarding trial objectives for the participant, hoped for/or expected knowledge to be gained and expectations of patient/participant commitment. The patient may feel that there will be loss of confidentiality, loss of focus on "their care," and potential for distracted concern of the physician regarding the trial. The fidelity of trust engendered by a perceived "fiduciary" responsibility to the patient may be challenged by the expectation for the research physician to address simultaneously the research and patient care needs [21].

In a careful ethical review of ancillary care responsibilities of medical researchers Richardson and Belsky developed a framework outlining the clinical care researchers owe research participants [22]. They posited two extremes for consideration. First that the researcher owed only that care needed for successful completion of the research and second that the researcher "ought" to respond as fully as a physician would for his/her patient. In this view, the researcher can be assumed to be acting as the patient's personal physician and the research participant as his/her patient. The authors feel that this confounds the roles of personal physician and clinical researcher and may exacerbate the therapeutic misconception. Participants may misconstrue the purpose of the clinical trial. The other perspective views the researcher solely as scientist and research participants solely as "subjects" thus potentially denying any need for care apart from the science, safety, and protection from injury. The authors then turn to a concept of "entrustment" that recognizes that the researcher does not have a fiduciary primary responsibility to the participant as patient but has a commitment to the research. With the researcher as a pure scientist model, the authors note that the legal metaphor is one of contracts. An entrustment model is proposed in which researchers responsibilities for care are limited in "scope and strength." Researchers are morally obligated to "engage" with participants as "whole people" and that participants must be treated as "ends" orienting the researcher's judgment. The authors' perspective is that "clinical researchers have limited entrustment responsibilities emerging form three principal conditions; the permission granted by participants, their vulnerability to researcher discretion and the trio of applicable duties of *compassion, gratitude and engagement*." The clinical scientist/researcher role is clearly not that of primary care physician even if the researcher's specialty is the disease or condition under investigation.

These types of patient concerns must be considered and thoughtfully and meaningfully addressed as part of the evaluation of a patient for becoming a participant.

At the time a patient becomes a participant in a clinical trial, the roles and relationship of the participant and physician change. After participant enrollment the research physician has additional obligations to the study [23]. These include personally conducting and supervising the research, informing patient of the investigational nature of the drugs being used, reporting adverse events to the sponsor, reading the investigators brochure, assuring that all colleagues assisting in the study are aware of their roles and obligations, maintaining accurate records, and that the IRB complies with Federal Regulations. There is no mention made of physician responsibilities to the patient. In addition, the investigator must ensure the proper and safe conduct of the research, assuring the validity and proper analysis and presentation of the data in order to assure success of the clinical trial. The expectations of participants may be influenced by a variety of exposures to research reporting. Generally, research that is reported as "breaking news" highlights successfully completed studies which may be labeled as "breakthroughs." This seems to be particularly important for chronic conditions and potentially fatal illnesses such as advanced cancer [24].

Cancer remains one of the leading causes of mortality in the United States, so cancer-related studies may be of special note in the public domain. There have been public and individual perceptions that research will lead to cure of disease and prolongation of life in circumstances for which there is not sufficient information to justify these expectations.

THE ROLE AND RESPONSIBILITIES OF IRB AND THE PATIENT AS PARTICIPANT (PRINCIPLES—AUTONOMY, BENEFICENCE, JUSTICE)

As a patient enters the realm of clinical research, their care during study participation is supervised not only by their research team but also by regulatory agencies of the United States, specifically the Department of Health and Human Services (HHS) and the Food and Drug Administration. These two oversight bodies are responsible for the composition, actions, and proper functions of IRB.

IRB were developed as a consequence of legislation designed to protect individuals participating in research from unwarranted risks, harms, and burdens of research. *The National Commission for the Protection of Human Subjects of Biomedical and Behavioral Research* (The Belmont Commission) developed the ethical principles which now serve, along with other internationally recognized ethics documents, as the basis of ethical research in the United States (CIOMS). This Commission was developed as a result of the public disclosure of the Tuskegee Experiment scandal in 1972. The Belmont Report was promulgated in 1979.

IRB are charged with reviewing all research receiving government funding to assure trial design appropriate to answer the research question(s), participant safety, appropriate research processes and procedures, and determine that the risks of trial participation are reasonable in relation to potential benefits for participants [25].

A number of commentaries have attempted to define what makes clinical research ethical. Emanuel et al. listed seven requirements: social or scientific value, scientific validity, fair subject selection, favorable risk benefit ratio, independent review, informed consent, and respect for potential an enrolled subjects [26]. The IRB is responsible for reviewing and evaluating the Informed Consent document and any other materials produced for study participants, including but not limited to advertisements and educational tools, etc.

The IRB membership characteristics are prescribed in the Code of Federal Regulations. IRB place special emphasis on invasive procedures (tumor biopsies) as well as genetic assessment of patient and tumor specimens and may require additional safety precautions/procedures for participant protection during study participation. All clinical research must be reviewed by the IRB for safety and risks at least annually and the IRB may require study modifications based on annual reports. Patients who become participants in clinical research have an additional safety net because of this regulatory and scientific oversight. The IRB has no role in the care of patients in clinical practice.

INFORMED CONSENT—FORM AND PROCESS—WHO NEEDS TO UNDERSTAND (PRINCIPLE—AUTONOMY, BENEFICENCE, NONMALEFICENCE)

Informed consent is a requirement for medical care in the context of patient/physician shared decision-making and patient-centered care. The Principle of Autonomy expects that the physician researcher assess the capacity and liberty of the patient/participant and allow time for deliberation and answering of questions before both researcher and patient agree upon clinical trial enrollment. The intent of informed consent is to assure that patients participate in research only if they have been made aware of the aims, methods, anticipated benefits and potential harms of the study, and the processes and procedures of study participation. The patient must decide if the proposed trial is consistent with his/her values, preferences, and interests. The patient must decide if their participation will be to their benefit. The role of physician researcher might also be clarified in this process, as should a realistic discussion of patient expectations for care.

Informed consent is the most prominent of ethical concerns when a patient becomes a participant in clinical research.

Participants may view research studies as treatment or "more advanced" than standard care, and in some circumstances, research may reach that expectation. There is a need for the researcher to explain the reasons that the patient may or may not be a candidate for clinical research and any additional "burdens" of the study for participants especially financial costs, time commitments, and special tests or procedures required due to research that may not have been otherwise expected. The research physician must have a comprehensive understanding of the study eligibility criteria and excellent communication ability in order to engender the patient's trust.

Detailed inclusions of content in written documents for providing informed consent in the clinical research context have been prescribed by Federal Regulation. These may appear to be more like legal documents than "informational" documents to the potential participant. The language and reading level of informed consent documents is of concern when dealing with complicated health issues (cancer, life, and death), ever more complicated clinical trial designs and greater expectations from participants (multiple tumor biopsies, extensive pharmacokinetics, repeated tumor-genetic analyses). It is imperative that the Informed Consent process be a **process**. This involves discussion of the trial with the participant or their surrogate, active listening on the part of the investigator, thoughtful responses to questions and a shared decision about participant enrollment. While signing the consent document is required, the "process" is not merely the signing of the consent document.

In some instances, physician communication has been reported to be "persuasive" and make use of "explicit recommendations" in the context of Phase I trials. Brown et al. note the informed consent process is complex and often difficult [27a]. While Jenkins et al. call for new education initiatives informed by research with patients and professionals to improve informed consent communication and processes [27b]. The quantity and quality of clinical communication among the oncologist, patient, and family were assessed as having important influence on patient's decisions to become research participants [28].

Patient comprehension of research may be impaired by lack of assessment of the potential participants' reading skills or health literacy. In a systematic review, Montalvo et al. noted continued therapeutic misconception and lack of understanding among research participants with regards to randomization, use of placebos, study benefits, and risks. The lack of standardized tools for assessing participant comprehension of research was evident [29].

Another systematic review evaluated participants' understanding of informed consent over a three decades time frame. The authors assessed understanding of participants in relation to the multiple parts of the consent document [30]. The proportion of participants who understood that they were free to withdraw from research at any time was the highest (75.8%), while 74.7% recognized the voluntary nature of their participation. Overall, the proportion of participants who recognized and understood the multiple components of the informed consent varied from 52.1% to 75.8%.

A number of attempts to evaluate methodology of the informed consent process have been undertaken. A Cochrane review of audiovisual informed consent information presentation concluded that the value of audiovisual interventions "remains largely unclear" and that many relevant outcomes need additional evaluation in randomized trials [31]. A systematic analysis of randomized controlled trials testing interventions to improve the informed consent process for research has been undertaken by Nishimura and associates [32]. Their review was consistent with the PRISMA guidelines and their methods were concisely described. The primary outcome of their study was participant understanding as assessed by quantitative testing. Understanding was divided into phases including immediate knowledge, retention and reading time. Secondary outcomes included participant satisfaction and study accrual rates. Fifty-four interventions were "tracked." Among these were multimedia, enhanced form, extended discussion and tests/feedback methods. Multimedia approaches demonstrated a nonsignificant increase in understanding scores while enhanced consent form had a significant increase in understanding score. The authors concluded that both *form* and *conversation* are important to the consent process. The role of multimedia remains to be determined as an aid to the informed consent process. Enama et al. reported a small study comparing two different IRB-approved consent forms for two Phase I clinical trials [33]. Study participants were randomized to a "standard" or "concise" consent document. The "concise" form had 63% fewer words than the "standard." Participants with the "concise" form scored as well as the "standard" participants in measures of comprehension. The authors noted that the "concise" form participants reported feeling better informed.

Health-care professional communication skills training are offered in a variety of settings and the impact of communications skills training has been assessed in a Cochrane review by Moore et al. [21]. Their review concluded that communication skills training of various types are effective in improving information gathering and supportive skills for health-care professionals. Long-term efficacy and which types of program were most effective could not be determined.

The need to emphasize that informed consent for research participation is a *communication process* and that it is not the reading and signing a document by the patient/participant cannot be understated.

Flory and Emanuel performed a systematic review of informed consent interventions designed to improve understanding of the information disclosed to research participants during the consent process [34]. They reviewed 32 studies including 42 trials that met their inclusion criteria. Multimedia interventions had improvement in understanding for 3 of 12 trials while 6 of 15 trials using "enhanced" consent forms had improvement in participant understanding. In three of five trials with extended discussion methods, additional time spent with the participant by a member of the research team or a neutral educator, understanding of the participant was improved. The authors concluded that multimedia and enhanced consent had limited impact on improving participant understanding compared to the standard informed consent process, but that extended discussion had a significant impact.

TYPES OF STUDIES AND STUDY PHASES (PRINCIPLE—AUTONOMY)

Phase I Clinical Trials

Understanding of the design and conduct of clinical trials is a challenge not only for patients/participants but also for many in the medical communities. While the standard nomenclature of trial phases remains, new trial designs, some based on tumor characteristics (genetic markers, protein expression, for example) as well as disease diagnosis, may further complicate the patient/participant understanding of the research goals.

The standard nomenclature of Phase I–IV studies may have little meaning for the patient/participant. The use of placebos, randomization, and the role of standard care may need explanatory conversations prior to and during clinical trial participation.

There are a many varieties of cancer research clinical trial formats. Traditional designations of Phase I–IV are being adapted to new technologies and scientific knowledge. Phase helps to describe the intent and purposes of the trials in general. Phase I studies in oncology have been of particular interest for patients, clinicians, research scientists, ethicists, and the public. Access to "new" anticancer agents has become a concern especially when no known beneficial treatment is available and promising new agents are being assessed continually in Phase I trials. Persons who are seriously ill with life-threatening disease want relief of the burdens of disease and to live. Hence, there may be a perceived conflict if research physicians performing Phase I studies and those individuals in existential circumstances do not have adequate understanding of the purposes of initial trials of anticancer agents. Phase I studies begin after preclinical observations of potential anticancer agents have been assessed in laboratory and animal experiments and shown to provide some type of anticancer effect. Phase I studies in human beings generally are *dose finding and agent toxicity studies* and often have correlative laboratory components such as pharmacokinetics and pharmacodynamics. The agent(s) may be given for the first time to a human being in the study. Frequently administration schedules of study drug are included in the research plan. The dose of study agent generally begins very low and is gradually increased as data about effects in humans are monitored. A safe dose or biologically active dose is sought for future study. For a participant in the early aspect of the study, the agent dose may be too low to be of clinical toxicity or benefit. At the highest dose used the agent may be too toxic or show no clinical effect on the cancer. Recent technology developments have added complexity to these studies which now may have serial tumor and normal tissue biopsy expectations, analysis of tumor molecular characterizations, and determination of "clinical benefit" for participants. Limited information indicates participants' acceptance of biopsies done for research purposes [35,36]. A separate informed consent for these procedures has been recommended and use has been included as part of IRB review for the studies at MD Anderson Cancer Center and other institutions. Phase I trials usually have a limited number of participants 15–20 and generally are of brief duration 12–18 months.

The role of informed consent in Phase I trials has been a major concern for clinical research physicians, ethicists, and patient/participants [37–39]. The potential participants in Phase I cancer clinical trials generally have progressive, incurable disease, and may have exhausted all treatment options of known or suspected benefit. They are therefore a vulnerable population in need of special care, support, and protection because of the threat to their life due to cancer. Special care must be taken in the evaluation and consent process for this group of patients not only to assure their understanding of the research but also to avoid adding to their suffering.

Therapeutic misconception and vulnerability are often cited as concerns in the informed consent process [38]. Pentz et al. differentiate "therapeutic misconception," which they define as misunderstanding the research purpose or how research differs from individualized care, and "therapeutic mis-estimation" as "incorrectly estimating the chances of research benefit as >20% or underestimating risk as 0%" [39]. An additional misconception may occur with the language used by researchers and the public in describing Phase I studies as "treatment" when there have yet to be demonstrated in the research setting specific treatment effects [40]. Miller et al. assessed hope and persuasion by physicians during informed consent for phase I trials in cancer [41]. The hopeful statements of the physicians were most often related to expectations of positive outcomes and the study providing options for the patient. The discussion of no treatment or a palliative/supportive care option was not mentioned (68%) and the state of disease as incurable was less frequently mentioned (85% of participants). Jenkins et al. reported a study of communication in the informed consent process for Phase I studies. Using questionnaires, physicians and patients' discussion of the studies and information given and received were evaluated [42]. The levels of agreement regarding key discussion information areas and how well this was done were evaluated. Discussion of prognosis was often omitted, while alternate care plans to Phase I trial entry had been reviewed and explained. The authors concluded that "fundamental components" of communication and information sharing were often missing. Brown et al. used audio recording to document discussions between research physician and potential participants in Phase I trials. (Informed consent for the recoding was obtained from each participant.) They then assessed, from the transcribed documents the communication themes for their analysis. Among these were orienting, educating patients, describing uncertainty and prognosis, persuading, decision-making, and making treatment recommendations. The authors' final conclusion was "patent centered communication that valued patient preferences while preserving the oncologist's agenda can be a helpful approach to these discussions" [27a]. The question that was not addressed is whether these patients facing existential threat and being recruited to Phase I studies have adequate and legitimate choices or preferences in regards to clinical trial enrollment.

Participants tend to overestimate the potential benefits and underestimate the potential toxicity associated with their participation.

Reasons for patients participating in Phase I cancer trails have also been evaluated by Catt et al. [43]. The authors used a 19-item measure of acceptance/decline for the clinical trial. They evaluated participants' hope, expectations of benefit, altruism, concerns, and general perceptions of the trial information. They observed that most patients were optimistic and had few concerns about the experimental aspect of phase I trials. Ninety percent consented to participate and 51% thought this was their "only treatment option," while 21% expected some medical benefit from their participation. The authors concluded that this group may be a self-selected group with optimistic expectations of personal benefit suggesting that achieving "genuine Informed consent" and avoiding therapeutic misconception may be difficult for these patients. Cox et al. performed a literature review of communication and informed consent for Phase I trials [44]. Based on their review, they concluded that patients consenting to Phase I trials have limited understanding of trial purposes and unrealistic expectations of benefit and risks during the trial. It was also not clear that the participants were aware of their right to withdraw from the trial. The authors recommended "clear and practical guidelines and training packages designed to ensure that all details of Phase I trials are effectively communicated."

The challenge for patient and clinician researcher is to have an open dialog that adequately describes the processes and procedures of the study, the time commitment and logistics of participation, and the risks and limited potential for benefit for the participant. The investigator must make clear to potential participants that the clinical benefit is expected to be very low and potential for toxicity is generally higher. A discussion regarding a patient's wishes for care, preferences for activity and living style, and alternatives to clinical trial enrollment need to be considered in the decision process. Adequate quantification of risk and benefit cannot be known until study completion. Participants in Phase I studies enter an uncertain future. Participants' expectations of benefit are associated with patient characteristics and the method of investigators' asking about expectations for benefit in the trial [45,46].

Phase II Clinical Trials

Phase II clinical trials generally are designed to seek additional information agent safety, efficacy for a specific cancer type, or biology and provide objective assessment of antitumor effects. These trials have more participants than Phase I studies to help establish agent dosing, schedule of administration, additional agent toxicity (not observed in the limited Phase I trial), and any observed antitumor effects. Generally, these trials, because they focus on anticancer beneficial effects, are referred to as *treatment trials*. Because these trials often have additional correlative studies and serial specimen acquisition the consenting process and risk assessment explanatory information may seem onerous for both investigator and participant. In addition to explaining the trial objectives, the investigator may also have a trial design that requires randomization to two or more study arms. The study design may assign treatment to differing doses or schedules of the same agent or to two or more differing agents or placebo. The use of placebos in cancer research has been difficult. Primarily, it has been held that a placebo will provide no treatment benefit and that it would be unethical to deprive a research participant of potential benefit of an active anticancer agent. The use of placebos in Phase II and III clinical trials has been examined by Daugherty and colleagues [47]. They examined the ethical, scientific, and regulatory perspectives of placebo use in randomized controlled cancer trials. With their review, they concluded that placebo-controlled studies are scientifically feasible and ethically justifiable. From the regulatory perspective, they may be necessary to meet standards for drug approval. Specific criteria were recommended and trial designs for such studies reviewed. Placebos are justified if the following criteria could be met: diseases with high placebo response rate, conditions that wax and wane, conditions with spontaneous remission, conditions with uncertain/unpredictable clinical course, when existing therapies have minimal benefit or severe adverse effects and in the absence of effective therapies. The authors base their conclusions on the concept of "clinical equipoise" that assumes an honest null hypothesis in the study design and justification and that participants should not receive a treatment known to be inferior to that which is otherwise available in clinical practice.

Randomized controlled trials add additional complexity for potential participant and investigators in adequately obtaining and assuring participant informed consent.

Phase III Clinical Trials in Cancer Patients

Phase III clinical trials are designed to assess benefits of treatments of promise from Phase II trials in a larger number of participants. These trials often have two or more active treatment arms and one investigational arm. That is, two or more treatments known to be beneficial for the disease category under study, with known or expected toxicity profiles and expectation of benefit for participants and one treatment of suspected but not proven efficacy. Phase III trials frequently assess "best" treatment in the comparison. Wright and associate evaluated reasons for patients entering Phase III cancer trials [48]. Patient, physician, and clinical research associate attitudes were assessed by questionnaires and analyzed by logistic regression methods. The patient's perception of personal benefit was the most important factor for deciding to participate. Decision support from the physician and the clinical research associate were also significant factors for participants. There were 19 patient-related factors, 1 physician-related factor ("the trial is asking an important clinical question"), and 4 clinical-research-associated factors identified. The most significant of the clinical research-associated factors was the amount of time spent discussing consent [48].

GENETIC AND GENOMIC STUDIES IN CLINICAL TRIALS (PRINCIPLES—AUTONOMY, BENEFICENCE, FIDELITY TO TRUST)

Next-generation sequencing and genome-wide association studies are examples of rapidly developing technology that have been useful in searching for the causes and etiology of cancer. There are two possibilities for genetic information discovery. The first participant may enroll in a clinical trial with genetic studies as part of the trial. Here, the informed consent process and document may describe the potential for actionable genetic findings and the policies and procedures to be followed as part of the trial if such are found. The second possibility is that of a participant offering specimens and data for large genome-wide repository use and an incidental finding of actionable cancer genetic risk are found. The process and procedure to be followed in this circumstance may or may not have been included in the informed consent process and document.

For participants in clinical trials, there are multiple ethical concerns regarding use of genetic and genomic research: informed consent for genetic and genomic studies in individual patients, the use of these data for study purposes, and the reporting of results of genetic/genomic studies to research participants. The rapid development of techniques to acquire genetic data regarding cancer risks related to genetic alteration/aberrations has resulted in recognition of multiple germline mutations associated with increased risk of cancer. In the past, these were given specific syndrome names but many are now known by the gene aberration or location. An example is the Hereditary Breast/Ovarian Cancer Syndrome now abbreviated to BRCA1/BRCA2-associated cancers. These genetic discoveries have resulted in clinical utility allowing the characterization and selection of patients for cancer genetic screening, risk assessment, and prophylactic interventions. This clinical use has also resulted in clinical trials selecting patients with these genetic changes to become participants in interventions to correct or abrogate the deleterious gene effects or treat diagnosed cancers. The application of genomic information for selecting patients for clinical trials and "tailoring" interventions to control cancer growth based on tumor-derived genetic information, referred to as "targeted therapy," has raised ethical and legal concerns for participants in clinical trials, patients who may become eligible for genetic-based trials and family members of those with genetic alterations/aberrations which may predispose them to cancer.

The development of large public databases of genetic information and the collection and use of these data raises concerns for participants regarding privacy, confidentiality, and future uses of individuals' genetic samples or data. Disclosure of genetic information demonstrating a predisposition to familial cancer syndromes is pertinent to the question of sharing of information with family members and the conditions and circumstances which might permit or require this type of disclosure.

A systematic review of informed consent for human genetic and genomic studies has identified a variety of ethical concerns with the emergence of large databases of genetic information [49]. The review noted particular concern with privacy and confidentiality. The investigators review was consistent with PRISMA. Their findings indicated that there was the potential for use of genetic data to stigmatize an individual or group of individuals. Release of data leading to identification of specific individuals was the number one privacy concern noted. The future use of data and uncertain future with regard to discrimination were also important. The role of the IRB in assuring detailed informed consent documents and processes especially in regard to privacy protections, future uses of study materials and data, and the processes to be used after completion of the study and/or participant withdrawal from study are ongoing concerns. The future use of participant tissues provided special consideration, the authors noting that the primary informed consent for collection and banking of specimens would be unlikely to provide adequate information regarding future research possibilities for participants.

The Genetic Information Nondiscrimination Act was passed by congress and enacted in 2008–09 [50]. This prohibits discrimination in health coverage and employment based on genetic information and provides definitions of genetic information.

The ethical and legal issues regarding return of genetic results to study participants when genetic abnormalities are discovered as incidental findings have been assessed by Lolkema et al. [51]. They noted a need to establish guidelines for disclosure of genetic information while preserving individual privacy. A framework for disclosure was reviewed based on a hierarchy of benefits regarding the genetic information, for example, finding of TP53 abnormality. For genetic data for which actionable positive individual benefits can be known, the authors recommend informing the participant. A decision tree for participants of next-generation sequencing studies is provided. Importantly, counseling regarding use of the genetic information and potential for disclosure or nondisclosure is noted as part of the informed consent process. The decision tree includes physician and participant roles and decision, the amount of data to be disclosed (none, only somatic mutations, or cancer-specific aberrations of known significance). Patient specific questions help guide the process.

The most detailed analysis regarding the questions of returning genomic results to relatives of research participants is the detailed work of Wolf et al. [52]. The authors conclude that researchers generally have a duty of reporting to participants but not their relatives. Nevertheless, the authors recommend that researchers consider responding to relative requests with appropriate authorization from the participant or the participant's legally authorized representative. Detailed pathways for sharing of participant's results with relatives are provided and include living adult, deceased adult, living child, and deceased child participant pathways.

Other considerations in obtaining and using research specimens are the processes and procedures for collecting and storing these specimens. An analysis of biorepository development has suggested that ideally a tissue banking model or prospective collection model of tissue/sample acquisition ought to be used [53]. This will meet the needs of researchers and provide quality assurance for the integrity of specimens and associated clinical data.

CONFLICT OF INTEREST IN RESEARCH

Conflicts of interest in research have the potential to undermine public trust in medical research and challenge the integrity of biomedical researchers. This has become an important topic and the focus of recent legislation with financial reporting requirements for pharmaceutical firms, physicians, and research scientists [54–56]. A conflict of interest is defined as an interest which threatens or appears to threaten one's ability to carry out one's primary duties and/or obligations in a fair and impartial manner (objectively). Generally, this has been interpreted in financial terms and focused on researchers, but other's interests may also pertain including IRB members, institutional leadership, Federal agencies, and pharmaceutical representatives. For researchers financial arrangements that have been considered to include direct payments for research, speaking on behalf of research sponsors, serving as paid consultants or on paid advisory boards for research sponsors, royalties, and stock ownership in the company sponsoring the research. The American College of Physicians has responded with a position paper addressing conflicts of interest such as gifts and services as well as financial arrangements between physician and industry. This position paper specifically addresses industry sponsored research. The Ethics and Human Rights Committee states that physicians have an ethical obligation to disclose their commercial ties to prospective research participants and "guard against bias in publication of research outcomes" [57]. The publicity and concerns regarding potential conflicts of interest has process to mitigate or "manage" these concerns. Institutions now have conflict of interest policies and committees. IRB also have policies, procedures, and detailed conflict of interest reporting requirements. Medical journals have initiated disclosure rules for researchers submitting research manuscripts for publication [58].

Importantly, the views of research participants regarding conflicts of interest in cancer clinical research trials have been assessed. In a survey study of potential research participants, Kim et al. collected data form 5478 persons regarding the importance of knowing conflict of interest information, whether disclosure should be required and whether the disclosure would potentially impact on willingness to participate [59]. They noted that respondents felt knowledge of conflict of interest was extremely or very important for the majority of participants. More than two-thirds felt disclosure should be made in the informed consent process and document. The participants in the survey indicated that they would be more concerned about individual conflicts than institutional conflicts.

A more direct intervention with research participants has been reported by Hampson et al. [60]. They interviewed 253 participants already enrolled in cancer trials in the United States about their attitudes regarding financial conflicts of interest among researchers and medical centers. The authors reported that more than 90% of participants had no concerns about financial influences of researchers or institutions. A minority of participants wanted disclosure of the reporting oversight system for researchers and for researchers' financial interests.

The concern regarding the potential for bias in research conduct or reporting has generated extensive discussion. Most recently, a challenge to the ideas of conflict of interest has suggested that focusing solely on the perceptions of conflicts of interest may be disservice to researchers and potential research participants [61,62].

Funding for cancer research comes from a variety of sources—the National Cancer Institute, the HHS, the Department of Defense, the pharmaceutical industry, foundations and voluntary not-for profit health organizations. The pharmaceutical industry support has been growing and has been of most concern in regards to potential or perceived conflicts of interest and research participants. Rosenbaum in her New England Journal of Medicine articles questions whether the conflict of interest regulatory process and procedures may be overzealous and inhibiting scientific development in the realm of new therapies and new knowledge of disease [61,62].

CONCLUSION

Ethical concerns when a patient becomes a participant in clinical research are varied and may be difficult to understand, manage, and mitigate. Informed consent is particularly challenging aspect of the research enterprise.

The need for the investigator to have a comprehensive understanding of the trial objectives, processes, and procedures for the research staff and participants and the skill to communicate these to the participant are essential. The knowledge regarding best practices is tenuous and further study is warranted. Action will be required to assure investigators are aware of and competent to implement ethical guidelines for beneficence and nonmaleficence.

A patient cannot become a participant if there is no access to available, potentially beneficial clinical research. A principle of justice for ethical research would suggest the need for greater efforts by the research community and the public to improve access for research.

Since some populations are at risk for exclusion from studies the generalizability of some study results may be severely limited and population specific. The health of communities may be impaired by lack of availability of studies.

Lack of knowledge regarding study design and intentions can influence enrollment in clinical trials. The misunderstanding of the intent of research and conflation of research with clinical care by both researcher physicians and patients, is of concern in establishing appropriate roles of research physicians and trial participant. The ethical obligations of research physician are not the same as that of that attending personal physician.

New areas of research and ever expanding biological knowledge bring new challenges for the public, research scientists, and potential research participants. Genetic and genomic applications to health, disease prevention, screening, and interventions, have become routine. Ethical frameworks for adjudicating patient/participant privacy and confidentiality need to be established as each new arena of science opens.

The fiduciary responsibilities of physicians to patients must be balanced by avoidance of meaningful conflicts of interest to assure patients that the physician is working in their best interests.

The researchers must assure that their work focuses on the development of new knowledge to "improve diagnostic, therapeutic, and prophylactic procedures and the understanding of the etiology and pathogenesis of disease", because it is essential for the results to be applied to human beings "to further scientific knowledge and to help suffering humanity."

References

[1] <www.hhs.gov/ohrp/humansubjects/guidance/belmont.htm>.
[2] World Medical Organization. Declaration of Helsinki. Br Med J 1996;313(7070):1448–9.
[3] Farrer L, Marinetti C, Cavaco YK, Costongs C. Advocacy for health equity: a synthesis. Milbank Q 2015;93:392–437.
[4] UyBico SJ, Pavel S, Gross CP. Recruiting vulnerable populations into research: a systematic review of recruitment interventions. J Gen Med 2007;22:852–63.
[5] Bolen S, Tilburt J, Baffi C, Gary TL, Powe N, Howerton M, et al. Defining "success" in recruitment of underrepresented populations to cancer clinical trials; moving toward a more consistent approach. Cancer 2006;106: 1197–204.
[6] Kwiatkowski K, Coe K, Bailar JC, Swanson GM. Inclusion of minorities and women in cancer clinical trials, a decade later: have we improved? Cancer 2013;119:2956–63.
[7] Eglestron BL, Pedrazo O, Wong YN, Dunbrack Jr RL, Griffin CL, Ross EA, et al. Characteristics of clinical trials that require participants to be fluent in English. Clin Trials 2015;12(6):618–26, pii:17407745592881 (Epub ahead of print).
[8] Heller C, Balls-Berry JE, Nery JD, Erwin PJ, Littleton D, Kim M, et al. Strategies addressing barriers to clinical trials enrollment of underrepresented populations: a systematic review. Contemp Clin Trials 2014;39:169–82.
[9] Ford JG, Howerton MW, Lai GY, Gary TL, Bolen S, Gibbons MG, et al. Barriers to recruiting underrepresented populations to cancer clinical trials. Cancer 2008;112:228–42.
[10] Mills EJ, Seely D, Rachlis B, Griffith L, Wu P, Wilson K, et al. Barriers to participation in clinical trials of cancer: a meta-analysis and systematic review of patient-reported factors. Lancet Oncol 2006;7(2):141–8.
[11] Comis RL, Miller JD, Aldige CR, Krebs L, Stoval E. Public attitudes toward participation in caner clinical trials. J Clin Oncol 2003;21:830–5.
[12] Todd AM, Laird BJ, Boyle D, Boyd AC, Colvin LA, Fallon MT. A systematic review examining the literature on attitudes of patients with advanced cancer toward research. J Pain Symptom Manage 2009;37:1078–85.
[13] Lai GY, Gary TL, Tilburt J, Bolen S, Baffi C, Wison RF, et al. Effectiveness of strategies to recruit underrepresented populations into cancer clinical trials. Clin Trials 2006;3:133–41.
[14] <www.cdc.gov/tuskegee/timeli>.
[15] Shavers VL, Lynch CF, Burmeister LF. Knowledge of the Tuskegee Study and its impact on the willingness to participate in medical research studies. J Natl Med Assoc 2000;92:563572.
[16] George SL. Research misconduct and data fraud in clinical trials: prevalence and causal factors. Int J Clin Oncol 2015; published online. http://dx.doi.org/10.1007/s10147-015-0087-3.
[17] Ellis LM. The erosion of research integrity: the need for culture change. Lancet 2015;16:752–4.

[18] Denicoff AM, McCaskill-Stevens W, Grubbs SS, et al. The National Cancer Institute-American Society of Clinical Oncology Cancer Trial Accrual Symposium: summary and recommendations. J Oncol Pract 2013;9:267–76.

[19] Project of the ABIM Foundation, ACP-ASIM Foundation, European Federation of Internal Medicine. Medical professionalism in the new millennium: a physician charter. Ann Intern Med 2002;136:243–6.

[20] Cassell EJ. The nature of suffering and the goals of medicine. N Engl J Med 1982;306:639–45.

[21] Moore PM, Rivers Mercado S, Grez Antigues M, Lawrie TA. Communication skills training for healthcare professionals working with people who have cancer. Cochrane Database Syst Rev 2013;28(3):CD003751 http://dx.doi.org/10.1002/14651858 CD003751 PMID 23543521.

[22] Ricahrdson HS, Belsky L. The ancillary-care responsibilities of medical researchers. Hastings Center Report 2004;34(1):25–33.

[23] Department of Health and Human Services Form 1572. <www.fda.gov/downloads/RegulatoryInformation/Guidances/UCM214282.pdf>.

[24] Abola MV, Prasad V. The use of superlatives in cancer research. JAMA Oncol 2015;1–2 http://dx.doi.org/10.1001/jamaoncol.2015.3931. [Epub ahead of print].

[25] Code of Federal Regulations Title 45 public welfare. Department of Health and Human Services Part 46 Protection of Human Subjects; <www.hhs.gov/ohrp/policy/ohrpregulations.pdf>.

[26] Emanuel E, Wendler D, Grady C. What makes research ethical? JAMA 2000;283:2701–11.

[27a] Brown R, Bylund CL, Siminoff LA, Slovin SF. Seeking informed consent to Phase I cancer clinical trials: identifying oncologists' communication strategies. Psycooncology 2011;20:361–8.

[27b] Jenkins VA, Anderson JL, Fallowfield LJ. Communication and informed consent in Phase I trials: a review of the literature from January 2005 to July 2009. Support Care Cancer 2010;18:1115–21.

[28] Albrecht TI, Eggly SS, Gleason MEJ, Harpeer FWK, Foster TS, Peterson AM, et al. Influence of clinical communication on patients' decision making on participation in clinical trials. J Clin Oncol 2008;26:2666–73.

[29] Montalvo W, Larson E. Participant comprehension of research for which they volunteer: a systematic review. J Nurs Scholarsh 2014;46:423–31.

[30] Tam NT, Huy NT, Thoa LTB, et al. Participants' understanding of informed consent in clinical trials over three decades: a systematic review and met-analysis. Bull World Health Organ 2015;93:186–198H.

[31] Synnot A, Ryan R, Prictor M, Fetherstonhaugh D, Parker B. Audio-visual presentation of information for informed consent for participation in clinical trials. Cochrane Database Syst Rev 2014;5:CD003717.

[32] Nishimura A, Carey J, Erwin PJ, Tilburt JC, Murad MH, McCormick JB. Improving understanding in the research informed consent process: a systematic review of 54 interventions tested in randomized control trials. BMC Med Ethics 2013;14:28 published online 2013 http://dx.doi.org/10.1186/1472-6939-14-28.

[33] Enama ME, Hu Z, Gordon I, et al. Randomization to standard and concise informed consent forms: development of evidence-based consent practices. Contemp Clin Trials 2012;33:895–902.

[34] Flory J, Emanuel E. Interventions to improve research participants' understanding of informed consent for research: a systematic review. JAMA 2004;292:1593–601.

[35] Lemech C, Dua D, Newmark J, et al. Patients' perceptions of research biopsies in Phase I oncology trials. Oncology 2015;88:95–102.

[36] Gomez-Roca CA, Lacroix L, Massad C, et al. Sequential research-related biopsies in Phase I trials: acceptance, feasibility and safety. Ann Oncol 2012;23:1301–6.

[37] Horng S, Emanuel E, Wilfond B, et al. Descriptions of benefits and risks in consent forms for Phase I oncology trials. N Engl J Med 2002;347:2134–40.

[38] Dubov A. Moral justification of Phase I oncology trials. J Pain Palliat Care Pharmacother 2014;28:138–51.

[39] Pentz RD, White M, Harvey RD, et al. Therapeutic misconception, misestimation and optimism in participants enrolled in Phase I trials. Cancer 2012;118:4571–8.

[40] Miller FG, Rosenstein DL. The therapeutic orientation to clinical trials. N Engl J Med 2003;348:1383–6.

[41] Miller VA, Cousino M, Leek AC, Kodish ED. Hope and persuasion by physicians during informed consent. J Clin Oncol 2014;32:3229–35.

[42] Jenkins V, Solis-Trapala I, Langridge C, Catt S, Talbot DC, Fallowfield LJ. What oncologists believe they said and what patients believe they heard: an analysis of Phase I trial discussions. J Clin Oncol 2011;29:61–8.

[43] Catt S, Langridge C, Fallowfield L, Talbot DC, Jenkins V. Reasons given by patients for participating, or not, in Phase I cancer trials. Eur J Cancer 2011;47:1490–7.

[44] Cox AC, Fallowfield LJ, Jenkins VA. Communication and informed consent in Phase I trials: a review of the literature. Support Care Cancer 2006;14:303–9.

[45] Weinfurt KP, Seils DM, Lin L, et al. Research participants' high expectations of benefit in early-phase oncology trials: are we asking the right question?. J Clin Oncol 2012;30:4396–400.

[46] Cheng J, Hitt J, Koczwara B, et al. Impact of quality of life on patient expectations regarding Phase I trials. J Clin Oncol 2000;18:421–8.

[47] Daugherty CK, Ratain MJ, Emanuel EJ, Farrell AT, Schilsky RL. Ethical, scientific and regulatory perspective regarding the use of placebos in cancer clinical trials. J Clin Oncol 2008;26:1371–8.

[48] Wright JR, Whelan TJ, Schiff S, et al. Why cancer patients enter randomized clinical trials: exploring the factors that influence their decision. J Clin Oncol 2004;22:4312–18.

[49] Khan A, Capps BJ, Sum MY, et al. Informed consent for human genetic and genomic studies: a systematic review. Clin Gene 2014;86:199–206.

[50] <www.ggenome.gov/Pages/PolicyEthics/Geneticdiscrimination/GINAInfoDoc>.

[51] Lolkema MP, Gadella-vanHooijdonk CG, Bredenoord AL, et al. Ethical, legal and counselling challenges surrounding the return of genetic results in oncology. J Clin Oncol 2013;31:1842–8.

[52] Wolf SM, Branum R, Koenig B, et al. Returning a research participant's genomic results to relatives: analysis and recommendations.. J Law, Med, Ethics 2015;43:440–63.

[53] Grizle WE, Bell WC, Sexton KC. Issues regarding collecting, processing and storing human tissues and associated information to support biomedical research. Cancer Biomark 2010;9:531–49.

[54] Thompson DF. Understanding conflicts of interest. N Engl J Med 1993;329:573–6.

[55] Conection.asco.org/magazine/exclusive-coverage/ssunshine-act-reoerting-letter-ceo. Sunshine act reporting: a letter from CEO Allen Lichter, M.D. ASCO connection August 1, 2013.

[56] Toroser D, DeTora L, Cairns A, et al. The sunshine act and medical publications: guidance from professional medical associations. Postgrad Med 2015;127:752–7.

[57] Coyle S, For the Ethics and Human Rights Committee. American College of Physicians–American Society of Internal Medicine. Ann Intern Med 2002;136:396–402.

[58] Drazen JM, deLeeuw PW, Laaine C, et al. Toward more uniform conflict of interest disclosures—the Updated ICMJE conflict of interest reporting form. N Engl J Med 2010;363:188–9.

[59] Kim SYH, Millard RW, Nisbet P, Cox C, Caine ED. Potential research participants' views regarding researcher and institutional conflicts of interest. J Med Ethics 2004;30:73–9.

[60] Hampson LA, Agrawal M, Joffe S, et al. Patients' views on financial conflicts of interest in cancer research trials. N Engl J Med 2006;355:2330–7.

[61] Rosenbaum L. Reconnecting the dots—reinterpreting industry–physician relations. N Engl J Med 2015;372:1860–4.

[62] Rosenbaum L. Beyond moral outrage—weighing the trade-offs of COI regulation. N Engl J Med 2015;372:2064–8.

CHAPTER

11

Training Current and Future Leaders

Jolyn S. Taylor and Diane C. Bodurka

The University of Texas MD Anderson Cancer Center, Houston, TX, United States

INTRODUCTION

A review of the topic of the ethical challenges of oncology would be incomplete without discussing the topic of leadership. A leader plays a pivotal role in determining how the values of medical ethics of autonomy, beneficence, nonmaleficence, and justice are applied [1]. While the medical community has long recognized the need to actively recruit and train future leaders within the field of medicine, such efforts have not been consistent or wide reaching. The Association of American Medical Colleges (AAMC) selected leadership as the focus of the presidential address during the 1997 annual meeting [2]. The Accreditation Council for General Medical Education (ACGME) developed a workshop for chief residents to develop leadership skills [3]. And within the specialty of oncology, the American Society of Clinical Oncology (ASCO) has created a year-long training program to develop leaders [4]. But why do we, as medical professionals, need to be concerned with leadership development initiatives? In this chapter, we will discuss why the medical profession must prioritize training current and future leaders, what the term "leadership" means, and how we can train medical professionals to become more effective and ethical leaders.

WHY DOES MEDICINE NEED LEADERS?

According to the Merriam–Webster dictionary, a leader is "a person who has commanding authority or influence." [5] Bennis and Nanus offered another perspective on what makes a leader, especially an ethical leader, when they stated that "managers are people who do things right and leaders are people who do right things." [6] Ciulla suggested an ethical leader is someone who influences others while respecting a follower's autonomy [7]. It is also the responsibility of a leader to pursue justice for the common

Ethical Challenges in Oncology.
DOI: http://dx.doi.org/10.1016/B978-0-12-803831-4.00011-7

good of his or her followers, to be beneficent, and to do no harm [8]. The surgeon Dr. Wiley Souba defined leaders as people who act as follows:

> "Listen to the environment and their people, exemplify and embody core ideology, applaud others, deal with problems, empower, enable and inspire people, [are] results-orientated, [and] serve their people." [9]

Whether on a large or small scale, the chief medical officer in a hospital, a charge nurse on a floor, or an attending physician on rounds, a leader should strive to be both effective and ethical [7]. The medical profession needs to promote leadership development training not only to encourage ethical actions within healthcare teams, but also because more effective leaders result in superior healthcare outcomes. Additionally, leadership training improves healthcare efficiency by increasing the overall quality of the workforce, enhancing organizational educational development, reducing employee turnover, and focusing strategic priorities [10].

At what organizational levels would medical professionals benefit from leadership development? At the level of healthcare executives, Lyons et al. reported that two-thirds of the physician chief executive officers they surveyed were required by their employers to pursue further formal training in order to develop their leadership skills. In part, this stems from a recognition that physician executives are expected to function as leaders above and beyond their roles as healthcare professionals. Among those surveyed by Lyons et al. almost one-half of the respondents acknowledged that their primary responsibility was to first act as a leader within their healthcare organization and then as a medical professional. This is in contrast to prior studies which reported that less than one-fifth of physician executives felt this way. Respondents noted that their primary reason for expressing a desire to take on more of a leadership role within their organizations was to be a "part of the healthcare solution" [11]. Leadership development training is essential to creating healthcare professionals capable of adapting to the ever-increasing challenges the medical profession faces.

At the trainee level, we also see the importance of developing leaders. Leadership development not only elevates the quality of trainees but also the level of trainee education. Itani et al. queried trainees from a surgical residency to quantify to what degree they perceived the value of leadership skills as they related to their career development. Ninety-two percent of respondents identified leadership skills as important for developing a career within the healthcare professions [12]. Medical trainees must balance clinical responsibilities, personal stressors, teaching responsibilities, and maintenance of their own educational needs. Learning how to effectively balance all of these demands, while being confronted with ethical dilemmas, can be achieved through leadership development training. The areas of leadership with the largest gaps between what the trainees felt was necessary and what they felt capable of achieving included communication, conflict resolution, leadership training and theory, time management, and effectively giving presentations [12]. Leadership training is needed to assist our medical trainees in reaching their full potential.

Leadership development training can benefit professionals at every level of an organization, not merely the executive and trainee levels. Souba and McAlearney emphasized in their works that strong leadership skills can make an important, effective impact at all levels within a healthcare system [9,13]. Leadership opportunities occur at the organizational, collegial, managerial, and personal level. Dine et al. identified four areas within medicine where effective leadership is especially important, including managing a team, establishing a vision, communicating effectively, and developing personal attributes such as wisdom, inspiration, vision, and integrity [14]. Improvement in these four areas would be beneficial if applied at any level of a healthcare organization.

The former president of the AAMC, Dr. Jordan Cohen, highlighted the importance of devoting resources in healthcare to train and to educate the future leaders of the field during his presidential address in 1997 [2]. He suggested that the institution of medicine needs to develop training programs to better prepare medical professionals to supply the "army of leaders" required to guide the medical profession through today's increasingly murky ethical world. In his view, a leader in the medical profession must be as dedicated to the study of ethics as he or she is to the pursuit of outcomes. Medicine cannot wait for a natural-born leader to rally the troops, but should instead recognize that professional ethics must be part of medical training in order to yield leaders to further the profession. As Cohen stated in his presidential address:

> "We need leaders to continuously challenge the ways we select and prepare students to become ethical caregivers. We need leaders to battle the reactionary forces limiting the achievement of diversity in the medical profession...We need leaders to transform our educational paradigm from a preoccupation with process to a devotion to outcomes...We need leaders to create educational environments dominated by self-directed, active learning, bolstered by electronic access to useful information resources, and people by faculty role models dripping with medical professionalism." [2]

Experts on the ethics of leadership suggest that altruism is associated with being a strong leader [7]. While one could argue that individuals who choose medicine as a profession are inherently altruistic, possession of this trait alone is not sufficient to create effective leaders [7]. We need leaders to guide our profession using the principles of medical ethics of autonomy, beneficence, nonmaleficence, and justice [1]. As the pace of advancement increases in medicine and in oncology, we are faced with ethical dilemmas of ever-increasing complexity. The need for well-trained leaders in the healthcare profession increases with every medical breakthrough. Medical professionals, especially oncologists, are tasked with making difficult decisions involving allocating resources to research and patient care, screening for genetic mutations, and broaching discussions regarding goals of end-of-life care. We, as a profession, must designate the development of current and future leaders from within the healthcare profession as a priority.

WHAT IS LEADERSHIP?

If we accept the necessity of developing strong leaders within the field of oncology, the logical corollary is to ask how we expect these men and women to lead. What are the different types of leadership? How should leaders lead?

Teams in oncology medicine may be as small as a clinical team composed of nurses, physicians, trainees, and the patient or as large as an entire division or cancer center. While each leadership situation is unique, the different styles of leadership can be generally categorized into a spectrum, including the laissiez-faire style, the authoritarian style, and the democratic style [8]. Table 11.1 summarizes these three leadership styles.

TABLE 11.1 Summary of Leadership Styles

Leadership Style	Characteristics of the Leader	Typical Healthcare Professionals
Laissiez-faire	**1.** A "nonleader" **2.** Avoids decision making **3.** Passes responsibility to others	Students or those trainees in the first or second year of a training program
Authoritarian	**1.** Provides structure and goals for followers **2.** Motivates through reward and punishment **3.** Does not foster creativity among followers	Trainees after the first or second year of a training program and most practicing healthcare professionals
Democratic	**1.** Interacts with followers as equals **2.** Acts as a role model who inspires followers by creating a shared vision and purpose **3.** Considered to be highly ethical	Healthcare professionals considered by peers to be very effective leaders such as those elected to lead academic departments or professional organizations

In order to illustrate the variations in these three leadership approaches, consider the following clinical scenario: Mrs. Smith is a 50-year-old woman with recurrent ovarian cancer. Her maternal grandmother developed breast cancer at age of 40, her mother died from ovarian cancer at age of 60, and her maternal aunt was diagnosed with breast cancer at age of 41. Mrs. Smith has two daughters who are 24 and 20 years old. You are a gynecologic oncologist and are meeting her for the first time. You notice from her prior medical records that she has not had genetic testing for the BRCA gene mutation and with her family history, she is a candidate for screening [15]. You would like to test Mrs. Smith because you know that patients with a BRCA gene mutation benefit from therapy with inhibitors of poly (ADP-ribose) polymerase (PARP) and it could inform your clinical decision making [16–18]. You also know that if Mrs. Smith does have the BRCA gene mutation, then her daughters are at risk of also having the mutation. Conversely, if she does not have the BRCA gene mutation then her daughters are not at risk. Women who carry the BRCA mutation can choose to reduce their lifetime risk of developing breast and ovarian cancer by undergoing more frequent screening exams and prophylactic surgeries [19,20]. Therefore, having this information could help her daughters make informed decisions regarding their health. When you ask Mrs. Smith if anyone has discussed testing for the BRCA gene with her she becomes upset and states, "I read online about BRCA gene testing. I will not be part of someone's science experiment."

Laissiez-Faire Leadership

The laissiez-faire leadership style has been referred to as a "nonleadership" style [8,21]. This style is characterized by a lack of clarity around articulation of a vision or goal for the team and little or no opinion expressed by the leader regarding decisive issues. A person utilizing this leadership style avoids making decisions him or herself and passes that responsibility on to the followers [22–24]. Laissiez-faire leadership is thought to lead to ineffective team dynamics as team members have no clear sense of direction and report higher amounts of dissatisfaction. Ultimately, if a leader consistently utilizes the laissiez-faire style of leadership, it will likely result in negative outcomes [8,21–24].

Response to Mrs. Smith: As someone early in his or her training, a medical intern may follow a laissiez-faire leadership style. Laissiez-faire leaders may recognize that, according to the principle of autonomy, Mrs. Smith has the ability to decline any medical recommendation regarding her body, including testing for the BRCA gene mutation. They may choose not to pursue the topic any further despite the possible benefit that could be derived from knowing her BRCA status. They may pass the responsibility of the possible negative ramifications of this decision on to the patient. Or they may assume that BRCA gene testing will be addressed by another medical provider. Not discussing the issue of BRCA gene testing requires less effort; however, this violates the principles of beneficence and nonmaleficence. In addition, there is a violation of the principle of justice in that there is no further discussion as to how this decision may affect her daughters' ability to receive appropriate medical care.

Authoritarian Leadership

In the authoritarian approach, the leader engages his or her subordinates to a greater degree than in the laissez-faire approach [8,21]. This leadership approach can also be categorized as "directive." This implies that while the leader engages actively with the group and is involved with problem solving, the leader expects that the team will follow his or her directions in order to solve the problem [23,25]. An authoritarian leader acts under the assumption that his or her followers require substantial direction provided by the leader in order to accomplish a task. In keeping with this assumption, an authoritarian leader tends to give direction and structure to followers' work and sets goals for them [8,21]. This leadership style is also known as "transactional" as the leader is, in a sense, entering into a transactional relationship with followers. If the followers perform as the leader requests, then the followers are rewarded according to previously agreed-upon terms. On the other hand, followers are also punished if they fail to act as the leader requested. Transactional leadership styles fall within the realm of what most people would consider to be the traditional boss who rewards or punishes employees in response to their work [25–27].

Within the broader category of authoritarian leadership, there are several subcategories of leadership style. Management by exception (passive) is a leadership style whereby the leader allows followers to function independently, mostly without oversight or guidance except when errors occur. Corrective action is taken when addressing errors, but the overall behavior is reactive and retrospective in nature and followers gain little in terms of self-development from interactions with the leader. Leaders who practice the management-by-exception (active) style, as opposed to management by exception (passive), pay close attention to any and all deviations from a plan of action and tend to micromanage followers. While this style is more proactive and may prevent errors from occurring instead of simply reacting to errors already committed, it allows little space for creativity and innovation and it emphasizes the negative consequences of making errors. The final subcategory of authoritarian leadership style is the contingent reward style. With this style, the leader clearly identifies what rewards, or punishments, can be earned by followers if they meet, or fail to meet, certain performance goals. This style attempts to motivate team members by using a reward system. The contingent reward leadership style typifies the transactional leadership model [22–27].

Authoritarian leadership style can be effective in clearly identifying team goals, but is not well-suited to deal with rapid change, such as is often encountered in the medical profession, or to promote innovation. As a result, this management style may lead to hostility, dependence, discontent, and loss of individuality among followers [8,25–27]. Leaders who tend toward an authoritarian leadership style have been characterized by their followers to be more egocentric, Machiavellian, critical, demanding, aggressive, or controlling and are more likely to fall short of becoming effective leaders [28].

Response to Mrs. Smith: A senior resident may choose an authoritarian leadership style. While authoritarian leaders recognize that Mrs. Smith has the autonomy to make a decision regarding her health, they engage in further discussion on the topic to ensure that she is making an informed decision. Following the principles of beneficence and nonmaleficence, they would provide Mrs. Smith with information regarding who is eligible for BRCA gene mutation testing and why it would beneficial, as well the possible ramifications to her daughters of testing. In this way, they can be assured that Mrs. Smith is properly informed to make a decision instead of relying on the information she has obtained from an internet search. Authoritarian leaders would clearly enumerate the goals they have for providing care for Mrs. Smith. They would describe how knowing whether or not she has a BRCA gene mutation would affect providing her medical care. Furthermore, authoritarian leaders use directive language more often and they may tell Mrs. Smith that they would like her to have the BRCA gene testing performed. If Mrs. Smith still declines testing or appears uncertain, they may attempt to enter into a transactional relationship with her. They may use a phrase such as, "I will take better care of you if you have the BRCA testing performed." While accurate in some ways, it also implies that medical care will be exchanged for patient compliance. Mrs. Smith's word choice in the initial statement of not wanting to be part of a "science experiment" already suggests hostility and mistrust. An authoritarian leadership style could worsen this. Upon confrontation with rising patient hostility, authoritarian leaders may struggle with finding other ways to communicate effectively with the patient.

Democratic Leadership

Democratic leadership is the third broad category of leadership style and is the most labor-intensive approach from a leader's perspective, but is also believed to yield more effective team results [8,21]. Leaders who choose to use a democratic leadership style are more likely to treat followers as if they believe them to be capable of performing the work on their own, but elect to work together with followers in order to jointly achieve a superior result. This leadership style has a leader working together with his or her team as equals [8]. Democratic, or participative leadership as Bass refers to it, involves active listening by the leader and engaging followers in the decision-making and planning process. Using a democratic leadership style results in higher levels of overall group member engagement, satisfaction, motivation, innovation, and production of results [8,23,25].

The broad category of democratic and transformative leadership includes several subcategories—individualized consideration, intellectual stimulation, inspirational motivation, and idealized influence. A leader who utilizes the leadership style of individualized consideration makes an effort to become familiar with his or her followers, treats them as individuals instead of as a collective entity, actively listens to questions or concerns, and promotes self-development among followers [22–27]. The next leadership style, intellectual stimulation, incorporates the positive qualities of individualized consideration and goes a step farther with the leader making an effort to stimulate followers to think through issues and problems for themselves and develop their own abilities. This can lead to an organizational climate of readiness to adapt to changes quickly and adeptly [22–27]. The next level of effectiveness among transformation leadership styles is the inspirational motivation style with which the leader motivates followers to more efficient productivity and higher levels of effort by creating a sense of communal priorities and effort along with a unified shared vision. Even more effective transformational leaders utilize the idealized influence leadership style whereby the leader serves as a role model for followers, is considered to be a highly moral and ethical person, is described as inspirational and a visionary who can mitigate crisis situations, and who uses power for the greater good [22–27]. Leaders using the idealized influence leadership style create a shared sense of purpose among followers. Followers of highly effective transformative leaders often describe those leaders as having substantial "charisma." Democratic or transformative leaders are typified as supportive, sensitive, nurturing, and considerate [28]. The drawback to more democratic leadership is that the work process can be less efficient, requiring more time and effort from the leader in order to achieve a goal [8]. Overall, however, democratic or transformative leadership is considered to be the most effective leadership style.

Response to Mrs. Smith: A democratic leadership style may be used by a clinician who has been recognized by peers to be highly ethical and an effective leader, such as a chair of an academic department. Democratic leaders would respect that Mrs. Smith has the autonomy to decide against BRCA gene testing; however, they would want to ensure that she is making an informed decision. They may determine Mrs. Smith's level of understanding using language that is collaborative and seek to engage Mrs. Smith. An example would be, "Could you tell me what you have learned from the internet regarding BRCA gene testing?" Democratic leaders would also ask Mrs. Smith to explain what she perceives her goals of treatment to be instead of enumerating their own goals for her. If her treatment goals included obtaining the best possible treatment response, democratic leaders could then explain how PARP inhibitors have improved outcomes in patients with BRCA-associated ovarian cancer. They may conclude by asking if this information changed Mrs. Smith's decision regarding BRCA gene testing. Finally, addressing the ethical principle of justice, democratic leaders would explain the possible ramifications of Mrs. Smith's BRCA gene testing for her daughters and ask if this was something which had been discussed with Mrs. Smith or her daughters before. Even if Mrs. Smith continues to decline BRCA gene testing, because democratic leaders respectfully engage others as equals, they would be able to maintain effective communication with her.

HOW DO MEDICAL PROFESSIONALS BECOME EFFECTIVE LEADERS?

To quote Hippocrates, "wherever the art of Medicine is loved, there is also a love of Humanity." Medical professionals strive to do what is best for their patients and for members of their teams. However, it is not always clear how a medical professional can become an ethical and effective leader while obtaining optimal outcomes. Given our previous discussion of the different varieties of leadership style, how do medical professionals learn to effectively utilize the different leadership styles?

Mentorship has been proposed as one way to develop leaders. Mentors within a medical specialty are aware of pertinent educational opportunities, act as models of successful leadership behavior, offer guidance at key career decision points, and provide feedback and support relevant to that particular specialty [11]. A report by the AAMC regarding the general professional education of physicians proposed that the solution to the current lack of adequate leadership education for the American physician should include exposure to role models within the medical profession. The report further advocated for rigorous evaluation of faculty teaching skills and frequent monitoring of the trainee experience under the faculty in order to ensure constructive mentoring was taking place. Little detail was given, however, with respect to how the faculty were to acquire the leadership skills necessary to act as mentors. The report, created in 1984, suggested that it may be necessary to hold leadership seminars for faculty, but offered no additional or further detailed suggestions [29].

Another more standardized method of obtaining leadership development training is to pursue an advanced degree such as a Master of Business Administration (MBA). Lyons et al. found that among physician executive officers, two-thirds of their healthcare systems required or offered to cover the cost of obtaining an MBA or other master-level degree meant to further develop their managerial and leadership skill set [11]. Such programs can be time intensive and lead to scheduling conflicts with clinical responsibilities of the healthcare professional. Though obtaining a formal degree or certificate would assist healthcare professionals in developing leadership skills, it is a solution that only a select few could realistically pursue.

Another method of teaching leadership development to healthcare professionals which seems to be growing in popularity is attendance at a short, intensive workshop within the home institution. Stoller et al. described a 1-day intensive workshop with internal medicine trainees in their first year of training. The workshop reviewed examples of leadership, orchestrated group activities with facilitator-led debriefings afterward, and included interactive lectures and table discussions among trainees questioning their perceptions of what makes a good leader. The workshop increased the trainees' valuation of leadership development training as well as their self-reported feelings of competency as leaders. All of the participants reported that spending time developing their leadership skills was valuable. Because the developers of the workshop mandated that all participants be excused from clinical duties in order to focus exclusively on the leadership development workshop, they effectively communicated the message that developing leadership skills deserves the requisite attention. Stoller et al. wrote that they believe this to be an invaluable step in changing the culture within the medical profession to view leadership development as a priority. Medical professionals with leadership training were seen as more likely to uphold the principles of medical ethics (autonomy, beneficence, nonmaleficence, and justice) and challenge accepted norms for the greater good, develop and share a clear vision, incorporate and recognize all members of a large multidisciplinary team in achieving a common goal, and serve as role models. Through this single day of training, the instructors were able to not only effectively explain to medical professionals the purpose and value of receiving leadership training, but also, through the varied learning styles of interactive sessions, to significantly increase the participants' perceptions of their own abilities to act as leaders within the medical profession [30].

Hemmer et al. also described a leadership development workshop for medical trainees. The group, similar to Stoller et al., found that if leadership training was incorporated within existing educational sessions, the importance and impact of the information was lost on trainees. Instead, they created a separate, year-long curriculum focused primarily on improving trainee leadership and management. This curriculum focused on the following five areas:

1. The difference between leadership and management
2. Interpersonal skills
3. Legal, marketing, and financial issues within a specialty
4. System-based quality improvement through structured programs such as Six Sigma or Lean
5. Informatics skills [31]

Only senior residents and fellows were invited to participate. All clinical responsibilities of the participants were excused during the six sessions that occurred throughout the year. The course included interactive lectures, computer laboratory sessions to test analytic skills being taught, required readings on the topics of economics and leadership in medicine, and a team-building exercise culminating in a presentation to all course participants. A pre and postcourse quiz was administered and demonstrated a 33% improvement in postcourse understanding of leadership and managerial skills. The trainees who completed the course overall responded with positive feedback. Trainees specifically identified the teaching sessions on how to develop leadership skills relevant to their specialty to be the most useful and engaging [31].

Instead of spreading a course out over a year, Klein et al. described an intensive 5-day retreat for trainees held at the beginning of their medical careers covering the important components of leadership development. During these 5 days, the principles of medical ethics were addressed through the following topics: honesty, integrity, reliability, responsibility, respect for others, compassion, empathy, self-improvement, self-awareness, knowledge of limits, communication, collaboration, altruism, and advocacy. There was a special focus placed on the ethically difficult situation of dealing with caring for patients at the end of life, a common occurrence among oncology patients. The retreat used didactic sessions, interactive lectures, role playing, self-scored questionnaires on ethical scenarios, and team-building activities such as a scavenger hunt and preparing a meal for the group to teach leadership skills. The ethical aspects of leadership were then revisited during monthly departmental conferences in order to continue to reinforce the importance of addressing these issues. While the study did not include an objective measure of success of the program, the participants provided positive feedback for the course [32].

Doughty et al. also described an intensive 3-day leadership workshop directed toward senior residents. The program directors first established a safe environment for sharing, then didactic sessions reviewed different education approaches to make leaders more effective teachers. Next, facilitators provided an overview of group dynamics, passed out a self-administered personality and decision-making assessment tool (Myers–Briggs Type Indicator) and a self-assessment of conflict management style. The workshop concluded with small group activities involving role playing of challenging ethical scenarios for leaders. The majority of the participants felt that the conference was beneficial in providing them with the necessary tools to act as leaders within their field. The authors also performed a follow-up survey 6 months after the leadership development workshop was completed to assess whether this initiative had lasting benefit. The participants reported there were several lasting benefits from the course including an improvement in their understanding of different personality types, conflict resolution, giving negative feedback, and assessing their own strengths and weaknesses as leaders [33].

Wipf et al. wrote about their experience developing a 6-hour course for intermediate and senior-level internal medicine trainees [34]. The 6 hours of course material were divided up over three workshops held weekly for 3 weeks and focused on three areas of leadership development, including teaching, communication, and management of clinical teams. These areas were chosen based on trainee survey responses identifying them as areas they perceived as weaknesses in their medical training. Clinical responsibilities interfered only minimally with the leadership development workshop. Instruction regarding leadership styles was performed through open-discussion lecture series, including video vignettes, focused seminars on Neher et al.'s five-step "microskills" of teaching, and role-playing sessions to practice newly learned leadership skills [34,35]. The trainees who completed the course rated it very positively and considered it to be highly beneficial [34]. The authors suggested that giving the participants time to learn in an environment free from clinical responsibilities validated the importance of leadership training and optimized learning by removing competing responsibilities.

In contrast to the above theory, Souba stated that teaching leadership to medical professionals did not necessarily have to occur as part of leadership workshops. While Souba noted that leadership education could occur in the form of formal education, he also felt that leadership training could occur through observation and experience. Mentors, role models, and even examples of ineffective leaders are important, just as taking on new challenges and making mistakes can improve your leadership skills. He argued that the most proficient leaders are "constantly learning" from their environments [9].

Though Souba suggested that leadership development could be occurring constantly within the environments of healthcare institutions, McAlearney investigated healthcare professionals' perceptions of barriers preventing this "constant development" [9,13]. She interviewed 35 employees and analyzed 55 healthcare organizations in order to propose a more effective strategy for developing leadership education for healthcare professionals [13]. McAlearney found six common reasons for inadequate leadership within healthcare including:

1. An industry lag in prioritizing leadership
2. Difficulty making leadership be representative of the population the healthcare network serves
3. Professional conflicts between medical specialties or groups
4. Inadequate time to devote to leadership training
5. Inadequate technical capabilities of the institution to develop leaders
6. Inadequate funding for leadership training [13]

She then suggested how healthcare institutions can overcome these challenges [13]. First, existing leadership must make a commitment that leadership development is a priority and change the organizational culture and view of leadership development programs [13]. The leadership development program must be created with buy-in from departments and groups who will utilize the services [13]. This in turn will lead to more engaged participants [13]. The healthcare system and existing leadership must make it clear that by pursuing leadership development of employees they are recognizing the value of those employees [13]. Such recognition can lead to greater employee satisfaction and efficiency [13].

In a later paper, McAlearney conducted 200 interviews of healthcare professionals and found that leadership development programs achieve their results by using skills-based training, 360 degrees feedback, action learning, and mentoring [10]. She also stated that an important aspect of any leadership development program is that it has a broad enough reach to develop skills in professionals within multiple organizational levels [10]. Though it is important to include all levels of professionals in leadership training, it is also important to divide these larger groups into more focused areas of training. These more focused areas may be specific to departments or groups which focus on certain topics such as quality improvement and cost containment and should tie into the larger organizational stated goals, strategic plans, and the principles of autonomy, beneficence, nonmaleficence, and justice. Finally, the healthcare organization must relate leadership development activities to employee satisfaction surveys so that employees experience greater job enjoyment resulting in higher employee retention [10].

One size should not, however, have to fit all. House and Howell highlighted that in order for leadership training to be accepted by a person and become a part of his or her routine, the training must take into account each person's unique personality type [28]. An example of this comes from outside the healthcare industry. Kirkbride advocated a multistep leadership development program using the Multifactor Leadership Questionnaire by Avilio and Bass to assess the leadership styles current employees already possessed [24,26]. The first step was to hold explanatory sessions to educate the participants about what to expect from the Multifactor Leadership Questionnaire and to provide an overview of the different leadership styles. Next, the participants completed the survey and were introduced to the concept that leaders often utilize the full range of leadership styles, from the laissez-faire style to the democratic style, depending on the situation. After this, that the facilitators provided feedback from the Multifactor Leadership Questionnaires and offered individualized action plans and coaching sessions to participants for how they could improve their leadership styles. Though this training initiative recognized that all leadership types may be used at some point by effective leaders, it emphasized that the majority of time should be spent acting as democratic and transformational leaders. Therefore, this initiative aimed to assist participants in understanding how frequently they currently utilized democratic and transformation leadership techniques and provided suggestions about how to shift a larger percentage of their leadership interventions toward a more democratic and transformational style [26].

Also using the Multifactor Leadership Questionnaire, Horwitz et al. assessed the leadership styles of surgical trainees across various stages of training to identify where to focus leadership development initiatives. Compared to the national average of leadership behaviors, surgical residents were more likely to utilize authoritarian and transactional leadership styles such as management by exception (active and passive) styles and were less likely to utilize democratic and transformational leadership styles such as the individualized consideration style. Horwitz et al. explained that these tendencies toward management by exception (active), or micromanaging, are frequently utilized by leaders because the consequences of incorrect action are extremely high in medical fields and leaders who are in charge of clinical teams are more likely to monitor trainees they are responsible for very closely. Overutilizing the management by exception (passive) style and waiting for followers to make mistakes before engaging with the team members could come from the ingrained hierarchy of medical training. Horwitz et al. suggested that the underutilization of the individualized consideration leadership style could be related to inadequate support among trainees and within clinical teams as well as a result of feelings of high independence among team members [22].

Women chose similar leadership styles compared to men in this cohort, unlike in other professions where women have been found to be more likely to utilize transformational leadership styles compared to men [22,25,36,37]. The uniformity of leadership style between the sexes could be a result of the years of similar medical training. In this study, women only differed from men in being less likely to utilize management by exception (active) [22,25,36,37].

In addition, there were minimal differences between leadership styles of senior compared with junior trainees. The more senior a trainee was, the less likely he or she was to choose a laissez-faire leadership style. This was the only significant difference in leadership style between the junior and senior trainees. This suggested that senior-level trainees did not develop or improve their leadership skills during their medical training. Senior trainees were not more likely to utilize transformation leadership styles compared with the junior trainees [22].

The finding that transformational leadership was underutilized could be used to tailor leadership development initiatives to trainees as they advance in their careers. Horwitz et al. noted that use of transformational and democratic leadership among trainees was associated with higher scores of perceived efficiency, desirable patient outcomes, and trainee satisfaction. Unfortunately, these styles were not frequently utilized. These findings underscore the necessity of providing leadership development training to medical professionals in order to increase utilization of transformational and democratic leadership and to create more effective healthcare leaders [22].

CONCLUSION

In summary, developing more effective and ethical current and future leaders within the healthcare profession is of the highest priority. While all three leadership styles, laissez-faire, authoritative, and democratic, have value, the most effective leaders utilize more democratic leadership styles compared with the other two. Democratic leaders are also considered to be highly ethical and able to uphold the principles of autonomy, beneficence, nonmaleficence, and justice. Developing the leadership skills of healthcare professionals requires a commitment of time and resources from healthcare institutions. It may, in some cases, require a change in the culture and the perception of the value of leadership development programs within that institution. There are several proposed methods for developing the leadership skills of healthcare professionals, including mentorship, formal academic coursework such as certificate programs or master degrees, focused workshops, and intensive assessments tailored to individual personality types [9,10,22,26,29–34]. No single method of teaching leadership development has been found to be superior to the others [9,10,22,26,29–34]. Leadership development workshops were well-received and considered successful across all trainee levels, emphasizing that leadership development training is beneficial at all stages in a healthcare professional's career. Recurring themes of these workshops included the concepts that successful leadership development training requires the exclusive focus on the subject material and removal of distraction from clinical duties, interactive learning sessions with role playing of difficult scenarios should be included in leadership training, and team-building exercises helped promote both personal and team leadership development and growth. In all of the examples of leadership development activities, healthcare professionals who participated felt that developing leadership skills was important to their future careers. As the medical profession becomes increasingly complex and ethically challenging, the success of our future will depend on how effectively we lead ourselves.

References

[1] Beauchamp TL, Childress JF. Principles of biomedical ethics. Oxford university press; 2001.
[2] Cohen JJ. Leadership for medicine's promising future. Acad Med 1998;73(2):132–7.
[3] 2016 Accreditation Council for General Medical Education (ACGME) Multi-Speciality and Pediatric Leadership Skills Training Programs for Chief Residents [cited 23.9.2015]. Available from: http://www.acgme.org/acgmeweb/tabid/166/meetingsandconferences/workshops.aspx.
[4] American Society of Clinical Oncology Leadership Development Program [cited 23.9.2015]. Available from: http://www.asco.org/professional-development/leadership-development-program.
[5] Merriam–Webster nd. Leader Merriam-Webster.com [cited 27.9.2015]. Available from: <http://www.merriam-webster.com/dictionary/leader>.
[6] Bennis W, Burt N. Leaders. New York: Harper & Row; 1985.
[7] Ciulla JB. Ethics and leadership effectiveness. The nature of leadership 2004;302–27.
[8] Northouse PG. Introduction to leadership: Concepts and practice. 3rd ed. Western Michigan University: Sage Publications, Inc.; 2015.
[9] Souba WW. The job of leadership. J Surg Res 1998;80(1):1–8.
[10] McAlearney AS. Using leadership development programs to improve quality and efficiency in healthcare. J Healthc Manag 2008;53(5):319–31.
[11] Lyons MF, Ford D, Singer GR. Physician leadership: How do physician executives view themselves? Physician Exec 1996;22(9):23–6.
[12] Itani KM, Liscum K, Brunicardi FC. Physician leadership is a new mandate in surgical training. Am J Surg 2004;187(3):328–31.
[13] McAlearney AS. Leadership development in healthcare: A qualitative study. Organ Behav 2006;27(7):967–82.
[14] Dine JC, Kahn JM, Abella BS, Asch DA, Shea JA. Key elements of clinical physician leadership at an academic medical center. J Grad Med Educ 2011;3(1):31–6.
[15] Force USPST. Risk Assessment, Genetic Counseling, and Genetic Testing for BRCA-Related Cancer in Women: Clinical Summary of USPSTF Recommendation. AHRQ Publication No 12-05164-EF-3. 2013.
[16] Kaufman B, Shapira-Frommer R, Schmutzler RK, et al. See more at: Olaparib monotherapy in patients with advanced cancer and a germline BRCA1/2 mutation. J Clin Oncol 2015;33(3):244–50.
[17] Gelmon KA, Tischkowitz M, Mackay H, et al. Olaparib in patients with recurrent high-grade serous or poorly differentiated ovarian carcinoma or triple-negative breast cancer: A phase 2, multicentre, open-label, non-randomised study. Lancet 2011;12(9):852–61.
[18] Oza AM, Cibula D, Benzaquen AO, et al. Olaparib combined with chemotherapy for recurrent platinum-sensitive ovarian cancer: A randomised phase 2 trial. Lancet 2015;16(1):87–97.

[19] Warner E, Plewes D, Hill KA, et al. Surveillance of BRCA1 and BRCA2 mutation carriers with magnetic resonance imaging, ultrasound, mammography, and clinical breast examination. JAMA 2004;292(11):1317–25.
[20] Domchek SM, Friebel T, Singer CF, et al. Association of risk-reducing surgery in BRCA1 or BRCA2 mutation carriers with cancer risk and mortality. JAMA 2010;304(9):967–75.
[21] Lewin K, Lippitt R, White RK. Patterns of aggressive behavior in experimentally created "social climates". J Soc Psychol 1939;10:271–99.
[22] Horwitz IB, Horwitz S, Daram P, et al. Transformational, transactional, and passive-avoidant leadership characteristics of a surgical resident cohort: Analysis using the multifactor leadership questionnaire and implications for improving surgical education curriculums. J Surg Res 2008;148(1):49–59.
[23] Bass BM. Two decades of research and development in transformational leadership. Eur J Work Organ Psychol 1999;8(1):9–32.
[24] Avilio BJ, Bass B. Multifactor Leadership Questionnaire. 3rd ed. Menlo Park, CA: Mind Garden Inc.; 2004.
[25] Bass BM. The Bass handbook of leadership: Theory, research, and managerial applications. New York, NY: Free Press; 2008. 458–97.
[26] Kirkbride P. Developing transformational leaders: The full range leadership model in action. ICT 2006;38(1): 23–32.
[27] Burns JM. Leadership. New York: Harper & Row; 1978.
[28] House RJ, Howell J. Personality and charismatic leadership. Leadership Q 1992;3(2):81–108.
[29] Geyman JP, Deyrup, James A. Subgroup analysis of teamwork skills within physicians for the twenty-first century/report of the project on the general professional education of the physician and college preparation for medicine. J Med Educ 1984;59:169–72.
[30] Stoller JK, Rose M, Lee R, Dolgan C, Hoogwerf BJ. Teambuilding and leadership training in an internal medicine residency training program. J Gen Intern Med 2004;19(6):692–7.
[31] Hemmer PR, Karon BS, Hernandez JS, Cuthbert C, Fidler ME, Tazelaar HD. Leadership and management training for residents and fellows: A curriculum for future medical directors. Arch Pathol Lab Med 2007;131(4): 610–14.
[32] Klein EJ, Jackson JC, Kratz L, Marcuse EK, McPhillips HA, Shugerman RP, et al. Teaching professionalism to residents. Acad Med 2003;78(1):26–34.
[33] Doughty RA, Williams PD, Seashore CN. Chief resident training: Developing leadership skills for future medical leaders. Am J Dis Child 1991;145(6):639–42.
[34] Wipf JE, Pinsky L, Burke W. Turning interns into senior residents: Preparing residents for their teaching and leadership roles. Acad Med 1995;70:591–6.
[35] Neher J, Gordonn K, Meyer B, Stevens N. A five-step "Microskills" model of clinical teaching. J Am Board Fam Pract 1992;5:419–24.
[36] Ochieng Walumbwa F, Wu C, Ojode LA. Gender and instructional outcomes: The mediating role of leadership style. J Manag Dev 2004;23(2):124–40.
[37] Eagly AH, Johannesen-Schmidt MC, Van Engen ML. Transformational, transactional, and laissez-faire leadership styles: a meta-analysis comparing women and men. Psychol Bull 2003;129(4):569.

CHAPTER

12

Ethics in Cancer Patient Education

Lorene Payne

The University of Texas MD Anderson Cancer Center, Houston, TX, United States

INTRODUCTION

Ethics is an integral aspect of all interactions between clinical professionals and patients, including those involving patient education. Although patient education does not usually involve the dramatic circumstances often associated with ethics, the entire education process does require application of ethical principles.

The literature encourages educators and clinicians to incorporate ethics into the process of patient education, yet recognizes that circumstances of the current medical care delivery systems often do not support clinicians in a way that allows adequate staffing or support for teaching. Redman summarizes the "...dilemma facing nurses: education is clearly essential to patient safety and flourishing...and yet, the dominant culture of medicine, which derives allocation of resources and privileges, pays only lip service to patient education" [1].

Physicians, too, recognize the ethical challenges related to patient education. While noting that our technology-rich environment provides health information messaging in multiple methods, it is recognized that much is written at a level beyond the average person's ability to understand or apply it. This limited ability to understand has been clearly linked to poor outcomes [2]. "The paradox is that people are awash in knowledge they may be unable to use" [3].

The ethical principles of autonomy, beneficence, justice, and nonmaleficence find direct application in the arena of oncology patient education. This chapter will describe those principles, explore applicable ethical codes which derive from them, and apply them to the practical world of patient education for the oncologic population. As we explore the basic ethical principles and codes, the invitation is extended to embrace the pronouncement "...patient education ought to be thought of as basic to ethical practice..." [1].

Ethical Challenges in Oncology.
DOI: http://dx.doi.org/10.1016/B978-0-12-803831-4.00012-9

ETHICAL PRINCIPLES

Our discussion will focus on four ethical principles presented in alphabetical order, not necessarily in order of importance: Autonomy, beneficence, justice, and nonmaleficence [4].

Autonomy

Autonomy is the recognition that each person has the right of self-determination, the right to choose their own preferences in care and treatment. It implies the patient is free from the control of others and is acting instead in accordance with their own values and judgments. Autonomy requires a capacity for critical thinking, deliberation, and independent decision making.

The principle of autonomy makes excellence in patient education imperative. Patients must be able to acquire information, understand and process it, and determine the best course of action for themselves. It is through the clear explanation of choices, communication, and teaching materials that are understandable and informative, that patients can exercise autonomy to their benefit. Without fully understanding medical choices, the right to autonomy is hollow and could even lead to harm, as patients would base decisions on incorrect, incomplete, or misunderstood information. To truly practice autonomy in medical decisions, the patient must receive exceptional education. The teaching must provide understandable explanations and materials, allow, encourage, and answer questions, thus providing them with a full understanding to make informed decisions and partner in their care.

For oncology patients who are unable to exercise autonomy because they are comatose or cognitively compromised due to disease process or treatment, autonomy may be problematic and presents a different set of ethical challenges for those acting on behalf of the patient. However, the educator's duty to provide complete teaching to those responsible for acting on the patient's behalf still exists.

Beneficence

Beneficence requires healthcare professionals to take actions that benefit others, providing for their good. It requires compassion and understanding of the patient's value system: determination of "good" is highly individual and dependent on each person's preferences.

In patient education, beneficence can apply on both an individual and community basis. Individually, posting teaching materials within an electronic patient portal is one example of beneficence. In the public health arena, making cancer prevention teaching accessible to communities through both outreach presentations and on the web are examples of educational beneficence.

Justice

Justice in cancer patient education refers to assuring that all members of society have access to medical care and education. Within a just medical care system, resources would be available to all equally.

A major issue relating to justice in patient education is that teaching materials are often written at a grade level above most people's level of understanding. Studies show that one-third of American adults have low health literacy, and a predominance of patient education materials are written at a level too high for most to understand, especially when dealing with a new diagnosis [5]. This is a violation of the principle of justice.

In this era of increasing use of electronic delivery of patient education, consideration must be given to facilitating access to computers or mobile devices, and/or provision of comparable learning opportunities for those who are not computer literate. Providing materials in any format cannot be effective unless people have access to them, understand them, have resources to place them in perspective, and are sufficiently broad in scope to be relevant to the patients accessing them.

Nonmaleficence

The principle of nonmaleficence is to do no harm. This has been an expectation of medical care throughout modern history. Related to patient education an example of nonmaleficence would be the unethical practice of bias in teaching materials that only presents one option, say the regimen of only one pharmaceutical company with which the prescriber has a financial relationship, in spite of other regimens with better clinical outcomes. Another example of nonmaleficence is inadequately evaluating patient understanding after an education session, resulting in "training" that actually was not mastered by the patient and may result in harm (Box 12.1).

BOX 12.1

ETHICAL TEACHING

JG is a 62-year-old woman of Mexican American heritage. She is receiving chemotherapy for treatment of Sarcoma and a central line has been placed. She and her husband attended a class to learn to flush the line and change the dressing. In class, they watched a video and a demonstration after which they successfully completed a supervised flush and dressing change. They were therefore checked off and sent home with an instruction packet. Twelve days later, JG presented in the emergency center with redness, swelling, and a foul smelling discharge at the site. Although the instruction packet they received in class included signs of infection, their English language skills were not sufficient to understand this written content and they explained they had been "too embarrassed to ask what the handouts said."

Principles violated: Nonmaleficence, considered her attendance at the class accomplished learning, but actually set her up for harm since the instructor did not assure her understanding of all teaching. Also, violated justice in that the patient and husband did not receive materials that met their needs in a language they could understand.

Actions that would have met ethical teaching: Thorough assessment of the patient would have revealed the limited English proficiency skills so the home-care packet could have been provided in Spanish. Use of Teach Back evaluation of the signs of infection would have revealed the lack of patient understanding and allowed correction before the patient went home.

CODES OF ETHICS

These four ethical principles have been among the guiding forces behind the development of codes of ethics within the healthcare professions. Each of the major professional groups has developed its code reflective of these principles and expects its membership to uphold them within their practice. Patient education is either specifically addressed, or implied, in each of these codes summarized in this section.

Health Education Code of Ethics

The National Commission for Health Education Credentialing (NCHEC) promulgates the Health Education Code of Ethics. The purpose of this group is to support the highest standards of educational excellence among health educators. It addresses health education needs at all levels from individual to community. Introducing its code of ethics, the group states that it provides "...a basis of shared values..." [6] and reminds health educators to aspire to high standards of ethical behavior. There are six articles, with each focusing on a different level of responsibility to

- promote good health among the public
- promote ethical conduct within the profession
- recognize professional boundaries and competence
- respect the rights, confidentiality, and needs of all, and adapt educational methods to meet diverse needs
- contribute to the profession through research and evaluation
- train health educators so as to benefit the educators and the public

Code of Ethics for Nurses

The American Nurses Association (ANA) [7] is the largest single group of healthcare professionals. Nurses are responsible for direct patient care including teaching. The code of ethics prepared for nurses focuses primarily on direct patient care but also encompasses expectations that apply to patient education. Among the nine provisions, the aspects that apply to education include:

- Respect for the dignity and worth of each individual
- Commitment to the patient at all levels including family, community and populations
- Promotion of health and safety
- Practicing in an ethical manner
- Research and scholarly contributions to the profession

Code of Ethics for Educators

The Cancer Patient Education Network (CPEN) prefaces its code of ethics with "Adherence to ethical standards is a key criterion in all aspects of cancer patient education..." [8] and among the provisions includes:

- Maintain personal integrity and competence
- Disclose potential conflicts of interest
- Promote and implement applicable standards

Code of Ethics for Public Health

The leadership council of the American Public Health Association (APHA) and the Public Health Leadership Society recognized that the ethical needs and demands of public health are unique to that milieu, as public health focuses on population groups and healthcare systems rather than individual patients and practitioners. Their code of ethics delineates the "...distinctive elements of public health and the ethical principles that follow..." [9]. There are 12 principles identified in the code: Those applying to education include:

- Inclusion of community members in the process of program creation
- Accessibility to services for all members of the community
- Incorporation of a variety of approaches to meet the needs of a diverse population
- Assurance of professional competence of employees
- Engagement in collaborations and affiliations that build effectiveness

EDUCATION PROCESS AND APPLICATION OF ETHICS

The foundation of ethical principles combined with codes of ethics describes expectations of the framework within which educators' performance and process is practiced. The education process is the sum of a series of educator, patient, and provider preparations and interactions carried out over time. It is not one episode or a single event. It follows a logical, cyclical process that begins with assessing patient needs (either individually or as a population), planning appropriate educational interventions, implementing the teaching plan, and then evaluating the results which begins the cyclical process of assess, plan, implement, and evaluate again.

Assessment

The results of needs assessment establish teaching needs, knowledge gaps, and required teaching materials design. Optimal teaching plans cannot be made until needs assessment is complete. Whether formal or informal, whether focusing on the individual patient or targeting the needs of a diverse community, the needs assessment identifies what content will be covered in a teaching session or program.

A formal assessment of adult health literacy, National Assessment of Adult Literacy (NAAL), was completed in 2003 and ethical educators will factor the results into their practice, their communications with patients and family members, and creation of educational materials. Health literacy is defined as the ability to obtain, process, and understand basic health information and services to make appropriate health decisions [10]. It is essential for understanding cancer topics and concepts, mastering skills, and self-care principles.

To establish the respondent's level of capability, examples of actual health literacy needs were tested. Survey participants were asked to perform common literacy tasks in three categories: Prose, document, and quantitative. Prose tasks revealed the person's ability to find and understand information presented in "continuous" reading such as instructional materials in which prevention and wellness information might be presented. Document tasks addressed how well the participant could find information and use it from "noncontinuous" sources such as health insurance forms and tables. Quantitative testing involved health numeracy skills such as understanding medication instruction (Table 12.1).

Responses were scored at the highest level as Proficient, then Intermediate, Basic, or the lowest Below Basic. The NAAL assessment established that health literacy levels of the US adult population are surprisingly low. Only 12% function at the Proficient level, 53% are Intermediate, and 35% are at either the Basic or Below Basic level (Tables 12.2 and 12.3).

Combining people in both the Basic and Below Basic groups reveals those considered to have low health literacy.

The NAAL revealed significant demographic differences in health literacy. Ethnically, 28% of the White population demonstrated low health literacy, 57% of Black Americans, and 65% of Hispanics. Those with less educational background were lower in health literacy skills and those over 65 years of age, regardless of educational background, scored lower in health literacy. People who spoke a language other than English in childhood also scored lower (the NAAL assessment was administered in English). Since many of these demographic groups are increasing through the aging population and immigration, the challenges of low health literacy are likely to increase in coming years.

The high incidence of low health literacy has prompted experts to recommend all healthcare providers and educators consider low health literacy another Universal Precaution [12]. This is similar to the advent of the use of gloves for all contact with blood and body fluid in the 1970s when the standard was adopted in response to AIDS. It was recognized that each encounter by the provider may expose them to the pathogen as it

TABLE 12.1 National Assessment of Adult Literacy (NAAL) - Literacy Tasks

Prose literacy scale →	**Prose tasks** Require the ability to search, comprehend, and use information from continuous texts such as news articles and instructional materials. Examples of prose tasks include locating information from a newspaper article, comparing different points of views in editorials, and interpreting the theme of a poem.
Document literacy scale →	**Document tasks** Require the ability to search, comprehend, and use information from noncontinuous texts such as job applications, maps, and food labels. Examples of document tasks include using a schedule to select a train, filling out appropriate information on a form, and locating a street on a map.
Quantitative literacy scale →	**Quantitative tasks** Require the ability to identify and perform computations using numbers embedded in printed materials. Examples of document tasks include balancing a checkbook, completing an order form, and calculating the interest on a loan.

From White, S. and McCloskey, M. (forthcoming). Framework for the 2003 National Assessment of Adult Literacy *(NCES 2005-531). US Department of Education. Washington, DC: National Center for Education Statistics. Available from: https://nces.ed.gov/naal/fr_tasks.asp.*

TABLE 12.2 Descriptions of Health Literacy levels, results from the NAAL

Health Literacy Level	Task Examples	Percentage
Proficient	Using a table, calculate an employee's share of health insurance costs for a year.	12%
Intermediate	Read instructions on a prescription label, and determine what time a person can take the medication.	53%
Basic	Read a pamphlet, and give two reasons a person with no symptoms should be tested for a disease.	21%
Below Basic	Read a set of short instructions, and identify what is permissible to drink before a medical test.	14%

Source: US Department of Education. National Center for Education Statistics (NCES). The Health Literacy of America's Adults: Results from the 2003 National Assessment of Adult Literacy. Available from: http://nces.ed.gov/pubs2006/2006483.pdf. 2006.

TABLE 12.3 Literacy by Race

Ethnic Group	Below Basic	Basic	Intermediate	Proficient
White	9%	19%	58%	14%
Black	24%	33%	41%	2%
Hispanic	41%	24%	31%	4%
Other	13%	21%	54%	12%

Source: US Department of Education. National Center for Education Statistics (NCES). The Health Literacy of America's Adults: Results from the 2003 National Assessment of Adult Literacy. Available from: http://nces.ed.gov/pubs2006/2006483.pdf. 2006.

was impossible to identify without testing which patients actually carried the HIV responsible for the disease. Therefore, every person was treated as if they were communicable: The universal precaution. Similarly, it is impossible to tell without testing which of our patients has low health literacy. So prudent, ethical practitioners will treat all patients as if they may have trouble in understanding.

Ethical practice expects that professionals involved in patient education must make themselves aware of these national assessment findings and incorporate mitigating measures into their practice. Specific measures are discussed in the next sections of the education process—planning and implementation. However, there are other considerations of the assessment phase in addition to the health literacy assessment findings. It must be determined individually what the patient and family consider their learning needs and what aspects they already understand. Omitting this step and presuming instead that the clinical staff knows what to teach is a violation of the respect and individuality of the adult learner. It may also be less efficient and less effective (Box 12.2).

To establish individual needs, ask the patient directly, using open-ended questions, what they have been told so far. Asking with respect, as opposed to questions phrased in a manner that puts people on the defensive, is best practice. Inviting, nonjudgmental questions such as "What have you been told about the biopsy results?" establishes a level of understanding without insulting. You will learn their exact understanding and can either

BOX 12.2

ASSESS

The nurse was preparing a chemo-naïve patient to begin oral chemotherapy. His teaching plan was to cover the drugs, schedule for administration, required laboratory tests, and safe handling of the chemo. Within the first few minutes, he noticed the adult daughter was attending to the instruction, but the patient did not seem to be listening. Stopping the planned instruction, he instead asked the patient "What is your biggest question or concern about starting chemotherapy?" With a relieved expression the patient asked "Am I going to be able to keep working through chemo? If I lose my job I have no insurance and won't be able to continue treatments." Until those concerns were addressed, it was not possible to actually absorb the chemo teaching.

Ethical actions demonstrated: The nurse was observant and attentive during teaching and reacted respectfully after noting the patient was not listening. The use of an open-ended question respected autonomy and invited dialogue which revealed the patient's teaching need. Meeting the patient's immediate needs then allows attention to the planned content.

provide more targeted information, if needed, clarify any misconceptions or move on to other topics.

Individual assessment should also seek to identify cultural and spiritual considerations and preferred language that may affect the education plan. The increasingly diverse population of the United States assures that the healthcare educator will be teaching people with a variety of values, beliefs, and usual practices that need to be considered in the teaching plan. For example, if teaching dietary methods for minimizing mucositis complications from chemotherapy, determining what types of foods and spices are common in the diet and preferences of the patient is imperative. Or, during provision of end-of-life care, the healthcare team needs to assess spiritual needs and practices and role expectations for caregivers, among other things.

It is not always practical to complete an individual patient teaching needs assessment. For instance, when topics are covered in a classroom setting or in a web-based training module, there is no opportunity for individual needs assessment. Another example where individual patient assessment is not indicated is assessing a community for public health prevention teaching. Instead of focusing on individual needs, seeking expert opinion on content and studying demographic trends and incidence of specific cancer types directs the education plan.

In addition to assessing direct patient need, healthcare providers may be queried about topics needed to fill gaps in patient teaching materials. For instance, trends toward genetic testing demand updated teaching materials with attention to numeracy literacy mitigation to assist patients to understand concepts of risk and benefit.

Often methods of needs assessment are informal in nature. An example of this is when clinicians note new pharmaceuticals, diagnostics, or treatments for which no teaching materials exist. The fact they bring this to attention of educators responsible for materials creation is an informal needs assessment. Or, the patient or family member will pose a question that itself becomes the needs assessment for that encounter.

Planning

Once needs are assessed, planning begins to meet those needs. Ethical practice requires that all planning incorporates steps to mitigate low health literacy and assure materials are accessible to most people. Steps to mitigate low health literacy in written materials used for teaching include [13]:

- Use simple words, avoid medical, or technical jargon
- Use at least a 12-point font
- Cover only two or three concepts per written piece
- Write at seventh grade level or below
- Organize the flow of ideas to enhance understanding
- Use "you" instead of "the patient"
- Examples and stories may animate the message
- Present the needed action, what the patient needs to do
- Clarify how the patient benefits from understanding/compliance
- Use bulleted points, avoid long sentences, and text-heavy paragraphs
- Use white space
- Choose images that support the message and are culturally balanced
- Include a family member in the teaching session

Consider during planning which support materials are most appropriate. Simple oral communication may be used, but when written materials complement the message the patient's understanding and recall may be enhanced. Some content is best supported through use of videos, models, or images. Studies show that incorporating a variety of strategies results in improved effectiveness of patient teaching [14]. Also, plan the setting which should respect the person's privacy, be comfortable, and minimize distractions.

One of the ethical considerations during the planning phase is to maximize accessibility to a variety of educational strategies. Indeed, in addition to ethical expectations, assuring accessibility is mandated through the Americans with Disabilities Act [15] which identifies reasonable accommodation for those with limitations. An example of compliance with Americans with Disabilities Act is providing transcripts for educational videos or using other methods to compensate for individual difficulties (Box 12.3).

Another ethical consideration when planning includes consideration of the origin of supplemental materials used in the teaching session. Materials used should be free of bias, whether created within a healthcare institution, or provided through an outside vendor. Education materials need to provide information and stimulate an understanding of what actions the patient can take; materials should not be marketing materials for a company or product. Part of the planning process may include setting ethical standards within the medical practice or institution that protect against unethical marketing materials masquerading as patient education.

BOX 12.3

MAXIMIZE ACCESSIBILITY

Mrs. P, a 72-year-old with chronic myelogenous leukemia wanted to attend a local conference for patients and their families but was fearful that her difficulty hearing would result in an inability to actually understand what was presented. After contacting conference organizers, a solution was found when they agreed to provide computer-assisted real-time (CART) transcription of proceedings as a caption on the bottom of the conference screen.

Ethical principle exemplified: Justice, as the technology improved accessibility to the proceedings to all attendees, even those with hearing impairment.

Certainly not all commercially available patient teaching materials are unethical, even if they contain the logo or name of the company providing them. For instance, if the physician has prescribed a unique injectable low–molecular-weight heparin to a postop patient on discharge and the pharmaceutical company has injection practice pads and syringes that provide good skills practice, using those materials does not violate ethical principles. However, if an oncology patient is now experiencing hyperglycemia secondary to high-dose steroid treatment, general diabetes teaching materials should be generic to avoid implying an endorsement of one company over another.

Increase in the use of the internet provides an opportunity to increase access for patients to information. Between one-third to one-half of adults use the internet and social media for healthcare information reasons [16]. Planning for web-based training and education includes consideration of reading level but also the layout, color, and ease of searching. The number of clicks needed to find desired information is critical to its accessibility. Another consideration is the meaning that color and images hold for people of different cultures.

In addition to following similar directives for written materials to provide beneficence to patients, suggestions for health literate web-based education [17] include:

- Making home page simple and inviting
- Using images that enhance content
- Providing engaging links, including audio and visuals
- Placing content in the center of the screen
- Labeling links clearly
- Assuring navigation buttons work properly
- Simplifying browsing features
- Including resources for further explanations

Another aspect of planning is to consider how success will be measured. Consider formal metrics that may be used for evaluating teaching encounters. Planning for what you will measure before implementing the teaching allows capture of preeducation metrics.

Will the program result in a decrease in readmission after hospital discharge? Will the teaching materials contribute to a decrease in the infection rate? Can an improved score for patient satisfaction be expected after the teaching? Will the education increase the numbers of people who receive screening for cancer? Identifying possible metrics during the planning phase allows an outcome measure to establish the value of the education. A baseline of the identified metric can be obtained and then compared with postteaching data.

Finally, choosing and preparing instructors is another aspect of planning education. Each individually licensed healthcare provider and health educator may serve as instructor in some circumstance and with the individual patient teaching that predominates in the healthcare field. All providers ethically have a responsibility to update their professional knowledge of teaching principles and mitigation steps for low health literacy and barriers to access.

Implementation

Implementation involves carrying out the teaching plan or program with nonmaleficence. It may take place in a variety of settings but generally will either be face-to-face or via a technology-based method. We will consider the ethical considerations of each of these two general implementation methods.

Face-To-Face Teaching

When personally teaching an individual or group of individuals, carry out the plans made during the planning phase. The first consideration is to provide for the respect and privacy of each person as per the planning phase (Box 12.4).

Assure that individuals have privacy to receive information, especially sensitive results, and the opportunity to ask questions without intimidation.

BOX 12.4

RESPECT

Families anxiously congregated in the surgical oncology waiting room. The nervousness was etched on their faces and each person's body posture broadcast their concern. There was a collective holding of breath whenever a surgeon came through the door, and pairs of eyes watched with concern as the family member was escorted from the room to hear the results of the surgery and the fate of their loved one. In the hall outside the surgical waiting room were a series of small consult rooms where the surgeon and family could discuss the results of the surgery in private.

Principle demonstrated: The right of the patient and family to confidentiality and respect for privacy.

Ethical practice that respects each person requires that the oral communication of the teaching experience be delivered in plain language with simple terms and avoiding jargon. This assures that even those with low health literacy can understand what is being said. Delivering information incrementally and allowing for reiteration between ideas or concepts increases chances that the message has been received. This approach is referred to as the "Teach Back" method. Although it is a common practice to ask the patient "Do you understand?" or "What questions do you have?," neither of those questions determine what the patient actually understood. Patients often respond in a way to save the practitioner time, or to avoid the appearance of ignorance.

Teach Back is a method of evaluation that invites the patient to repeat the information in their own words. It allows educators to assure they have spoken clearly and explained well. It is not a test of the patient and requires some practice by the educator to assure it is phrased as such. Questions such as "I want to be sure I didn't miss anything. Can you remind me what I've said so far?" or "Since your wife couldn't be here for the class, what are you going to tell her that you learned?" or "Please tell me what you might see if the wound is infected." Each of these approaches requires patients to restate concepts and instructions, allowing you to hear what they actually understood. Follow-up discussion and explanations can then focus on what is unclear or misunderstood before moving on to other topics. Adopting Teach Back into your daily practice and patient communication demonstrates professionalism and ethical care [14].

Another aspect of implementation that follows ethical guidelines is the use of multiple modalities to increase the likelihood of reaching those with different learning styles. Visual learners commit information more readily from presentation slides, videos, and images, while auditory learners benefit from hearing the information via the oral presentation or podcasts. Kinesthetic learners recall best if they have an opportunity to physically handle materials. Using multiple teaching strategies increases patient understanding. Research has shown that 67% of patients receiving education that incorporated multiple teaching modalities had better outcomes [14]. Therefore, a class preparing patients and their caregivers to take care of an indwelling Foley catheter after discharge would ideally include a presentation and discussion of the equipment while speaking the steps out loud, followed by participants practicing how to empty and measure the bag, and including written step-by-step handouts to take home.

Teaching Through Technology

The existence of the internet expands potential teaching outreach and thus increases the opportunity for many people of diverse backgrounds and locations to find educational information, thus supporting the principle of Justice. It also challenges the educator to create materials for a target population that is not homogenous. The majority of the work for online education is completed at planning and creation; implementation becomes simply a matter of making it available and monitoring the site for activity and feedback. As has been discussed, the internet is a common source for health information and an ethical educator will assist people to know which sites are authoritative.

Institutions are recognizing that the internet experience can be personalized, identifying patterns in searches and website visits and suggesting similar pages to target populations.

This affords an exciting possibility to target patient education information as content is customized into the web experience and meet the ethical principle of justice and the current regulatory demands of Meaningful Use. Meaningful Use regulations incentivize the use of electronic health records to individualize health information (among other requirements) [18].

Other technologies providing interactivity have been embraced by the public and offer autonomy for monitoring compliance with teaching and possibly capturing outcomes. There are gadgets for reminders of medication administration for instance. Wrist-worn devices capture self-care practices such as the number of steps taken in a day and even vital signs. Patients can enter their compliance with smoking cessation and there may be opportunities to reach out for clinical trials participants [19].

One ethical consideration of web-based education is that patients' privacy must be protected during communications and interactions first under the Health Insurance Portability and Accountability Act (HIPPA) of 1996 and also through the Health Information Technology for Economics and Clinical Health Act (HITECH) passed in 2009. Thus, secure servers and encrypted messaging protect patient–provider communication. When patient portals are used, secure communication between healthcare providers and patients must be assured. Suggestions to protect patients' rights to privacy include the use of secure servers and password protection. Recent studies are finding reassuring low rates of violations of privacy [16].

Evaluation

Ethical teaching includes evaluation to assure beneficence and guard against nonmaleficence. Acknowledging the patients' past experiences and existing knowledge base, it is respectful to provide a mechanism for patient and family member involvement during the formative evaluation stage of planning education programs and materials. This also supports justice, in that involving a diverse patient population increases the likelihood that the resulting programs and educational materials will meet the needs of the population.

Ideally, the planning phase includes identification of evaluation metrics to use and will be carried out after conclusion of the teaching. After teaching, summative evaluation may be as simple as using Teach Back. Alternatively, the evaluation step will compare pre and postresults and outcomes to demonstrate the efficacy of the teaching plan and value of education to the cancer treatment and outcomes. Implementing formal evaluation methods of teaching plans and programs contribute to the body of knowledge of best practices in education which is an expectation of some of the professional ethics codes.

Omitting evaluation may potentially be an example of nonmaleficence, in that an expectation of benefit may be presumed but may not have been realized. For instance, a treatment plan may include pain control measures and, after sharing handouts about pain control, the clinician may chart that teaching was complete for pain self-management. If the patient later presents with complaints of uncontrolled pain, subsequent clinicians may presume the patient is following the initial algorithm and advance to a stronger drug or dose which may not have been necessary if proper evaluation showed that the patient was not following initial pain control methods properly.

SUMMARY

Ethical principles of autonomy, beneficence, justice, and nonmaleficence, combined with the codes of ethics of numerous professional organizations apply to cancer patient education in multiple ways, through the entire education process, and within the many ways a teaching encounter may be experienced.

The Institute of Medicine's Global Forum held in 2013 dedicated a workshop to explore ethical issues, focusing on the desirability of creating a code of ethics that would apply to all healthcare professionals. They promote the idea that society would benefit from such a transdisciplinary code rather than individual professional groups focusing on their independent roles. They reason that inclusion of the public together with the entire team of healthcare professionals in describing the new code could create improved trust and reciprocity: Expectations between healthcare providers and between individuals and society. "A transdisciplinary code of ethics, applicable to all health professionals and created with public input, would be the first step toward generating a social contract that can meet the contemporary needs of health professionals and the patients and communities they serve" [20].

As patient education is provided in the healthcare arena, all educators are bound by the basic ethical principles to provide ethical care to the populations they serve.

References

[1] Redman B. Ethics of patient education and how do we make it everyone's ethics? Nurs Clin N Am 2011;46:283–9

[2] Berkman ND, Sheridan SL, Donahue KE, Halpern DJ, Crotty K. Low health literacy and health outcomes: An updated systematic review. Ann Intern Med 2011;155(2):97–107.

[3] Koh, H, Rudd RE. The Arc of Health Literacy [Internet]. 2015. [cited 10.9.2015] JAMA. Published online August 06, 2015. http://dx.doi.org/10.1001/jama.2015.9978. Available from: http://jama.jamanetwork.com/article.aspx?articleid = 2426088.

[4] American Nurses Association (US). Short Definitions of Ethical Principles and Theories Familiar Words, what do they mean? Silver Spring, MD. [cited 3.8.2015]. Available from: http://www.nursingworld.org/MainMenuCategories/EthicsStandards/Resources/Ethics-Definitions.pdf.

[5] US Department of Health and Human Services. America's Health Literacy: Why We Need Accessible Health Information. [cited 22.8.2015]. Available from http://health.gov/communication/literacy/issuebrief/.

[6] National Commission for Health Education Credentialing (US). Health Education Code of Ethics. Whitehall, PA. [cited 18.8.2015]. Available from: http://www.nchec.org/code-of-ethics 6b. O'Connor & Andrews, 2015; Raman, 2015.

[7] American Nurses Association. Code of Ethics for Nurses. Silver Spring, MD. [cited 18.8.2015]. Available from: http://www.nursingworld.org/codeofethics.

[8] Cancer Patient Education Network. Code of Ethics. Charlottesville, VA. [cited 18.8.2015]. Available from: http://www.cancerpatienteducation.org/pdf/CPEN_CodeofEthics.pdf.

[9] Public Health Leadership Society of the American Public Health Association. Principles of the Ethical Practice of Public Health. [cited 18.8.2015]. Available from: http://phls.org/CMSuploads/Principles-of-the-Ethical-Practice-of-PH-Version-2.2-68496.pdf.

[10] Health Resources and Services Administration. About Health Literacy. [cited 12.9.2015]. Available from: http://www.hrsa.gov/publichealth/healthliteracy/healthlitabout.html.

[11] US Department of Education. National Center for Education Statistics (NCES). The Health Literacy of America's Adults: Results from the 2003 National Assessment of Adult Literacy. [cited 31.7.2015]. Available from: http://nces.ed.gov/pubs2006/2006483.pdf. 2006.

[12] Agency for Healthcare Research and Quality. AHRQ Health Literacy Universal Precautions Toolkit. [cited 20.8.2015]. Available from: http://www.ahrq.gov/professionals/quality-patient-safety/quality-resources/tools/literacy-toolkit/index.html.
[13] Centers for Medicare and Medicaid Services. Toolkit for Making Written Material Clear and Effective. Baltimore, MD. [cited 15.9.2015]. Available from: https://www.cms.gov/Outreach-and-Education/Outreach/WrittenMaterialsToolkit/index.html?redirect = /WrittenMaterialsToolkit.
[14] Friedman A, Cosby R, Boyko S, Hatton-Bauer J, Turnbull G. Effective teaching strategies and methods of delivery for patient education: a systematic review and practice guideline recommendations. J Cancer Educ 2011;26:12–21.
[15] United States Department of Justice. A Guide to Disability Rights Laws. [cited 15.9.2015]. Available from: http://www.ada.gov/cguide.htm.
[16] Grajales III FJ, Sheps S, Ho K, Novak-Lauscher H, Eysenbach G. Social media: A review and tutorial of applications in medicine and health care. J Med Internet Res 2014;16(2):e13.
[17] US Department of Health and Human Services. Health Literacy Online. [cited 15.9.2015]. Available from: http://health.gov/healthliteracyonline/index.htm.
[18] HealthIT.gov. Meaningful Use Regulations. [cited 16.9.2015]. Available from: http://www.healthit.gov/policy-researchers-implementers/meaningful-use-regulations.
[19] Lim MSC, Wright C, Hellard ME. The medium and the message: Fitting sound health promotion methodology into 160 characters. JMIR mHealth eHealth 2014;2(4):e40.
[20] Wynia M, Kishore S, Belar C. A unified code of ethics for health professionals: Insights from an IOM workshop. JAMA 2014;311(8):799–800.

Further Reading

Agency for Healthcare Research and Quality (US). The SHARE Approach—Using the Teach-Back Technique: A Reference Guide for Health Care Providers. July 2014. [cited 12.8.2015]. Rockville, MD. Available from: http://www.ahrq.gov/professionals/education/curriculum-tools/shareddecisionmaking/tools/tool-6/index.html.

Glossary

(from NIH website http://www.niehs.nih.gov/research/resources/bioethics/glossary/index.cfm)

Autonomy 1. The capacity for self-governance, i.e., the ability to make reasonable decisions. 2. A moral principle barring interference with autonomous decision making.

Beneficence The ethical obligation to do good and avoid causing harm.

Confidentiality The obligation to keep some types of information confidential or secret.

Ethics (or morals) 1. Standards of conduct (or behavior) that distinguish between right/wrong, good/bad, etc. 2. The study of standards of conduct.

Health literacy The ability to obtain, process, and understand basic health information and services to make appropriate health decisions.

Honesty The ethical obligation to tell the truth and avoid deceiving others.

Justice 1. Treating people fairly. 2. An ethical principle that obligates one to treat people fairly.

Nonmaleficence A principle of bioethics that asserts an obligation not to inflict harm intentionally.

Privacy A state of being free from unwanted intrusion into one's personal space, private information, or personal affairs.

Teach Back Effective clinician–patient communication in which patients are asked to use their own words to explain what the clinician taught. It checks the clarity of explanation provided by the educator and level of understanding received by the learner. Teach Back allows the educator to know what the patient actually heard from the instruction.

CHAPTER

13

Commercialism in Cancer Care

Michael S. Ewer

The University of Texas MD Anderson Cancer Center, Houston, TX, United States

INTRODUCTION

In its simplest form, the ideal healthcare delivery system of any country should provide the highest level of benefit to the greatest number of patients at the lowest manageable overall cost. This ideal should be achieved while respecting patient autonomy and privacy, and through mechanisms that foster empathy, caring, and respect within the patient–provider relationship. Those seeking care should have complete confidence that the relationship will be transparent, honest, appropriately balanced, void of conflicts of interest or biases, and will be undertaken with the highest level of integrity. Uncertainties regarding diagnoses and treatment interventions must be disclosed, and the balance between risk and benefit of employing the goods and services offered must be shared with and perhaps more importantly, understood by the healthcare consumer. Medicine as it is practiced today in most developed nations including the United States is rife with barriers to this ideal. Mechanisms of compensation almost always involve third parties, and teams of professionals often provide care. Providers share protected health information within a healthcare entity or disclose it to other entities or providers who are all involved in some aspect of care delivery, payment for the provided goods and services, or simply managing the complex health care of a patient and the entities that provide it. Further barriers to providing optimal and ethical care exist as well and relate to fact that health care in the United States has evolved into a huge, largely for profit enterprise that presently consumes in excess of 17% of the gross domestic product of the United States economy—considerably more than is spent on a per person basis for health care in other countries. One should not imply that the business of health care is universally in conflict with the goals of most providers, i.e., to offer ethical and optimal care, but when commercial expediency conflicts with such goals, there are possible ethical concerns, conflicts of interests may be present, and the ideals of personal, confidential, affordable, and ethical care may become increasingly difficult to achieve.

It is within this framework that the impact of commercialism in health care must be examined. While all who seek health care are affected, cancer patients are especially

Ethical Challenges in Oncology.
DOI: http://dx.doi.org/10.1016/B978-0-12-803831-4.00013-0

vulnerable to the deleterious sequelae that may arise as commercialism becomes an increasing factor for both those who seek and those who provide health care. We recognize the positive role that commercialism has played in modern cancer care; new classes of therapeutics have been made available to our patients through research initiatives that have been fostered by private industry. It is not merely the new pharmaceuticals that have improved care; new implantable devices, less-invasive surgical techniques, and advances in supportive care for the cancer patient have all furthered joint initiatives of scientists, caregivers, and industry. We now have treatments or cures for many forms of cancer, the diagnosis of which previously implied certain death. The perception, however, that to achieve the benefits of modern cancer therapy, care must proceed without restriction or waiver along paths paved through commercial interest and profit motives, and that any change in the relative roles of academia and industry could impair further progress in the care of the cancer patient may well be incorrect.

Notwithstanding the huge advances that have been made, we have not fully appreciated how they should be utilized, integrated, and paid for so that they are available to the largest number of patients who can truly benefit from their availability at a cost that society can afford. Some resources are limited both by availability and by cost, and some forms of treatment that may benefit some do not benefit others; there are uncertainties that give rise to numerous clinical trials that attempt to answer questions related to oncologic efficacy, but concerns related to cost and mechanisms that integrate cost with benefit are unpopular. Such trials give rise to allegations of rationing of health care, or worse yet, "death panels." The concept of placing a cap on the price of one year of quality life, as has been done in other countries such as Great Britain, is repugnant to many in this country and the United States Food and Drug Administration (FDA) does not take cost into consideration when approving new pharmaceuticals or devices. While it is difficult for many patients to comprehend these realities, especially when they believe that their survival could be compromised, is it in the public interest to capitalize on such realities through marketing gimmickry or price manipulation? The question as to whether or not media influence or marketing have gone too far in our society to the extent that they result in more harm than good is an issue that is both highly complex and aggressively debated. Having gone down our present path of extensive commercialism as we provide health care makes it exceedingly difficult to reverse our course, yet some understanding may help place the benefits and risks of expanded commercial involvement in health care into perspective. This chapter will explore some of the commercial relationships as they affect cancer patients and identify barriers to optimal health care that have evolved with the rise of commercialism. The question of whether the desire to earn and retain money has done so to the detriment of our patients afflicted with cancer must be asked, and while the answer may, at least in some respects, be troubling, ambiguous, and uncertain, it is important that these issues enjoy public discussion and comment.

SIMPLIFIED MODEL OF PRESENT HEALTHCARE FINANCING

To understand where in our healthcare system commercialism exerts influence, the various interactions between patients, providers, and payers must be appreciated. Fig. 13.1

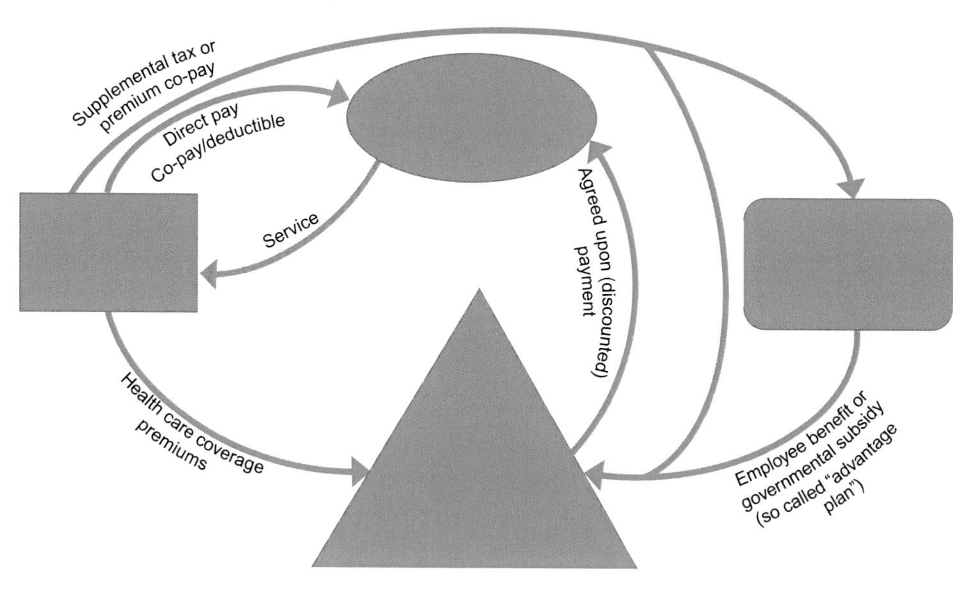

FIGURE 13.1 Simplified schematic of the relationships between healthcare consumer (patient), payers, and providers.

shows a highly simplified cartoon that depicts these relationships. While they apply to most complex medical encounters, they are especially relevant for cancer patients, many of whom seek more than one provider, are seen in complex facilities, and ultimately pay for the rendered services through third-party or governmental payment programs.

What is clearly apparent is that a simple model comprised of a single provider and a single consumer who seeks treatment for cancer and who pays the provider directly for the services rendered no longer exists but for the most basic or simple of encounters. While the patient and the disease diagnosis represent relative constants, other coalitions and relationships come into play and are both complex and volatile. The primary provider, e.g., initially may be the primary physician, later the cancer center or specialty practice, and at the end of life, a hospice-focused facility. But providers also include pharmacies, device manufacturers, and ancillary service providers, each of whom must coexist with, interact with, and to some extent, compete with one another for healthcare dollars. Additionally, providers must adhere to an ever-growing and ever-more costly and complex administrative burden due to changes in healthcare regulation. Furthermore, each of these providers must deal with associates: pharmacies with drug wholesalers or manufacturers and device manufacturers with support providers and marketing staff being but the simplest examples.

Those that ultimately pay the providers may be private or governmental entities, supplemented by individual copays and secondary insurance plans. Private entities are most commonly insurance companies through individual or group nonemployee-related plans, or through insurance company-based plan administrators of employee benefit plans. The distinction is relevant in that the latter enjoy some legal protections under the federal Employee Retirement Income Security Act (ERISA) of 1974. Among the governmental plans, we have Medicare, a federal program introduced in 1965 that covers those over the age of 65 as well as some younger persons with included disabilities such as end-stage kidney disease and amyotrophic lateral sclerosis. Medicaid, a means-tested program that provides care to those with low income and limited resources is managed by the individual States. Other governmental programs include the Veterans Health Administration and the Indian Health Services. Some governmental programs incorporate nongovernmental administration of the plans. These may be thought of as hybrids in that they have components of the traditional governmental coverage supplemented by private insurance. Such plans, often referred to as "advantage plans," offer expanded coverage but may limit the extent of covered providers. Private entities operate and administer such plans on a for-profit business basis; their products are both highly marketed and associated with significant direct (within the third-party payer's budget), or indirect (paid for by the provider who must obtain coverage certification, preapproval, and denial appeals) administrative costs. To some extent, these payers have assumed the role of rationers or gatekeepers in that they are not required to provide payment for all healthcare-related costs that may be desired by the patient or recommended by the physician. They may limit or exclude payments to entities that have been identified as out of network for a variety of business or financial considerations. Marketing by such entities is undertaken in an attempt to expand the pool of healthy participants and to ensure that there are adequate number of covered patients so that devastating claims can be managed without financial devastation. Denial of payment for noncompliance with precertification requirements or for providing services by entities that are not in network are commonplace. Such denials may leave the provider with obligations to deliver care with little or no possibility of payment and leave the patient with restrictive or otherwise compromised options. Providers accept negotiated discounted reimbursement payments as part of the grand scheme in which they function, thereby making the initial billed charges little more than a fiction for those whose bills are ultimately paid by third-party payers, but a burdensome reality for those who seek to pay directly, and who wish to pay a fair price for the services they receive [1].

The ultimate sources that fund health care in the United States are also complex but fall into three primary groups. Approximately half of funding comes from governmental sources that include Medicare, Medicaid, Veterans Administration Health Care, and programs such as the Indian Health Services and the Children's Health Insurance Program. Such funds are considered public and they are paid for through direct and supplemental tax programs; the population at large, therefore, funds these programs. Private health insurance including employer-provided plans pay approximately 34%, and consumers pay approximately 13% of the healthcare budget in the form of direct premium payments, copays, and deductibles [2]. Other plans, where both governmental and commercial components co-exist, often referred to as Medicare or Medicaid Advantage Plans, are widely marketed and may be disproportionately costly.

Healthcare providers find themselves in the midst of an ever-changing business environment. The costs of providing care escalate and administrative requirements become increasingly demanding and burdensome, adding greatly to costs beyond what are actually needed to provide optimal care. Revenues shrink as a result of negotiated discounted payment rates imposed on providers by third-party payers, and even major institutions function under threat of being excluded by third-party payers for reasons that are sometimes unclear and deemed arbitrary, and for which the possibility of a negotiated or litigated resolution is not economically viable [3]. Additionally, some entities find themselves in the midst of a competitive race to provide the latest in technological equipment or expertise and to provide services that may go beyond addressing the optimal medical needs of their clients. Adding such new technology may be hugely important from a marketing standpoint but the extent to which it truly impacts outcomes for cancer patients may not be as clear and obvious as marketing initiatives imply. Costs of new technology may be sufficiently large so that cost shifting is required, thereby escalating the charges of unrelated services. Nevertheless, aggressive marketing is undertaken, the expenses for which further impinge upon what otherwise might be available to address the true medical needs of our patients. Individual facilities often spend millions of dollars annually on marketing [4]. It is within this context that providers at all levels strive to achieve the goals of providing care that is efficacious, cost-effective, compassionate, and private. Many of these goals raise concerns that may be in conflict with the goals of commercialism; ethical considerations arise with regard to each of these goals.

PROVIDING ETHICAL CANCER CARE

As can be seen, how cancer care is provided and paid for in the United States has many components that are motivated, at least in part, by a desire for monetary profit. Providers, of course, in their need to survive in the present arena must remain competitive, and marketing to attract clients as well as negotiated rate payments that are paid by third-party payers are part of this equation. Providing ethical care in this environment, in part due to commercialism and the highly complex financial business model, may be challenging. Some aspects of ethical care have been compromised in instances where the need for efficiency, expediency, and fiscal reality conflict with broader ethical considerations. These realities have become so ingrained in how we provide and finance medical care in general and cancer care in particular that the potential erosion of patient autonomy is assumed to be a necessity and a natural result of how we provide and pay for medical services in our present healthcare environment.

The treatment and management of cancer patients and the involvement of commercialism offer clear examples of how we have been led down paths that may compromise the theoretical goals of an affordable, practical, and ethical healthcare delivery system as these goals are balanced with the commercial interests of the business community. The compromises that providers make on a daily basis have become so ingrained into our practices that a new norm exists. Often we strive to limit ethical compromise—we try to minimize the effects of these compromises on our patients—but we do not always succeed. We recognize that we must provide care on an ongoing basis, and to do so must take certain realities into account. Nevertheless, a system less focused on commercial aspects ultimately might achieve care that is less costly, less compromised, one that enjoys enhanced confidentiality, is less prone to conflicts of interest, and one that instills a greater level of confidence among healthcare consumers.

ECONOMIC AND ETHICAL IMPEDIMENTS TO AUTONOMY

Some of the compromises we make for our patients stem from a balance between an individual's desires with regard to health care and his or her rights to have these desires addressed accordingly. Yet in some instances these desires and rights must be balanced against the broader interests of society at large. We may think of this balance as the balance between **microethics**, i.e., the goal of the provider to be the zealous and uncompromising advocate of his or her patient and **macroethics**, i.e., the interests, realities, and priorities of society at large with respect to the allocation of limited medical resources. Yet even in the context of the macro versus micro consideration, commercial interests abound at many levels of the health care we provide.

This having been stated, we should not intimate that commercialization as it relates to cancer care is invariably harmful or largely unethical. Much has been learned and new goods and services have become available that have made a huge difference in both the quality of life and the survival times for many with cancer. Yet in our efforts to provide the best care we can, it is imperative that we understand that commercial achievements sometimes yield little direct benefit, have resulted in unmanageable cost, have caused sufficient invasions of privacy so as to undermine openness and honesty of the doctor–patient relationship, and have fostered an environment where trust and confidence in the healthcare delivery system may be compromised.

As we consider autonomy, our decisions have been largely impacted by commercial and economic considerations. In a more natural environment, consumers weigh the cost of products or services with the benefit of consumption and make decisions accordingly. In health care, the costs often are borne by parties other than the ultimate consumer. In that respect, there is more than one cost that must be considered: The first by the consumer him or herself, the second by society at large through taxation that ultimately pays a large part of healthcare expenses, and finally, the global payouts of nongovernmental third-party payers. For the individual, as the costs to him or her decrease, the desire to consume the product or service increases and probably does so geometrically. As the demand increases in the absence of cost considerations, pricing is free to rise without the usual limits imposed by the economic law of supply and demand. It is this economic reality coupled with an inability to place societal limits on this balance that have contributed to the extraordinarily high prices for some of the newer pharmaceuticals, many of which offer only modest, albeit statistically proven, benefit [5]. The cost of a growing number of anticancer pharmaceuticals is staggering to the extent that some feel the charges are so disproportionate to the benefit that they are unconscionable. When the balance between demand and minimal out-of-pocket cost to the patient is modified by a third-party payer's refusal to fund the requested product or service, patients often opt for a less financially burdensome alternative. As ethicists, we look at this aspect of health care in several ways: Is failure to provide desired option rationing? Do we have ethical obligations beyond the acceptance of the principle of autonomy to serve the broader societal interests and needs under macroethical considerations? These concerns, of course, are not limited to new pharmaceuticals. Older drugs that have traditionally been inexpensive may undergo substantial increases in price despite the absence of indicators that the costs of manufacture or distribution have risen. Mergers and acquisitions that have reduced competition are often felt to be the cause of such increases and have led to practices that some describe as price gouging [6,7].

In exploring the presumed role of commercialism, we must be aware that the United States spends upwards of 17% of our gross domestic product on health care, fully one-third more than other western countries, and does so without achieving significant healthcare benefits quantitated by metrics of life expectancy, estimations of quality of life, or reduced infant mortality [8]. While such end points are multifactorial and complex, clear benefit of enhanced spending in this country has not produced an unmistakable proportional advantage. Can we draw attention to specific elements of commercialism that affect health care in general and cancer patients in particular for which alternate paths may be considered in the future to provide a more ethical as well as a more affordable model for health care?

Commercialism, of course, did not enter the healthcare system in a single moment of guided (or misguided) policy but evolved stepwise. While many believe that policy changes to achieve a more favorable balance between ethical care of the cancer patient and commercialism are impossible, an understanding of these issues is an essential starting point. We will examine the ethics of marketing and a market-driven healthcare system, new drug development, and direct-to-consumer advertising of both goods and services. In doing so, however, we must be alert to the fact that to continue progress in treating cancer and to maintain the momentum conducive to innovation, there must be sufficient incentives to do so. We ultimately must strive to create an environment that provides the right amount of care delivered in optimal surroundings and to do so at a cost that represents a fair value. To do so may require a substantial reining in of commercialism with regard to health care in general and cancer care specifically.

THE BASIS FOR COMMERCIALISM IN HEALTH CARE

Remuneration for healthcare goods and services traditionally implied a direct payment to the provider. In today's healthcare environment indirect remuneration is much more common; indirect remuneration may be thought of as any mechanism by which payment is provided other than through a direct payment by the consumer for the product or service provided to the provider of that service. Providers of care, regardless of whether they are paid directly or indirectly, have found ways to market what they provide with the intent to increase revenue; the incentive to do so, even when ethical concerns abound, is great and the risks of not engaging in marketing strategies, even when ethical considerations are overtly manifest, may threaten the very economic viability of the entity.

In the direct remuneration model, the ethical dilemmas are easy to identify: The provider may advise care beyond what is deemed optimal for the specific purpose of increasing revenue. The provider's fiduciary duty to the consumer has been breached and the consumer, unfortunately, may not be sufficiently aware of the breach to raise questions. The ethical dilemma of breach of fiduciary duty becomes greater as the provider's role expands as may take place when a provider performs in-house laboratory services, imaging, or dispenses medications all of which may incrementally raise revenue for the provider. Modern oncologic care requires an extremely high level of such services; while it is difficult to determine if an individual imaging study or laboratory test constitutes an ethical breach or conflict of interest, the fact that many more such studies are performed in this country without incremental overall benefit suggests that motivation or incentive beyond best practice standards could be playing a role.

The relationship between commercialism and indirect healthcare payments is much more complicated; these relationships play a huge role in how we provide health care in this country. As we noted, services are commonly provided by an individual who is employed by an entity that contracts for adjusted or discounted payment with a health plan to pay an agreed portion of the sometimes arbitrarily defined cost. An additional portion may be paid by the person actually receiving the services in the form of copay, and the uncovered portion simply disappears under the designation of "contractual adjustment." Further complicating this is the fact that providers of some goods such as pharmaceuticals may offer assistance programs to some who might not be able to afford prescribed medication or copayment or for whom a third-party payer is either not identified or has declined to cover that particular product in the unique circumstances of the involved patient. Additionally, discount coupons may be widely available that may encourage patients to use drugs that may be more costly than alternatives. Such programs, while helpful to some, have been subject of criticism as copayments were introduced to encourage consumers to use less expensive alternatives. Such programs may mitigate criticism of excessive price increases by pharmaceutical marketers or may be used as a marketing ploy; some have referred to them as a sham [9]. Finally, premiums may be paid by individuals, employers, the government itself, or various combinations. Not surprisingly, within this circuitous route, the meaning of fair, transparent, and equitable pricing, i.e., ethical pricing seems largely to have evaporated.

POTENTIALLY UNETHICAL EXAMPLES OF COMMERCIALISM THAT AFFECT CANCER PATIENTS

Patient autonomy and the physician's duty to act in the best interest of the patient are fundamental aspects of ethical behavior. Our care models have evolved in ways to assure that these fundamental principles should be upheld, yet we have partial or total breakdown with regard to some aspects of care, many related to commercial interests. Several examples follow:

Example 1

A cancer patient is in need of an imaging study. She has a health plan that has an agreement to pay a discounted rate to the imaging center. A second patient with similar problems comes to the same facility as a self-pay patient and has the same imaging study performed using the same device and read by the same physician. The second patient may be asked to pay a substantially higher price, sometimes as much as three times the amount that is ultimately received for the identical services under the terms of the payment agreement with the third-party payer. The rationale for this discrepancy is that the facility or institution performing the study agrees to offer a discounted rate to the third-party payer in exchange for the expected large volume of patients for whom the third-party payer will be paying. When asked about the discrepancy in actual accepted payment, the facility notes that whatever they charge a self-pay patient is their base charge for the service and the discounted rate to the third-party payer is negotiated and contracted. It is based on the charge to the self-pay patient, discounted or capped according to contract. Failure to pay the full amount by the patient who pays directly may result in the ultimate sale of the full debt to a collection agency that then may actually pay less for the legal right to attempt collection than the third-party payer paid for the service and what the self-pay patient would have been able and happy to pay. These practices are widespread; they are totally legal under our present system but, as physicians treating cancer patients with widely diverse mechanisms of payment, we must ask questions beyond whether or not a business practice falls within the law; they are clearly not transparent and they are probably not equitable.

While commerce is rampant with instances of unequal pricing or even gimmickry designed to maximize profits, the question remains as to whether or not such practices are ethical when they are related to healthcare services. We have come to expect varying prices with regard to some services such as transportation and hotel accommodation, and some delight in the knowledge that they paid a lower amount than others for these services. However, in dealing with a cancer patient who requires a service, is the model of yield management, or greatly divergent pricing, ethical? Should legal barriers to such practices be implemented? The Affordable Care Act remained silent on pricing disparities, effectively endorsing the established and ongoing practice that allows patients to be caught in a trap of self-pay gauging? Parity of pricing for health care should be the norm, not the exception; disparity is considered by some to be wantonly unfair, unethical, and as a business practice, a practice that should be eliminated. It should be noted that some, in an attempt to level the playing field, have contracted with professional advocates or negotiators who may or may not ultimately achieve a negotiated discount for self-pay or residual amounts but who do so in exchange for a fee that might be better applied to furthering patient care at an affordable price [10]. While the agreed price for a good or service ultimately depends on many economic considerations, a more ethical approach could be achieved through fair, equitable, and ethical pricing. The adage, pay a fair price for what you buy (or charge a fair price for what you sell) should apply broadly for all healthcare goods and services, but especially in the care of cancer patients where costs are high and cost consistency should be both an expectation for patients and a public policy consideration of huge importance. Ultimately, in looking at the interests of the business stakeholders, providers, and consumers, we must ask if these practices, beyond whether or not they are ethical, may add significantly to the ultimate overall cost of medical care in the United States, and therefore, on that basis alone, be legislatively prohibited under public policy [11].

Example 2

In 2013, Americans spent an estimated $329 billion on prescription drugs; the costs of these products continue to rise and, at present, we spend on average approximately $1000 for these products for every man, woman, and child in the United States. One report noted that the major pharmaceutical companies spend more on marketing these products than they do on research and development of new agents and, in the extreme case of Johnson and Johnson in their effort to market a cardiovascular agent, the expenditures for marketing exceeded those for research and development by a factor of more than two [12]. In the broader sense, honest and legitimate marketing is neither illegal nor unethical; marketing has been deemed a free speech right. Marketing is intended to increase the demand for products of a specific type as well as to increase the market share of a particular product brand; in the case of pharmaceuticals, these strategies have been highly successful. Marketing is restricted in instances where there are public policy concerns; tobacco advertisements on radio and television ceased on January 1, 1971 and geographical limits have been implemented with regard to some products with the intent of protecting children.

The United States FDA has limited some off-label marketing practices but distribution of material reflecting use beyond those specifically approved by the FDA may be permissible under certain conditions and may have an indirect marketing goal [13]. Notwithstanding the constitutional right to free speech that includes advertising that is not false or misleading, there remains a fundamental difference between general marketing and the marketing of pharmaceutical products that raises both the issue of public policy concerns for individual patients and issues related to societal or macroethics.

From the standpoint of public policy, drugs should be used to improve health or to make life more enjoyable for those taking them. Along with the potential benefits, there are risks in using many of the pharmaceutical agents available. This implies that for each agent a level of optimal use may, at least theoretically, be defined. Greater use by large numbers implies a potentially higher level of adverse events than might be observed with optimal use; use below that level suggests that broader use might be appropriate. Our society's goal should be to achieve optimal levels of use. This goal, however, is likely to be in direct conflict with the marketing goals that are oriented to maximizing profit rather than achieving optimal use. One must at least raise the question as to whether marketing of pharmaceuticals directly to consumers is sufficiently detrimental to the population at large so that free speech protections should be curtailed on the grounds that a public policy concern exists.

Another concern related to marketing clearly has broad implications but has received little attention. Under normal market theory, an item will be supplied at a cost level that will maximize overall profit; if priced sufficiently high, it will not be purchased and if priced low, demand will increase, limited only by supply. With regard to pharmaceuticals, a compounding factor shifts this fundamental economic balance in that often the ultimate consumer is not the payer, and therefore demand can rise with little concern for cost. Pharmaceuticals in general, but especially those used in the treatment of malignancy as well as for the supportive care of the cancer patient, are costly and so the burdens of these costs, if not borne by the patient, will be paid for through government payments or by private insurance. In either case, these costs are ultimately met by society at large. This begs the question as to whether marketing of products to increase demand, where the major source of payment comes largely from governmental coffers or insurance premiums, can continue to be justified on the basis of our constitutional right to free speech. Interestingly, at the time of this writing, in addition to the United States, only New Zealand, a nation that has more stringent drug price controls than does the United States, allows direct-to-consumer advertising of prescription pharmaceuticals.

Furthermore, direct-to-consumer advertising, now legal in the United States is hugely expensive. The World Health Organization reported that Pfizer's advertising campaign for the agent Lipitor cost $250 million; total marketing costs were just under $5 billion in 2008 [14]. Again, this cost is derived from the sale of agents paid largely from public or pooled funds. We must ask if such huge expenditures are ethical in the broader macroethical societal domain and what better ways we might discover to improve general health with the resources now consumed with marketing.

Example 3

Continuity of care is especially important for cancer patients. In an environment where mergers and acquisitions exist among both providers and payers, continuity may be compromised causing patients to scramble in order to find a network where their established provider can be paid. Failure to do so results in the transfer of patients to alternate facilities or risks their acquiring huge debt to pay the sometimes inflated costs imposed on self-pay patients. Payers choose network providers based on economic considerations but when continuity of care within an entity is of paramount importance for patients, effectively forcing them to migrate to other providers may be hugely burdensome or, worse, potentially dangerous to the extent that they could be life threatening. Ethical mechanisms that facilitate continuity of care rather than financial denial can and should be established. Third-party payers should bear costs to the same extent that would be their obligation if in-network care were provided. Their contractual escape clauses that allows nonpayment and places the burden of payment on the shoulders of patients present an ethical dilemma; in some instances such practices are considered unconscionable.

THE NEED FOR ENHANCED REGULATION TO PRESERVE ETHICAL CARE FOR THE CANCER PATIENT

Free market forces in the face of untethered commercial interests with regard to the treatment of cancer patients have demonstrated a clear need for either self-regulation or administrative oversight to ensure ethical care. Self-regulation has failed to achieve our goals in this regard. The Affordable Care Act has been largely ineffective in moving us in the direction of diminished commercial influences. Many now are required to spend more in the form of copays and deductibles than before passage of the Act. With regard to the potential ethical concerns associated with commercialism in health care, our priorities should be directed toward offering patients the greatest number of options for care and to use the maximal proportion of our healthcare dollars for the benefit of patients. To do so, we must reduce costs that do not directly benefit patients; the primary targets for this are administrative and marketing expenditures. Healthcare services and their costs should be transparent, consistent, and reflect a level of fairness beyond mere compliance. These ethical goals should be a priority and, in the absence of self-regulation, legal, legislative, and judicial influence and oversight may be required. As providers, we should not be satisfied with the status quo—one where we spend more and get less than is the case in much of the rest of the civilized world. We must strive to put the percentage of our gross domestic product spent on health care back in line with that of other nations, and we must spend that money wisely treating those who suffer from diseases, be they acute or chronic, benign or malignant, curable or lethal; we must do so fairly, equitably, and ethically.

References

[1] McGrath T. My Daughter's $29,000 Appendectomy. Philadelphia Magazine.

[2] CBO's 2011 Long-Term Budget Outlook, Congressional Budget Office, June 2011.

[3] Houston Chronical March 13, 2016.

[4] http://yourbusiness.azcentral.com/average-marketing-budget-hospital-17444.html.

[5] University of North Carolina at Chapel Hill. "Costs for orally administered cancer drugs skyrocket: Patients may increasingly take on cost burden." ScienceDaily. ScienceDaily, April 28, 2016. <www.sciencedaily.com/releases/2016/04/160428132130.htm>.

[6] Painful pills: sudden price rises for long-established drugs lead to calls for action. The Economist, September 26, 2015 [accessed 02.05.2016] http://www.economist.com/news/business-and-finance/21665436-dramatic-rises-price-some-medicines-prompt-calls-action-hillary-clinton-

[7] Bill George. What's behind skyrocketing drug prices. CNBC—commentary-CNBC.com http://www.cnbc.com/2015/12/21/whats-behind-skyrocketing-drug-prices-commentary.html.

[8] D. Squires and C. Anderson, U.S. Health Care from a Global Perspective: Spending, Use of Services, Prices, and Health in 13 Countries, The Commonwealth Fund, October 2015.

[9] Michael Hiltzik. Why big pharma's patient-assistance programs are a sham. Los Angeles Times, September 25, 2015 accessed at http://www.latimes.com/business/hiltzik/la-fi-mh-pharma-s-sham-patient-assistance-programs-20150925-column.html.

[10] Christina LaMontagne. How to negotiate hospital bills and avoid medical bankruptcy. U.S.News & World Report Money. August 19, 2013 [accessed May 2] http://money.usnews.com/money/blogs/my-money/2013/08/19/how-to-negotiate-hospital-bills-and-avoid-medical-bankruptcy.

[11] Porter ME, Lee TH. The strategy that will fix health care. Harvard Business Review; October 2013.

[12] Ana Swanson. Big pharmaceutical companies are spending far more on marketing than research. The Washington Post [accessed 19.04.2016] www.washingtonpost.com/news/wonk/wp/2015/02/11/big-pharmaceutical-companies-are-spending-far-more-on-marketing-than-research.

[13] Stephen Feller; United Press International. FDA allows Amarin to market off-label use of Vascepa. March 9, 2016 [accessed 19.04.2016] http://www.upi.com/Health_News/2016/03/09/FDA-allows-Amarin-to-market-off-label-use-of-Vascepa/1121457535374/.

[14] Bulletin of the World Health Organization 2009;87:556-64 [accessed 19.04.2016] http://www.who.int/bulletin/volumes/87/8/09-040809/en/.

CHAPTER

14

A Perspective on the Cost of Cancer Care

Jeffrey S. Farroni

The University of Texas MD Anderson Cancer Center, Houston, TX, United States

INTRODUCTION

...better the hard truth, I say, than the comforting fantasy. —***Carl Sagan, Demon Haunted World, 1995.***

There is much concern over the cost of health care. It is an issue that cuts across the focal plane of the medical enterprise from the individual patient's care to federal policy and international commerce. Health-care spending in the United States surpassed $3 trillion in 2014 [1]; of that, about $170 billion was spent in the "war" against cancer (Cancer Prevalence and Cost of Care Prevention. National Cancer Institute. <http://costprojections.cancer.gov/>). Another way to put US health-care spending in perspective is to think of it as its own economy. From that point of view, the health-care ecosystem in the US surpasses the financial productivity of many countries, Table 14.1. We are beyond the threshold whereby health-care economics, access, and costs can be ignored. The ethical values entrenched in justice and the fair distribution of limited resources looms ever larger in our consciousness. The economic impact continues to increasingly be levied on the individual patient.

Patients face difficult choices of whether or not to seek treatment or fill a prescription out of economic concern. Such distress is not relegated to the realm of hypothetical. A survey of the general population conducted in 2014 indicated that 23% of respondents aged 18–64 had difficultly or were unable to pay for medical expenses [2]. In addition, 23% of all respondents experienced medical problems but did not visit a doctor or clinic; and 19% did not pursue a recommend treatment or follow care [2]. These data illuminate the impact of direct health-care cost on avoidant behavior should raise alarms on its own merit. Often the devastating financial aftermath of these personal costs is overlooked, e.g., bankruptcies, financial burden distribution to family/caregivers, loss of employment/benefits, etc.

Ethical Challenges in Oncology.
DOI: http://dx.doi.org/10.1016/B978-0-12-803831-4.00014-2

TABLE 14.1 Total Annual Health-Care Expenditures[a] in the United States is Roughly Equivalent to the Gross Domestic Product of the Following Countries[b]

Bulgaria	Cyprus	Macedonia	Lesotho
Uruguay	Afghanistan	Madagascar	Bhutan
Ethiopia	Nepal	Malta	Liberia
Lebanon	Honduras	Armenia	Cape Verde
Slovenia	Bosnia & Herzegovina	Tajikistan	San Marino
Tunisia	Gabon	Haiti	Central African Rep.
Lithuania	Iceland	Benin	Belize
Costa Rica	Mozambique	Bahamas	Djibouti
Turkmenistan	Cambodia	Niger	Seychelles
Tanzania	Georgia	Rwanda	St. Lucia
Serbia	Papua New Guinea	Moldova	Antigua & Barbuda
Panama	Botswana	Kyrgyzstan	Solomon Islands
Yemen	Senegal	Kosovo	Guinea-Bissau
Libya	Brunei	Guinea	Grenada
Ghana	Equatorial Guinea	Suriname	St. Kitts & Nevis
Jordan	Chad	Mauritania	Samoa
Dem. Rep. Congo	Jamaica	Sierra Leone	Gambia
Bolivia	Zimbabwe	Togo	Vanuatu
Côte d'Ivoire	Rep. Congo	Timor-Leste	St. Vincent & Grenadines
Bahrain	Namibia	Montenegro	Comoros
Latvia	Albania	Barbados	Dominica
Cameroon	Mauritius	Malawi	Tonga
Paraguay	South Sudan	Fiji	São Tomé & Príncipe
Trinidad/Tobago	Burkina Faso	Eritrea	Fed. States of Micronesia
Uganda	Mongolia	Swaziland	Palau
Zambia	Mali	Burundi	Marshall Islands
Estonia	Nicaragua	Guyana	Kiribati
El Salvador	Laos	Maldives	Tuvalu

[a]*Centers for Medicare & Medicaid 2010 estimated cost.*
[b]*GDP calculations based on International Monetary Fund's world economic outlook database estimates from 2010.*
United States spending on cancer-related costs alone is represented by the shaded area.

Cost is of particular concern for patients who suffer chronic illnesses that involve expensive interventions, such as cancer chemotherapy. It is estimated that 57%–68% of personal bankruptcies in the United States are due to medical-related expenses [3,4]. This phenomenon, which is not observed in other health-care systems around the world, is particularly insidious in that not only must a patient suffer the burden of illness, they must also bear the weight of potential financial ruin; a weight that can encumber a patient's family as well. The escalation of treatment cost places a moral weight on health-care providers. Do we ask the individual physician to ration interventions as a means to steward limited resources? Do we ask the physician to choose which patients are "worthy" or have sufficient value to warrant the most expensive and innovative treatments? The short answer is that we should not place such an obligation on the individual but that does not mean we must ignore policy solutions to address cost. Although there are many areas where costs may be managed, the focus here will be the price of treatment. The "comforting fantasy" posited by pharmaceutical manufacturers is that the high cost of chemotherapy is justified due to the tremendous regulatory barriers and expense in bringing them to market. This narrative is bolstered by claims that such exorbitant pricing is crucial to incentivize and support the high cost of innovation and represent the market value of these crucial interventions. The rhetoric is often evoked as a dire warning that attempts to reign in pricing will deny patients the next generation of wonder drugs. Before considering ways to reduce cost, a view of health care is considered that frames the rationale and justification for limiting the market. We must grapple with the ethical dilemma of balancing the need for critical, potentially life-saving therapies, with the potentially prohibitive cost that many patients face. This chapter explores pharmaceutical treatment cost through an ethical lens, including factors that contribute to cost, and actions that can be taken to push against the tide of well-funded interests who wish to maintain the status quo.

ETHICAL PERSPECTIVE ON TREATMENT COST

The Ethical Imperative

Part of the reason why ethics play such an important role in medicine is the inherent vulnerability that illness inflicts on us all [5]. The toll of disease can render us physically impaired or mentally incapacitated. We may rely, to varying degrees, on the devotion, kindness, and empathy of others for our care. It is in this context of vulnerability that we seek guidance from physicians who are endowed with the knowledge and skill to make us whole. Therefore, a heightened duty of care and responsibility is placed upon physicians, and the health-care team. Society expects physicians to act in the patient's best interest. A common manifestation of these obligations is embodied in principalism, i.e., respect for person (autonomy), beneficence (nonmaleficence), and justice [6]. Disease burden is particularly salient in the oncological setting. The "cancer" label metastasizes into one's sense of personhood to corrupt the notion of "self;" this is a core element of suffering [7]. Cancer subsumes one's identity from diagnosis, to "cancer patient," to cancer "survivor."

These labels mask the moral enterprise and value-laden decisions inherent to the treatment of cancer. Ethical dilemmas present themselves when values conflict or judgments diverge. An ethicist may be employed to help the patient or provider navigate these challenging issues. The traditional role of the ethicist is to engage in dialogue on ethical issues faced by stakeholders [8–11]. However, the typical practice tends to be reactive to ethical dilemmas where ethical issues are often perceived as problems needed to be "fixed." In this model, issues are "solved" as they arise; thus, the focus tends to be more on conflict resolution and compliance. As such, the traditional view tends to take an issue-centric framework to ethical inquiry. In this model, the stakeholders may ignore or not even seek ethical advice. Indeed, commentators have suggested that negative attitudes persist regarding the role of ethics and is viewed, along with regulatory compliance, as a barrier or obstruction to patient care or scientific progress (Bell J, Whiton J, Connelly S. Final report: evaluation of NIH implementation of Section 491 of the Public Health Services Act, mandating a program of protection for research subjects, OIG, Arlington, VA. <http://www.hhs.gov/ohrp/archive/policy/hsp_final_rpt.pdf>; 1998 [retrieved 05.07.2012] and Grob G. Testimony before the committee on government reform and oversight regarding institutional review boards: a time for reform. <http://www.oig.hhs.gov/oei/reports/oei-01-97-00193.pdf>; 1998 [retrieved 05.07.2012]). In part, the emergence of new forms of practice for the ethicist is an attempt to alter these perceptions and to change the way that ethical values are incorporated into practice settings [12].

There is no greater example of the intersection of patient-centered care and clinical research than the treatment of cancer. The National Cancer Institute has a budget of $5.21 billion for fiscal year 2016 (National Cancer Institute. NCI budget and appropriations. National Cancer Institute. <http://www.cancer.gov/about-nci/budget> [retrieved 27.06.2016]). This investment coupled with the clinical cost of cancer treatment (see Table 14.1 supra) makes cancer one of the highest health-care priorities. As such, much national effort is being focused to harmonize clinical research and patient care. The rhetoric is charged with the hope of innovation since "[c]ancer is a leading cause of death ... cancer research is on the cusp of major breakthroughs. It is of critical national importance that we accelerate progress towards prevention, treatment, and a cure—to double the rate of progress in the fight against cancer" (White House Cancer Moonshot Task Force. The White House Office of the Press Secretary. January 28, 2016. <https://www.whitehouse.gov/the-press-office/2016/01/28/memorandum-white-house-cancer-moonshot-task-force> [retrieved 27.06.2016]). These statements were backed up by a pledge of $1 billion for further research (FACT SHEET: Investing in the National Cancer Moonshot. The White House Office of the Press Secretary. February 1, 2016. <https://www.whitehouse.gov/the-press-office/2016/02/01/fact-sheet-investing-national-cancer-moonshot> [retrieved 27.06.2016]). The heavy emphasis on cancer treatment opens up an enormous niche for high-priced pharmaceutical interventions. The amalgamation of desperation, advocacy, and hope with free market, profit-motivated entities results in a milieu of moral and ethical quandary.

Much societal value is placed on the understanding, diagnosis, and treatment of disease and improving human health. The public expects researchers to be ethically accountable and socially responsible. Due in part to the value that research plays in the furtherance of public health, scientific inquiry is both a virtue and a social good and researchers are moral agents in its pursuit. Science and ethics must be interdependent and cooperative human endeavors. Enhancing research integrity requires institutional and systemic structures. Education, mentoring, consultation, and anticipatory guidance increase awareness of ethical behavior. Critical reflection on ethical principles and underlying moral values leads to improved ethical behaviors. Regulatory compliance is necessary but not sufficient to ground ethical research practices. Integrating ethical research norms throughout research practice leads to better scientific outcomes; they promote the aims of research.

Research investigators, clinicians, and their teams are likely to confront new ethical challenges within the complex interplay between the goals of basic/applied sciences, clinical medicine, and those of partnerships between academia and industry. As such, tensions may arise due in part to the divergent meanings associated with the evolving changes that confront clinical practice, research purposes, and resources [13]. Scientific progress is contingent upon principle investigator-centered competitiveness for funding and publications. Such pressures, coupled with the high-paced demand for data and outcomes, often give rise to conflicts of practices, values, norms, and rules. Needed is further ethical inquiry into the nature, goals, values, and guiding principles that organize this new research enterprise [14–16].

Ethics occupies a critical role in all aspects of research-based patient care from the patient bedside and through all phases of clinical research. Clearly, the ethical issues will differ depending upon where the consultation arises within the research enterprise. This is due to differences in environment, goals, settings and study design, the respective values, etc. The ethicist must pay attention to the increased pace of discovery and the more demanding challenges of having well designed postapproval studies, conscientious data analysis, and public transparency of results. The research enterprise itself has a developing culture of hurdles that need to be overcome, such as activities geared toward expeditious reporting of results targeted toward specific research outcomes.

Much of the commentary regarding ethics in clinical research focuses on the traditional research ethical "issues", e.g., consent, risk/benefit assessment, subject selection/enrollment, specimen use, etc. It is possible that burdens may arise during the conduct of research that influence ethical analysis whereby providing "solutions" to dilemmas that conform to their own goals [14]. The adoption of "typical" research ethics principles and that of emerging research should not be an end run to usurp ethical duties and responsibilities [17]. Additional duties and responsibilities that are not merely confined to the resolution of "issues" but rather extend beyond the investigator/subject or physician/patient relationship. In research practice, ethical values operate predominately in the realm of personal and professional relationships where the humanity of the subject as well as that of the practitioner is the primary locus of concern. Deliberately focusing the lens of inquiry on the relationships themselves has the capacity to open up the moral spaces in which human subjects research is practiced. The expectation of increased impact and more effective interventions provides the moral justification for their endeavors [11], but here the value of efficiency may require ethical constraint.

Many patients make an investment in hope upon receiving their cancer diagnosis. Hope manifests in numerous ways from confidence of a competent doctor to the anticipation of favorable tests results and the prospect of an effective treatment. Access to treatment may be elusive due to inadequate insurance, the exorbitant treatment-related costs that may do harm, to an unaffordable degree by the patient and an unaffordable support network. That is not to say successes are unobtainable (see Box 14.1).

BOX 14.1

HOPE AND PROMISE

Billy Tauzin has disclosed a medical journey that embodies the hopeful miracle. Mr. Tauzin has a long and distinguished career in the US House of representatives for Louisiana's third district. He experienced heavy bleeding on day, which culminated in the diagnosis of duodenal adenocarcinoma. He underwent the Whipple procedure by which portions of his stomach, intestines, and pancreas were removed. Unfortunately, his cancer progressed despite the technical success of the very complicated and risky procedure. The only option left for Mr. Tauzin was to enroll in a research study to evaluate this form of chemotherapy that had not been FDA approved for his specific condition.

The results surpassed his expectations and he has been cancer-free for over 10 years as of this writing. His treatment was so successful that he declared that "[i]t' a good thing we tried [this drug] because it worked like a miracle."

While Mr. Tauzin's story is inspiring, the promise of a miracle may be out of the grasp for many. These miracles do not come at a nominal cost. Patients may pay hundreds of thousands of dollars per year on pharmaceutical agents, unless drug cost is covered by the sponsor of a clinical trial. The desperate hope of a cutting edge therapy may be quashed if administration of the first dose is denied due to cost. Even with insurance, the out-of-pocket expenses can be tens of thousands of dollars. Such a financial burden can be devastating for a family. Should a patient be held hostage to high treatment cost? Should the specter of hope via innovative therapy even be offered, if the cost is well out of the patient's reach? What if lifetime therapy is initiated and the treatment costs continue to rise well above the standard rate of inflation? Part of the answer to these questions lies in the fundamental view one has to the accessibility of health care.

Health Care as a Human Right

Our ability to contribute to societal duties, e.g., engaging in commerce, raising families, participating in the democratic process, etc. is predicated to a large extent on our health status. Relegating a baseline level of health care to traditional markets forces places access and the quality of that care to the whim of externalities that may place other interests higher, e.g., profits. Health care deserves elevation to a human right because it is such a fundamental element of society and a critical element of our personhood. Many international declarations codify the fundamental right status of health care (Article 25 of the Universal Declaration of Human Rights; Article 12 of the International Covenant on Economic, Social and Cultural Rights; Article 24 of the Convention on the Rights of the Child; Article 5 of the Convention on the Elimination of All Forms of Racial Discrimination; Articles 12 & 14 of the Convention on the Elimination of All Forms of

Discrimination Against Women; Article 25 of the Convention on the Rights of Persons with Disabilities).

Even if the view prevails that health care is indeed a market, it warrants special consideration in terms of how it ought to be regulated. The difference between the health-care market and the market for, say, automobiles, is choice. This sentiment was echoed in an early Supreme Court challenge to the affordable care act when Justice Ginsberg noted that "[u]nlike the market for almost any other product or service, the market for medical care is one in which all individuals inevitably participate. Virtually every person residing in the United States, sooner or later, will visit a doctor or other health-care professional" [National Federation of Independent Business v. Sebelius, 567 U. S. ____; 2012 (Ginsberg RB, concurring)]. At some point in our lives, we will access the health-care enterprise. A national health interview study conducted by the CDC in 2009 of US adults revealed that 98.9% of *all* respondents had visited a doctor at some point in their lives [18]. Just the fact of being alive places us in that market. This involuntariness imbues a special duty by society to ensure not to exploit patient care for profit.

There are certain essential core functions of government that shouldn't be left to market forces. These include such things as the military, public utilities, and the infrastructure that drives our society. We have observed the consequences of turning a blind eye to the invisible hand of market forces. For example, privatization of domestic prisons, the outsourcing of military functions abroad and energy deregulation has resulted in higher costs, decreased accountability, and erosion in the public trust.

Our founding documents posit self-evident truths including "certain unalienable Rights, that among these [is] Life" [The Declaration of Independence, para. 2 (U.S. 1776)]. Framing the issue of cost, in part, needs to be viewed in the context of an imperative to provide equitable access to health care. However, the general interpretation of our Constitution is that the government may not deny one of life without due process of law but that there is not an affirmative obligation to provide health care or ensure the health of its citizenry. Prisoner health is one exceptions where there is a constitutional requirement to provide health care (*Estelle v. Gamble*, 429 U.S. 97; 1976). Granted, the requirement sets a very low bar in terms of the duty imposed by prison officials (The Supreme Court determined that prison officials have to exhibit, at a minimum, greater than deliberate indifference to an inmate; a standard below ordinary negligence. *Id.* at 104.). The US stands in contrast to most other countries. Countries with "younger" constitutions almost universally address health care, whereas "older" ones generally do not [19]. For example, the South African Constitution contains the right to access health care in its Bill of Rights [The Constitution of South Africa Ch. 2 § 27 (1997)]. Access to care has been codified internationally in that "[e]veryone has the right to a standard of living adequate for the health and well-being of himself and of his family, including food, clothing, housing and *medical care...*" [The United Nations. 1948. *Universal Declaration of Human Rights*. § 25.1 (emphasis added)]. A majority of countries do indeed have some form of right to health care, however, the US falls in the 45% of countries who have no such protections [19]. A majority of the US population receives medical care through the use of private insurance, provided mostly by one's employer.

The history of health-care reform in the United States is tumultuous with early efforts to provide national health coverage failing in the New Deal era. However, employer-centered insurance coverage began to flourish during World War II. Companies were clamoring for workers during the labor shortage that occurred created by the war. The Congress had implemented federal wage restrictions; thus, health benefits were an end run around this legislation to attract employees [*EBRI Health Benefits Databook*, first edition, 1999; *EBRI Databook on Employee Benefits*, fourth edition, 1997 (updated); and Marilyn J. Field and Harold T. Shapiro, eds., *Employment and Health Benefits: A Connection at Risk* (Washington, DC: National Academy Press, 1993)]. Workers were now incentivized to stay with a company. Over time, this has resulted in a strong private insurance market; a corporatized system whose ultimate profit-motivated fealty is to its shareholders and not necessarily with the health of the patient.

The market failure of health-care cost is complex and multifactorial from arbitrary pricing [20] to collusion [21]. The fundamental concept of health care as a market is that more competition would lead to reduced cost and better outcomes. This has not been the case for some pharmaceutical agents, e.g., the price of Gleevec has increased threefold, from about $2600/month to about $9200/month, to match the price of competitors (http://www.nbcnews.com/health/cancer/utterly-broken-drug-market-high-cost-surviving-cancer-n369261).

One would think that increased products in that therapeutic space would lower cost but the opposite has occurred. There was hope that health-care reform via the Affordable Care Act would require cost controls but provisions to insure this were removed to facilitate passage that was not part of the legislation per se. What it was intended to do was to is provide a larger customer base for the private insurance sector.

Other countries, e.g., the Netherlands, have put limits on the amount of profit a private company coming in the health-care market, striking a balance between making a profit and serving the interests of the patient. During the recent health-care debate in the United States, a single-payer option, as in many other developed countries, was barely considered. While the Affordable Care Act in the United States attempts to increase access, e.g., dependent coverage to 26 years old, eligibility for perexisting conditions and medicare drug benefits. The savings for cancer treatments alone is $75 million (from HHS memo 9/21/2112). It also aims to reduce costs by encouraging efficiency, e.g., the medical loss ratio (MLR) which mandates that an insurer cannot spend more than 20% on indirect cost thereby allocating 80% to providing benefits, has saved an estimated $2.1 billion on health insurance premiums (HHS, 9/11/2012). The savings has been derived from rate review and MLR that have saved $1 billion and $1.1 billion, respectively, to nearly 13 million consumers. The drive for the health-care marketplace to relegate care as a product is increasing over time as costs continue to rise. A private corporation is wrapped in a fundamental ethical dilemma in the inherent conflict between the fiduciary duties to shareholders to maximize profits vs the care of the patient. Cancer chemotherapy offers a marketplace, which highlights this dilemma.

DISPARITIES IN CANCER TREATMENT COST

Uncontrolled Cost of Care

We spend over twice per capita on health care than the next country in the Organization for Economic Cooperation and Development [22]. While spending much on an important public need is not unscrupulous per se, however that investment should be

TABLE 14.2 Estimated New Cancer Cases and Deaths in the United States for 2015

	New Cases	Deaths
All cancer types	1658,370	589,430
Genital organs[a]	329,330	58,610
Digestive system[b]	291,150	149,300
Respiratory system	240,390	162,460
Breast	234,190	40,730
Urinary	138,710	30,970
Lymphoma	80,900	20,940
Skin	80,100	13,340
Leukemia	54,270	24,450
Oral cavity	45,780	8650
Other	163,550	79,980

[a]*Includes cervical, ovarian, prostate, and testicular cancers.*
[b]*Includes esophageal, stomach, intestinal, colon, and pancreatic cancers.*
Data adapted from *Howlader N, et al. SEER Cancer Statistics Review, 1975–2012, National Cancer Institute. Bethesda, MD;* (Based on November 2014 SEER data submission, posted to the SEER web site, April 2015); available from: *http://seer.cancer.gov/csr/1975_2012/*

justified by acceptable incremental benefit. Our health outcomes are toward the bottom of the list despite the expense in terms of things like infant mortality and life expectancy [23].

A recent study found that ~44,789 people die each year as a result of not having insurance [24], which surpasses the annual number of deaths attributed to breast cancer (Table 14.2). It would be interesting to know how the public's perception of health-care access would change if the same level of awareness and advocacy were given to it as we give to breast cancer.

Cancer costs are shared by: private insurance (50%), Medicare (34%), out of pocket (8%), other public (5%), and Medicaid (3%). Over the past 20 years, the total medical costs of cancer have nearly doubled, they have shifted treatment costs away from the inpatient setting and toward the outpatient setting and the share of these costs are increasingly not paid for by private insurance and Medicaid has increased [25]. This translates to the burden of treatment cost being increasingly imposed upon the patient. Hence, the imperative exists to not only understand contributors to high cost, but also mechanisms to bring these cost within sustainable levels that are equitable in terms of affordability for the patient and economically viable for manufacturers. With these points in mind, why do companies invest so much in Food and Drug Administration (FDA) approval? What does FDA approval allow a company to do? How may a clinical trial be influenced by the high cost of development? Remedies such as the 12 Century Cures Act seeks to make the regulatory process more streamlined. However, some have critiqued this proposal as weakening the safety and efficacy standards of FDA review [26]. The importance of FDA approval is that it opens the gates to the health-care marketplace so that investments may be recouped.

The typical justification for high drug prices includes the cost of research development, it reflects its value to the customer, market forces set cost, and controlling for cost would disincentivize innovation [27]. It has been estimated that "out of 5,000 to 10,000 screened compounds, only 250 enter preclinical testing, five enter human clinical trials, and one is approved by the Food and Drug Administration" [28]. The pharmaceutical industry estimates that, as of the early 2000s, it cost $1.2 billion to bring a single drug to market [28]. This amount has been heavily criticized as an overestimation [29]. Pharmaceutical companies spend almost 20 times more on marketing than on research and development [30]. Other estimates place the cost of development around ~$50 million per new drug. This assessment is based upon a number of considerations including the incentive in place that motivated development. Companies are rewarded by expanding market share by introducing new agents that are very similar to the existing market with little clinical benefit, and often at higher cost. They often factor in the cost of marketing as well as indirect and subsidized cost. Much of the laboratory and preclinical research that provides the rationale for market development is funded by the public, e.g., the National Institutes of Health. It seems reasonable that critics of the private sector will decry exorbitant pricing when their product is sold on the shoulders of societal investment. Is it just to presume that companies have a reciprocal obligation to the public not to engage in usurious practices? This is not to say companies cannot be profitable but rather the more challenging question of what is a reasonable profit based upon the special duties imposed by caring for the sick.

Fair Price

> We agree with those who say the price we have set for Gleevec is high. But given all the factors, we believe it is a fair price, *Daniel Vasella, Novartis' CEO [31]*

The notion of morally conscious pricing, i.e., "just price," for cancer treatment has been posited whereby a drug is valued by not only its potential clinical benefit but also its societal and personal cost [27,32]. The concept of "just price" dates back to Aristotle, and later by Marcus Aquinas, but a recent interpretation of this doctrine is that if the market price was usurious or as a result of collusion then the proper authority may interfere in the transaction to impose a fair price [33]. A recent example of a perceived usury involves an agent that has been around for some time, pyrimethamine.

Pyrimethamine is a folate analog used for the treatment of toxoplasmosis; an infection caused by a relatively common global parasitic protozoan. Symptoms include head and body aches, fatigue, and fever, which can be quite sever, even life threatening, in the immunocompromised and vulnerable. Susceptibility to infection increases in areas where sanitation is poor or contamination is prevalent. Pyrimethamine was developed in the 1950s and proved to be an effective treatment. In 2015, the manufacturer of pyrimethamine was purchased by Turing Pharmaceuticals. Martin Shkreli, Turing's CEO, promptly increased the price from $13 a pill to $750 per pill. He cited the need to make the drug profitable to provide funds for further research and development of new agents as the rationale for the price hike. Critics immediately pointed out that this was unconscionable as it would put it out of the price range for patients. This also held Medicare hostage in paying such a high price, of which, the public will have to bear the brunt of the company's profit. Some have viewed his R&D claim as suspect, since by their estimation, pyrimethamine is a perfectly effective agent for this condition.

A powerful critique against this move by Turing Pharmaceuticals is the fact that pyrimethamine is on the World Health Organization (WHO) List of Essential Medicines. Starting in 1977, the WHO created a catalog of "essential medicines." An essential medicine is "intended to be available within the context of functioning health systems at all times in adequate amounts, in the appropriate dosage forms, with assured quality, and at a price the individual and the community can afford" (World Health Organization. *Essential medicines and health products: Essential medicines.* Available at: <http://www.who.int/medicines/services/essmedicines_def/en/> [retrieved 08.10.2015]).

This is not Mr. Shkreli's first pharmaceutical controversy. In 2010, while employed for a capital management firm, Mr. Shkreli sent a letter to the FDA regarding an inhalable form of human insulin for the treatment of diabetes mellitus. This citizen petition urged the FDA to not approve the drug due to not meeting study endpoints, insufficient safety and effectiveness determination, and misleading information provided by the manufacturer. He said that approval of this agent would be "dangerous and [would] not comply the FDA's mandate to secure public health interests" (Letter from Citizens for Responsibility and Ethics in Washington to the US Attorney for the Southern District of New York. Exhibit A. July 9, 2012. Available at: <www.citizensforethics.org/page/-/PDFs/Legal/Investigation/7-9-12_Shkreli_NY_US_Attorney_Letter.pdf?nocdn = 1> [retrieved 24.09.2015]). Although Mr. Shkreli did indicate that he does have a conflict of interest in the FDA's determination, he failed to convey the short position his firm had taken on the manufacturer. The insidious nature of this request, i.e., manipulating the clinical pipeline and public safety/benefit for economic gain, underscores unethical practices that the patient citizen may exert.

Could public outcry, social media shaming and political pressure exert a "just price" on Turing Pharmaceuticals? Within a couple of weeks, the company agreed to lower the price of pyrimethamine (but as of this writing they have not done so). The reaction to the company's conduct also incurred the scrutiny of the government, as Mr. Shkreli appeared before the Congress to testify on his pricing practices. Unsurprisingly, he pleaded his fifth Amendment privilege to avoid disclosing his rationale. Also, the company posted a $15 million loss in the quarter following this decision undercutting the financial motivation for the price hike in the first place. The case of Turing Pharmaceuticals has highlighted the challenges of a market-based system. Again, the issue is not whether or not a company may profit but rather balancing that profit with the obligations entwined with participating in an endeavor involving the most vulnerable among us. One way to address high cost is by changing the behavior of the actors involved in the transaction. The easiest, of which, is to empower one's own moral agency and change one's own conduct.

Changing Practice

Pharmaceutical marketing occupies a rather unique niche in US health care. Other than New Zealand, the United States is the only country that allows direct-to-consumer advertising for prescription drugs. The company—consumer relationship is complicated by the fact that there is a third-party gatekeeper standing in the way of a successful transaction. Hence, in these advertisements, you will always hear, "Ask your doctor if [drug] is right for you." This being the case, it is arguable that the physician is the true customer since the drug's success rests on the physician's judgment. It is this very judgment that may be a powerful instrument in managing exorbitant cancer drug cost.

Determining an intervention for a patient's condition involves balancing a variety of factors such as health status, disease progression, risks/benefits, patient preferences. Beyond those considerations are the inherent moral agency of the medical judgment. A challenging question to consider is what role cost should play in medical decision-making? The term "financial toxicity" has arisen and is used to describe a pharmaceutical agent who effectiveness does not merit it's high cost [34]. In other words, the minimal anticipated benefit is not justified by the potential financial harm.

Zaltrap (ziv-aflibercep) was hailed as an important new therapeutic option for patients with metastatic colon cancer; a disease by which innovative interventions are sorely needed (Renergen, Inc. Press release. August 3, 2012. Available at: <http://investor.regeneron.com/releasedetail.cfm?releaseid = 698064> [accessed 04.04.2016]). However, despite the approval of this new agent, some experts felt it was not much better than the prior standard of care. At twice the price of standard therapy, physicians at the Memorial Sloan-Kettering Cancer Center made the bold calculation that the month and a half of extended life expectancy was not worth this cost. They decided to take a stand against soaring drug cost by choosing not to prescribe Zaltrap for patients with metastatic colon cancer. In the immediate aftermath of their decision, the company decreased the cost by 50%.

Critics decried this move as medical rationing and deprivation of patient autonomy. Commentators claimed it was unethical to deprive patients the benefit of this agent, particularly in such a unilateral way that excluded an individual patient's treatment preferences. Others took issue with the fact that part of the physicians' analysis was based on economics. What is not debatable was the manufacturers willingness to reduce their seemingly arbitrary price and demonstrates the power providers have in advocating for their patients. This example illustrates an example of leveraging individual power for broader action through moral conviction.

Balancing Negotiating Power

If health care is indeed relegated to the caprices of the free market, then the playing field should be fair. Adam Smith's invisible hand would require an equitable ability to negotiate the best bargain for each stakeholder. Such is the case for many countries in that they are able to enter into negotiations with drug manufacturers and settle on the best price possible to the benefit of the patient. For example, tyrosine kinase inhibitors (a class of drugs used in the treatment of an increasing variety of cancers) cost about twice as much in the United States compared to other countries that are able to negotiate prices [35]. Such is not the case for the US government and it's important to peak behind the curtains of the political process to not only reveal other interests at play but also to demonstrate the importance of participation in the political process (see Box 14.2). The overall cost of the program was one of the contentious issues that was a barrier to the bill's passage. The initial estimate is that Part D

BOX 14.2

A FREE MARKET?

Efforts to curb prescription drug costs reached an apex in the early 2000s during the passage to the Medicare Modernization Act. The legislation included the creation of prescription drug benefits under Medicare Part D. At the time, the hope was to curtail spiraling drug costs. Medicare Par D included amending the Social Security Act to include a "noninterference" clause stating:

> In order to promote competition under this part and in carrying out this part, the Secretary—(1) may not interfere with the negotiations between drug manufacturers and pharmacies and PDP sponsors and (2) may not require a particular formulary or institute a price structure for the reimbursement of covered part D drugs.

At the time, a tsunami of pharmaceutical lobbyists flooded the capital. Inclusion of the noninterference clause was essential from their perspective. Congressman Tauzin (see Box 14.1), along with other members of the majority leadership, engaged in a nonholds barred campaign for passage of Part D. A vote on the act was held in the dead of night with voting open for several hours (15 min or so is the norm). This extension allowed the time necessary to whip the vote in favor of passing by a narrow two-vote margin. The consequence of this legislation was to bind the hands of the federal government from negotiating with pharmaceutical companies.

would cost $395 billion over 10 years. However, the actual cost was upward of $534 billion [36,37]. Soon afterward, it came to light that the actuary who calculated the cost was told by the Medicare administrator not to come forward with the adjusted cost estimate. The administrator was faced with dismissal from his job if the true cost projections were released prior to the vote for fear that it may not pass [38]. Members of the US House indicated that indeed they would not have voted for the bill if they knew the actual anticipated cost. Revelation of these shenanigans, at the very least, point to the significance of this law to, and support by, the pharmaceutical industry (see Box 14.3).

This is an important provision that it effectively prohibits the federal government from using its vast purchasing power and leverage into negotiating lower drug prices. Interestingly, this did not apply to the Veteran's Administration (VA) due the political unsightliness of holding veterans hostage to this policy. As a result, prescription drugs prices are significantly lower for the VA, while at the same time providing a reasonable profit for the manufacturers [39]. Medicare could benefit significantly from similar authority [40]. It is estimated that Medicare would be able to save $40–80 billion per year if it were able to negotiate [39].

Encouraging Innovation

The entrepreneurial spirit traces back to our country's founding. The ability to "secur[e] for limited times to authors and inventors the exclusive right to their respective writings

BOX 14.3

THE REVOLVING DOOR

The Pharmaceutical Research and Manufacturers of America (PhRMA) is one of the most powerful lobbying forms in Washington, DC. PhRMA has spent almost $300 million to lobby the Congress between 1998 and the first half of 2016, according to the Senate office of Public records (compiled by the Center for Responsive Politics).

Spending trends track important legislation. For example, leading up to the passage of the Medicare Modernization Act in 2008, PhRMA increased its annual spending to over $16 million (over five times what they spent just 5 years prior). An increase in lobbying expenditures was observed leading up to the passage of the Affordable Care Act in 2009. The peak lobby investment during that period averaged over $23 million, well above the mean of $15 million spent over the past two decades.

Soon after the passage of the Medicare Part D, a number of lawmakers went on to become pharmaceutical industry lobbyists. Rep. Tauzin, who had championed the inclusion of the Medicare Part D, the noninference clause, became president and CEO of PhRMA, earning a 42-million annual salary.

and discoveries" is an expressed power provided to the Congress by the US Constitution (U.S. Const. art. I, § 8). In creating the US Patent and Trademark Office, the Congress has enabled a mechanism for a limited, state-sanctioned monopoly over one's novel ideas and works. The exclusivity afforded by patent protection encourages innovation, development, and commercialization of one's ingenuity and labor. One custom that has arisen is for a pharmaceutical company with patent exclusivity to pay potential competitors not to develop a generic alternative to their product.

This practice of "pay-for-delay" has several downstream effects. Importantly for the manufacturer, it means a continued profit stream. For example, Novartis recently negotiated an agreement with Sun Pharmaceutical to delay production of imatinib (the generic form of Gleevec) for about 7 months when the drug went off patent in July of 2015 (WSJ, http://www.wsj.com/articles/SB10001424052702304908304579563560797460496). That seemingly small delay can result in an additional $2.7 billion in global sales. The unfortunate consequence is that patients are denied more affordable options and the health-care system continues to be burdened with high cost. There are bipartisan proposals in the Congress to limit this practice.

ADVOCACY OF THE PATIENT CITIZEN

End-user empowerment is another mechanism to address exorbitant prescription drug cost. The patient citizen is defined here as a patient or caregiver who embraces their civic role as a member of democratic society to engage in activism that seeks to right injustice in the health-care enterprise. Manipulating the machinations of democracy may be another way to address the cost of unaffordable medications. The example of Medicare Part D illustrates the importance of engaging the political process. The arc of Rep. Tauzin's story (Boxes 14.1–14.3) highlights the incestuous and ethically conflicting relationship between policy making and the private sector companies who stand to benefit from those very policies. Holding policymakers accountable at the ballot box is an important mechanism, we have to effectuate change. For example, there is a petition that implores the President, the Secretary of HHS and the Congress to adopt and/or support legislation of some of the remedies previously mentioned, e.g., empowering CMS to negotiate drug pricing [41] and intellectual property reform (Protest High Cancer Drug Prices so all Patients with Cancer have Access to Affordable Drugs to Save their Lives. Available at ⟨https://www.change.org/p/secretary-of-health-and-human-services-protest-high-cancer-drug-prices-so-all-patients-with-cancer-have-access-to-affordable-drugs-to-save-their-lives> [retrieved 08.10.2015]). Time will tell whether or not advocates for this cause will not only sign the petition but leverage it into political action.

Exerting one's rights to influence policy change in representative democracy can indeed can be effective; however, it is no panacea. In fact, one study suggests that policy at the federal level is driven by the economic elite [42]. The analysis of almost 1800 policies between 1981 and 2002 reveal that public support had negligible effect on whether or not a particular policy was adopted [42]. The policies adopted by the Congress tended to align with the interests of the economically powerful, e.g., Chamber of Commerce, corporate interests [42]. The issue of money, and its influence, in politics is only becoming more poignant in the aftermath of US Supreme Court decisions like *Citizen's United v. FEC* (*Citizens United v. Federal Election Comm'n*, 558 U.S. ___; 2010) and *McCutcheon v. Federal Election Commission* (Shaun McCutcheon, et al., *Appellants v. Federal Election Commission*, 572 U.S. ___; 2014). These rulings have had the effect of protecting campaign contributions as protected speech and prohibiting aggregated contribution limits. The power to influence legislation has grown proportionally with the means and resources, such as the case of the pharmaceutical industry whose pockets are deep. Between 2009 and 2014, the health-care sector has spent an average of $510 million to lobby the Congress, with about 35% of that money coming directly from the pharmaceutical industry (Compare to lobby expenditures between 1998 and 2002, which average $210 million. Raw data from The Center for Responsive Politics. Available at: <https://www.opensecrets.org/lobby/indus.php?id=H> [retrieved 08.10.2015]). The average patient, health-care provider, or citizen can easily feel overwhelmed when faced with such insurmountable influence. However, there are glimmers of hope such as the case of pyrimethamine's price reduction after ignited escalation.

The public's strong reaction to the commodification of health care and the exploitation of the inherent vulnerability of the sick culminated in Turing Pharmaceuticals to reconsider their pricing practices (http://www.epvantage.com/Universal/View.aspx?type=Story&id=596784&isEPVantage=yes¬Sub=false). As of this writing, it remains unclear what the lowered price will be but it does point to the success and the power of popular uprising. Particularly with a persistent media to highlight these unjust pricing practices. There remains hope that mechanisms such as the leukemia petition will highlight these issues and compel policymakers, legislators, and manufacturers to reexamine their business models under the light of ethical practice. While we should not deny a company of reasonable profits, we can hold them morally accountable for affordability with a regard to the lifesaving products they produce. Part of where this starts is what we can do in our role as an individual, a citizen and participant in the health-care enterprise (whether as patient or provider).

CONCLUSION

Treatment cost does represent a potentially life-altering barrier for some patients in receiving therapy. Increasing unaffordability of medical interventions may appear to be an insurmountable problem due its sheer magnitude and complexity. This chapter merely cracks the door to that complexity by way of revealing the interests at play, e.g., policy, lobbying, corporate fiduciary duties, etc. However, we are also reminded of our ethical obligations and moral agency to advocate for improving patient care.

One way to become more involved in effectuating change is reimagining and expanding our responsibility as providers or as patients. Engaging the political process, exerting one's authority for positive change and participating in positive activism, one may be able to impact seemingly insurmountable obstacles in the health-care ecosystem, e.g., treatment cost. Social and behavioral factors have a broad and profound impact on health across a wide range of conditions and disabilities. A better balance is needed between the clinical approach to disease, presently the dominant public health model for most risk factors, and research and intervention efforts that address generic social and behavioral determinants of disease, injury, and disability. Rather than focusing interventions on a single or limited number of health determinants, interventions on social and behavioral factors should link multiple levels of influence, i.e., individual, interpersonal, institutional, community, and policy levels [43].

Investments in research are predicated on the promise of future health impact that, if unmet, may ultimately undermine public trust and enthusiasm [44]. The emergence of translational programs imbues a new scientific paradigm that calls for a reexamination of clinical research *as a practice* (Table 14.3) [12]. This practice has three primary tenants: (1) an individual commitment beyond the patient–physician, researcher–participant, patient–self relationships that traditionally define the perspectives; (2) the integration of ethical norms throughout the heath-care enterprise; and (3) critical reflection on how to embody these principles into every day practice.

Perceived interpersonal collaboration processes (such as greater trust, cohesion, and communication) are correlated to increased productivity [45]. Intrapersonal characteristics, such as the propensity to endorse multidisciplinary values and behaviors, are predictive for research productivity [45]. Removing oneself from the feeling of powerlessness prevents us from successfully effectuating profound changes in our health-care system for the better.

TABLE 14.3 Operational Beliefs for the Individual Health-Care Provider

1. Commitment to patient care includes expanding the clinical ethics landscape to include patient-centered research, one's larger role in society, political participation, and activism
2. Integrating ethical norms throughout research and clinical practice leads to better health outcomes and the quality of patient care
   a. The public expects clinicians and researchers to be ethically accountable and socially responsible
   b. Scientific inquiry is both a virtue and a social good; clinicians, researchers, and patients are moral agents in its pursuit
   c. Clinical practice, research, and ethics are interdependent and cooperative human endeavors
   d. Enhancing research integrity requires institutional and systemic structures
   e. Education, mentoring, consultation, and anticipatory guidance increase awareness of ethical behavior
   f. Regulatory compliance is necessary but not sufficient to ground ethical research practices and clinical care
3. Critical reflection on ethical principles and underlying moral values leads to embodiment of these norms by stakeholders in the clinical research enterprise
   a. How can we maximize the chance of accomplishing our goals of increased discovery, training, and expedited research outcomes? Are we critically reflecting on these goals? What are the metrics and incentives for achieving these goals and how can we lower obstacles and potential conflicts?
   b. What do we mean by "culture change" as it pertains to our clinical research enterprise?
   c. How will the emphasis on public–private partnerships, intellectual property, and entrepreneurship change the research landscape and investigator incentives?
   d. Will the ethical values, duties, obligations, and relationships change in this model, and if so how?
   e. Will new models of ethical inquiry be needed to meet these emerging demands?

References

[1] Keehan SP, Cuckler GA, Sisko AM, et al. National health expenditure projections, 2014–24: spending growth faster than recent trends. Health Aff (Millwood) 2015;34(8):1407–17.

[2] Fund TC. Trends in coverage, medical debt, and access to care findings from the Commonwealth Fund Biennial Health Insurance Survey, 2014. <http://www.commonwealthfund.org/interactives-and-data/infographics/2015/jan/biennial-insurance-interactive>; 2015 [Accessed 17.11.2015].

[3] Himmelstein DU, Thorne D, Warren E, Woolhandler S. Medical bankruptcy in the United States, 2007: results of a national study. Am J Med 2009;122(8):741–6.

[4] LaMontagne C. NerdWallet health finds medical bankruptcy accounts for majority of personal bankruptcies. <http://www.nerdwallet.com/blog/health/2014/03/26/medical-bankruptcy/>; 2015 [Accessed 29.09.2015].

[5] Ten Have H. Respect for human vulnerability: the emergence of a new principle in bioethics. J Bioeth Inq 2015;12(3):395–408.

[6] Beauchamp TL, Childress JF. Principles of biomedical ethics. 7th ed. New York: Oxford University Press; 2013.

[7] Cassel EJ. The nature of suffering and the goals of medicine. N Engl J Med 1982;306(11):639–45.

[8] Cho MK, Tobin SL, Greely HT, McCormick J, Boyce A, Magnus D. Strangers at the benchside: research ethics consultation. Am J Bioeth 2008;8(3):4–13.

[9] Emanuel EJ, Wendler D, Grady C. What makes clinical research ethical? JAMA 2000;283(20):2701–11.

[10] Havard M, Cho MK, Magnus D. Triggers for research ethics consultation. Sci Transl Med 2012;4(118):118cm1.

[11] Sugarman J, McKenna WG. Ethical hurdles for translational research. Radiat Res 2003;160(1):1–4.

[12] Farroni JS, Carter MA. Translational ethics: a humanistic framework for research ethics. Int J Humanities 2013;10(3):73–84.

[13] Littman BH, Di Mario L, Plebani M, Marincola FM. What's next in translational medicine? Clin Sci (Lond) 2007;112(4):217–27.

[14] Maienschein J, Sunderland M, Ankeny RA, Robert JS. The ethos and ethics of translational research. Am J Bioeth 2008;8(3):43–51.

[15] Morin K. Translational research: a new social contract that still leaves out public health? Am J Bioeth 2008;8 (3):62–4. discussion W61–63.

[16] Shapiro RS, Layde PM. Integrating bioethics into clinical and translational science research: a roadmap. Clin Transl Sci 2008;1(1):67–70.

[17] Petrini C. Ethics in translational public health. J Public Health (Oxf) 2008;30(4):514–15.

[18] Pleis JR, Ward BW, Lucas JW. Summary health statistics for U.S. adults: National Health Interview Survey, 2009. Vital Health Stat 2010;10(249):1–207.

[19] Heymann J, Cassola A, Raub A, Mishra L. Constitutional rights to health, public health and medical care: the status of health protections in 191 countries. Glob Public Health 2013;8(6):639–53.

[20] Light DW, Kantarjian H. Market spiral pricing of cancer drugs. Cancer 2013;119(22):3900–2.

[21] Allen S, Farragher T. Partners, insurer under scrutiny. The Boston Globe. January 23, 2009.

[22] Anderson GF, Hussey PS, Frogner BK, Waters HR. Health spending in the United States and the rest of the industrialized world. Health Aff (Millwood) 2005;24(4):903–14.

[23] OECD. Health at a Glance 2013: OECD Indicators; 2013. <http://dx.doi.org/10.1787/health_glance-2013-en>.

[24] Wilper AP, Woolhandler S, Lasser KE, McCormick D, Bor DH, Himmelstein DU. Health insurance and mortality in US adults. Am J Public Health 2009;99(12):2289–95.

[25] Tangka FK, Trogdon JG, Richardson LC, Howard D, Sabatino SA, Finkelstein EA. Cancer treatment cost in the United States: has the burden shifted over time? Cancer 2010;116(14):3477–84.

[26] Avorn J, Kesselheim AS. The 21st century cures act—will it take us back in time? N Engl J Med 2015;372 (26):2473–5.

[27] Kantarjian H, Rajkumar SV. Why are cancer drugs so expensive in the United States, and what are the solutions? Mayo Clinic Proc 2015;90(4):500–4.

[28] America PRaMo. Key industry facts about PhRMA, The Pharmaceutical Research and Manufacturers of America website. <http://www.phrma.org/about/key-industry-facts-about-phrma>; 2013 [Accessed 11.01.2013].

[29] Light DW, Warburton R. Demythologizing the high costs of pharmaceutical research. Biosocieties 2011;6 (1):34–50.

[30] Light DW, Lexchin JR. Pharmaceutical research and development: what do we get for all that money? BMJ 2012;345:e4348.

[31] Vasella D, Slater R. Magic cancer bullet: how a tiny orange pill is rewriting medical history. New York: Harper Business; 2003.

[32] Kantarjian HM, Fojo T, Mathisen M, Zwelling LA. Cancer drugs in the United States: Justum Pretium—the just price. J Clin Oncol: Off J Am Soc Clin Oncol 2013;31(28):3600–4.

[33] de Roover R. The concept of the just price: theory and economic policy. J Econ Hist 1958;18(4):418–34.

[34] Silverman E. 'Financial toxicity:' who's really to blame for high cancer drug prices? Wall Street J 2014; October 7.

[35] Experts in Chronic Myeloid L. The price of drugs for chronic myeloid leukemia (CML) is a reflection of the unsustainable prices of cancer drugs: from the perspective of a large group of CML experts. Blood 2013;121 (22):4439–42.

[36] Pear R. Bush's aides put higher price tag on medicare law. New York Times 2004;30:1. January.

[37] Congressional Budget Office (CBO). Letter from Douglas Holtz-Eakin, Director, Congressional Budget Office, to the Honorable Jim Nussle, Chairman, House Budget Committee, regarding the comparison of CBO and administration estimates of the effect of H.R. 1 on direct spending. In: Congressional Budget Office editor; 2004.

[38] Goldstein A. Foster: White House had role in withholding medicare data. Washington Post 2004; March 19.

[39] Baker D. The savings from an efficient medicare prescription drug plan. <http://www.cepr.net/documerns/efficient_medicare_2006_01.pdf>; 2006 [Accessed 18.11.2015].

[40] Frakt AB, Pizer SD, Feldman R. Should medicare adopt the Veterans Health Administration formulary? Health Econ 2012;21(5):485–95.

[41] Tefferi A, Kantarjian H, Rajkumar SV, et al. In support of a patient-driven initiative and petition to lower the high price of cancer drugs. Mayo Clinic Proc 2015;90(8):996–1000.
[42] Gilens M, Page BI. Testing theories of American politics: elites, interest groups, and average citizens. Perspect Polit 2014;12(3):564–81.
[43] Zerhouni E. Medicine. The NIH roadmap. Science 2003;302(5642):63–72.
[44] Duyk G. Attrition and translation. Science 2003;302(5645):603–5.
[45] Stokols D, Hall KL, Taylor BK, Moser RP. The science of team science: overview of the field and introduction to the supplement. Am J Prev Med 2008;35(2 Suppl):S77–89.

Further Reading

Jones V. When chemo saves your life. Better Health 2009; <http://getbetterhealth.com/when-chemo-saves-your-life-an-interview-with-billy-tauzin/2009.01.29>; [Accessed 3.09.2015].

CHAPTER

15

Mergers, Acquisitions, and Consolidations in the Health-Care Industry: Does Bigger Have Ethical Consequences?

Robert (Bob) Brigham

The University of Texas MD Anderson Cancer Center, Houston, TX, United States

INTRODUCTION

Health-care delivery is a highly personal activity. In the context of modern health care, it implies a very special relationship between the patient and their caregiver at many levels. This personal relationship is shaped and influenced in ways beyond the specific patient care or scientific discovery goals of the provider and patient. The complexity, cost, and rapid rate of scientific advancement within health care create a complex business model that is affected by governmental, economic, social, and technological influences. As care providers in this milieu, it is important to understand the influencers that can affect practice, education, and research. Additionally, we must anticipate what the future may bring and consider how institutions should adapt to large forces buffeting the health-care landscape as core components of health-care delivery are managed; it is crucial to understand the effect that these various factors have on the core patient health-care experience. Focused oversight of every aspect of the patient–provider relationship provides unique opportunities, on the one-hand to foster the evolution of clinical care, and to further research and the advancement of scientific discovery, while at the same time providing services designed to optimize these goals in a friendly, collaborative, and nurturing environment.

This chapter will look at the evolving health-care environment, address marketplace considerations, and the impact these considerations have on mergers, acquisitions, and system consolidations. These factors have an impact on cancer-care providers as they

Ethical Challenges in Oncology.
DOI: http://dx.doi.org/10.1016/B978-0-12-803831-4.00015-4

strive to heal the sick, advance the science and share knowledge with both patients and other health-care stakeholders.

THE INDUSTRY LANDSCAPE

Among the more recent trends that have altered or that are anticipated in the future to alter the climate of the health-care industry is the continued and escalating extent of health industry mergers and acquisitions [1] (Fig. 15.1).

In 2015, the last year for which statistics are available, a brief overview provides meaningful details: Barnabas Health and Robert Wood Johnson Health System in New Jersey merged to form the state's largest hospital network. Supply chain giant, Cardinal Health, acquired the Harvard Drug Group for $1.1 billion, and Aetna, one of the larger stakeholders in the health insurance and plan management industry, paid 37 billion dollars to acquire Humana; the resulting merger was the industry's largest ever [2]. Such mergers and acquisitions are not limited to health insurance and management companies. In Pennsylvania, Genesis HealthCare, a postacute care provider bought 24 skilled nursing facilities, and the nation's second largest pharmacy health provider, CVS Health, bought Target's 1600 pharmacies across 47 states at a cost of $1.9 billion. Furthermore, in behavioral health, Acadia Healthcare acquired three new systems consisting of 500 beds in 17 facilities [3]. In the arena of electronic health records, industry giant, Cerner Corp merged with Siemens Health Services. VHA, a large supply chain company, merged with University HealthSystem Consortium (UHC), creating one of the largest health services companies in the United States. In cross-sector acquisitions, Intermountain Healthcare,

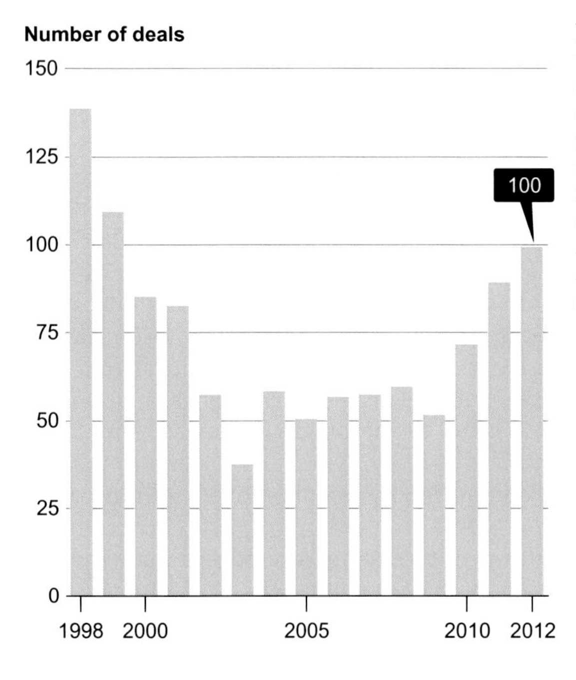

FIGURE 15.1 The trend of hospital mergers from 1998 to 2012. While the number of mergers was stable from 2002 to 2009, that number has increased dramatically over the past few years. The total number of mergers reflects only part of the story, in that the size of the individual mergers, e.g., the number of actual hospital beds provides a picture of a decreasing number of key players in the industry. *From American Hospital Association. Trends Affecting Hospitals and Health Systems. In: AHA trendwatch chartbook 2012. Washington, DC: AHA; 2012.*

took over full ownership of Amerinet, a large health-care group purchasing company, and Illinois Health and Science plans to purchase IBA Molecular, a radiopharmaceutical manufacturer [4].

In 2016, large institutions such as Stanford University, University of Pennsylvania Health System, Boston Children's Hospital, North Shore-Long Island Jewish Health System, Partners HealthCare, Ascension, Merck, and New York University (NYU) have all had their names associated with mergers and acquisitions mentioned in the news.

Sectors of the health-care industry, from providers, insurers, consultants, suppliers, medical device companies, and entities associated with technology have all investigated the climate of mergers and acquisitions that cross sector boundaries. Clearly bold integration and growth initiatives have dominated the health-care industry [5] (Fig. 15.2).

As this trend evolves, there is little suggestion that these mergers will abate in the foreseeable future. Not fully obvious in these statistics is the tendency within the industry to move from independent operations to part of a provider system, a trend that reaches back for most of the past decade. Fig. 15.3 depicts the shrinking percentage of stand-alone hospitals from 2003 through 2013 [6].

This phenomenon is similarly reflected in looking at trends of independent physician practices versus those practices that are in a formal relationship with a larger organization. A recent study from Health Affairs found that, among physicians filing Medicare claims, 35.6% worked in groups of more than 50 in 2011, an increase from 30.9% in 2009 [7].

In addition to formal mergers and acquisitions, new relationships are forming in the industry in an attempt for organizations to achieve the benefits of partnerships short of actual ownership changes. Mayo Clinic has unveiled its Care Network enabling members

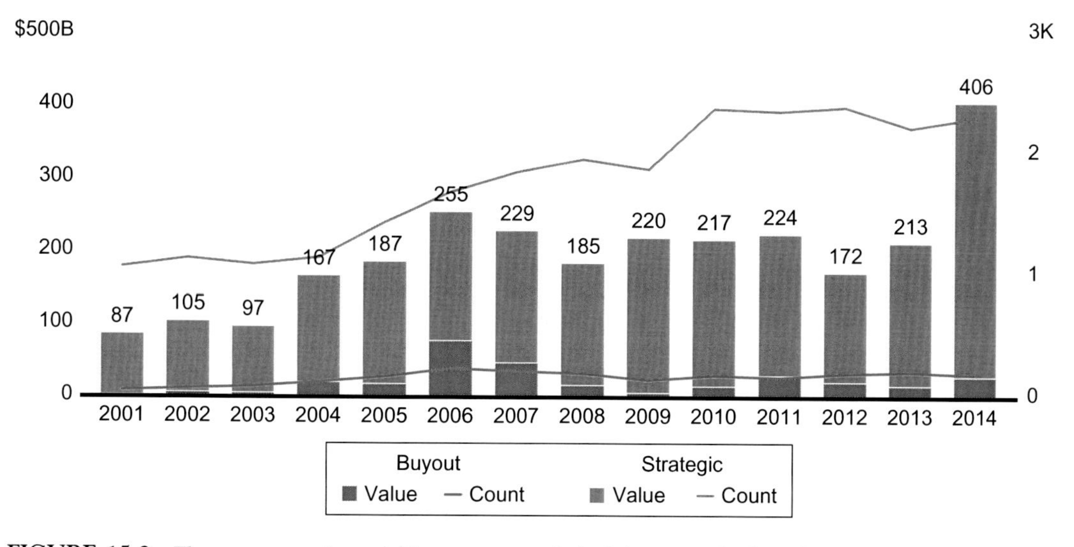

FIGURE 15.2 The mergers and acquisitions over a period of time must look at the size of the takeover; one reflection of which is "deal value" reflected in this bar graph. *From Bain & Company. Global healthcare private equity report 2015. Boston, MA: Bain & Company; 2015.* Notes: *Excludes spin-offs, add-ons, loan-to-own transactions and acquisitions of bankrupt assets; based on announcement date; includes announced deals that are completed or pending, with data subject to change.*

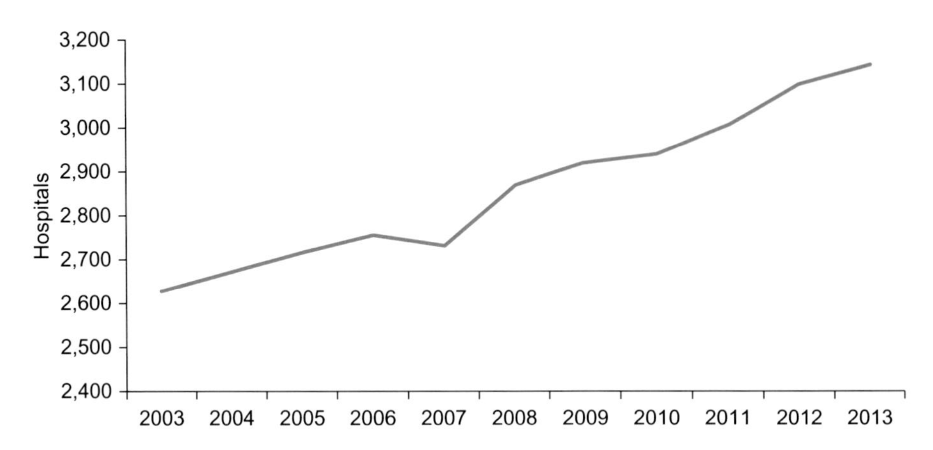

FIGURE 15.3 The change in independent versus system hospital facilities from 1999 to 2013. *From American Hospital Association. Number of hospitals in health systems, 2003–2013. In: AHA trendwatch chartbook 2015. Washington, DC: AHA; 2013. Available from: http://www.aha.org/research/reports/tw/chartbook/2015/chart2-4.pdf.*

to take advantage of Mayo Clinic expert resources on a subscription basis but with obligations to meet specified criteria. Similarly, The University of Texas MD Anderson Cancer Center in Houston has developed several levels of partnerships to enable them to expand their impact on cancer care. Consequently, new and innovative alliances are emerging in the marketplace with increasing frequency.

DRIVERS OF THE MERGERS AND ACQUISITIONS TREND

The confluence of any number of factors may trigger the decision-making of an organization to seek to improve its position by merger or acquisition. However, there are no guarantees that the actual transactions will realize the intended goals and there is reasonable debate in the industry as to the advantages and disadvantages of cooperation versus competition, smaller focused organizations versus large centralized systems, or the impact of these considerations on cost and quality. Adding to the uncertainties is the potential role of government oversight and regulation with regard to any of the industry's sectors. Health-care leaders assess the ever evolving and seldom clear future their organizations face; they must attempt to predict and navigate amidst multiple factors, many clearly beyond their control. Reasons to consider merger or acquisition may stem from any of the following influences as well as from other considerations that may arise within the continually evolving health-care environment.

1. *Managing costs in an environment of declining revenue.* Health-care providers face an unenviable business model where the costs of drugs, technology, labor, and services continue to rise but many of the revenue sources remain fixed. Medicare, for many organizations, is one of the largest payer groups and the reimbursement for services provided that are paid by Medicare is set independently of provider costs. Furthermore, increases in Medicare payments to providers are at a rate generally below what providers

are experiencing on the cost side. Hospital systems achieved an overall operating margin of only 3.1% in 2013, with 14.5% of them experiencing a negative operating margin [8]. Given that health-care industry consumption is close to 17% of the United States gross domestic product (GDP), there is little societal sympathy for health care system budgetary woes. To combat the problem of increasing costs with constrained income, some organizations consider a merger to gain economies of scale and to leverage shared services across a broader base of patient care activity. Infrastructure and administrative support costs such as those for information technology, billing and financial services, human resource management, and operational administrative work can be managed through a shared service model and allocated with the intent to enhance efficiency, avoid redundancy, and reduce work-related wasted resources. Costs are further reduced by a larger patient volume managed by the physical infrastructures, thus, reducing the per unit fixed cost. These considerations are not unique to the health-care industry, and are analogous to those also seen in transportation, hospitality, and food service. The Barnabas Health and Robert Wood Johnson Health System merger in New Jersey which brought together 11 hospitals, 30,000 employees and 9000 physicians reflects this motivation. In an organizational statement, they said, "Through sharing of resources and best practices, the merger will promote the highest quality health care delivery and also enable greater economies of scale" [9]. In another example, Utah-based Intermountain Healthcare acquired the group purchasing company, Amerinet, that served 83,000 members. The intent was to leverage the large-scale purchasing power to drive down costs and take advantage of even greater supply chain process efficiencies [10].

2. *Increase patient base and market share for greater opportunity in working with third party payers.* As revenues from current business become constrained, organizations search for ways to generate necessary income to sustain their mission. The ability to serve a greater and more clinically diverse number of insured patients can improve an organizations ability to gain higher reimbursement rates or ensure inclusion (or continued inclusion) in various health plans. This is aided by either a more comprehensive scope of services, or the consumer preference for inclusion in a larger network with greater access options, or both. The greater number of included patients also contributes to managing actuarial risk in population health management. Again, the Barnabas Health and Robert Wood Johnson Health System merger recognized this as the Barnabas Health CEO noted, "The new health system will effectively comprise of every clinical service from primary to quaternary and greatly strengthen our commitment to medical education and research. The merger also will provide a larger enough geography to be appropriate for the migration to population health" [11]. Similarly, the payer community has looked at ways to increase its market share to help secure a larger customer base and use its size to leverage cost-effective discount rates from the providers. Aetna CEO, Mark Bertolini, stated "The acquisition of Humana aligns two great companies. The complementary combination brings together Humana's growing Medicare Advantage business with Aetna's diversified portfolio and commercial capabilities to create a company serving the most seniors in the Medicare Advantage program and the second-largest managed care company in the United States" [12].
3. *Increasing organizational capabilities.* Many times organizations will seek mergers or acquisitions to gain new capabilities that could take much longer and cost significantly

more to produce with internal development. The acquisition of Siemens Health Services by electronic health record giant Cerner Corporation was priced at $1.3 billion to create a merged effort that would invest more than $650 million into research and development. Cerner's long history of competence in the development of health records and management of large data are now bolstered by Siemens' device and imaging expertise. Cerner CEO, Neal Patterson, stated "By combining client bases, investments in R&D, and associates we are in a great position to lead clients through one of the most dynamic eras in health care. A unique feature of this acquisition is we'll continue working with Siemens AG in an R&D capacity, in order to advance the interoperability of electronic health records with medical devices" [13].

4. *Diversifying services or products.* For many organizations, the need to meet market demands by vertical integration or diversification is essential to compete or grow. In health care, governmental or economic influences can trigger questions for organizational leadership regarding the need to expand services or integrate with other organizations to expand the comprehensiveness of their business model. Cardinal Health's $1.1 billion takeover of the Harvard Drug Group was undertaken, in part, to expand its telesales programs and packaging options [2]. The acquisition of Catamaran, a pharmacy benefits manager, by UnitedHealth Group (UHG), the nation's largest insurer, affords UHG the opportunity to become an innovative force in serving the pharmacy-care marketplace. The recent drive to enable participation in an Accountable Care Organization (ACO) has caused many health-care institutions to examine their range of services and to explore new partnerships that may enable them to enter into arrangements with partners at a cost-point that is competitive. There were approximately 500 ACOs at year end 2013 and The Center for Medicare Services announced the start of 123 new ACOs in January of 2014 [14]. The total number of ACO-covered lives is now estimated at close to 20.5 million. Insurers, physicians, hospitals, and networks have all been involved in exploring new partnerships in response to this phenomenon. In an example reflective of this, the physician groups of Mercy Health Muskegon (Michigan) and Advantage Health/Saint Mary's in Grand Rapids, Michigan aligned to form Mercy Health Physician Partners. David Blair, MD, president and CMO of Mercy Health Physician Partners in Grand Rapids stated, "This expansion provides for the creation of larger specialty teams, greater focused expertise and enhanced standards of care" [15].
5. *Access to capital.* The very narrow and often negative operating margins of many hospitals has increased the cost and difficulty of accessing capital for their operations. Regardless of whether it is a small rural hospital or a large urban hospital serving a poorly insured population, the provider's survival may be challenged without their engaging in some form of partnership. The recent trends in bond ratings reveal that the upgrades have fallen precipitously since 2011 and rating downgrades have climbed [16] (Fig. 15.4).

The Mayo Clinic Health System in the Midwest evolved over many years as smaller rural hospitals and physician groups sought the advantages of partnering with a strong organization whose bond rating and access to capital was an important enabler of those rural operations.

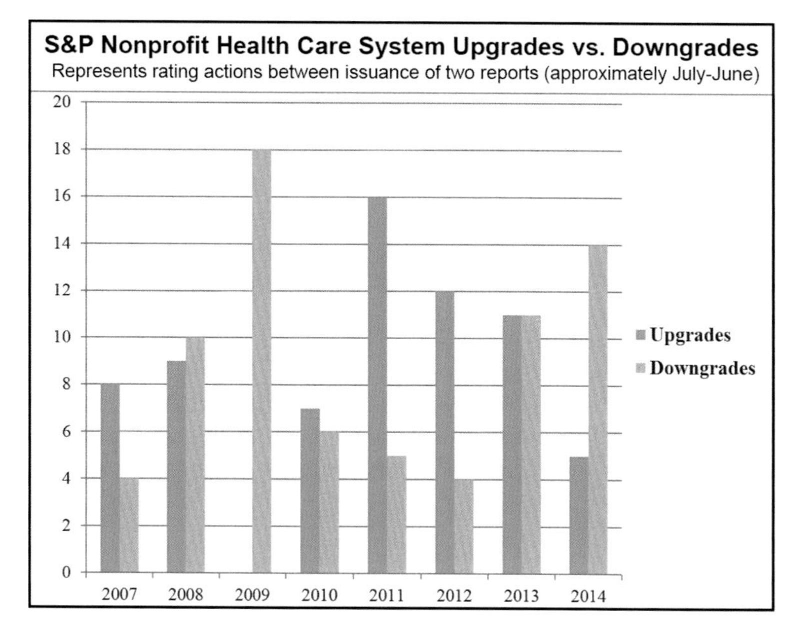

FIGURE 15.4 Comparison of not-for-profit bond rating upgrades and downgrades. *From Laidlaw K. The influential big three: credit rating agency forecasts. In: The capital issue. Lancaster Pollard; 2014. Available from: http://www.lancasterpollard.com/NewsDetail/tci-fe-credit-rating-agency-forecasts. *Represents rating actions between issuance of two reorts (approximately July-June).*

These examples represent some but not all of the factors that cause leaders in the health-care industry to consider new forms of partnerships. Regardless of the driving force, the activity surrounding new partnership models continues at a remarkable pace and is one of the contributing factors that will continue to shape the environment in which we provide healthcare.

RISKS ASSOCIATED WITH CONSOLIDATIONS

Caregivers are concerned that large systems with corporate interests may become less sensitive to the needs of individual patients or mission-focused research interests as these systems concentrate on larger population management and financial performance priorities. Physicians recognize the uniqueness of the individual patient and often are unsure as to whether the large corporate presence is sufficiently sensitive to those nuances so vital to the physician–patient relationship. Caregivers also have concerns as to resource allocation, and if sufficient space and human resources will be appropriately allocated to only the top priorities as identified by executive leadership.

There are concerns that the mindsets of business leaders are different than those of care providers and they may have a disproportionate influence in direction-setting in larger health-care organizations. While business minds may look for scalability and consistency to drive efficiency, physicians often deal in the reality of a unique and difficult case that does not fit well with a standard or generic approach. Such concerns raise ethical questions when a caregiver feels that his or her obligation to be the "zealous advocate" for the patient could be compromised. Furthermore, the drive to optimize the total organizational performance may conflict with an individual provider's specialized interests and thus create a conflict of commitment. These factors may create ethical challenges for care providers and leaders of smaller as well as larger clinical enterprises. Will a larger corporate enterprise erode the entrepreneurship, personal commitment, and motivation of practitioners? Will focused enterprises such as cancer centers be overwhelmed in their unique purpose as they compete for resources with noncancer priorities? These are very real concerns in the turbulent health-care setting and ethical questions often arise when impactful decisions may be beyond the control of an individual caregiver.

While mergers and acquisitions represent opportunities for improving operational performance, they are associated with recognized potential risks that can undermine the delivery of value through the many vertical and horizontal integration possibilities. Many industry analysts fear that consolidations will not be used to drive value but rather to create oligopolies to maintain high prices and protect respective industries from the pressures of competition. Some of these considerations along with their ethical overtones have been addressed in other chapters in this volume. These concerns do have an ethical component and may impact the very nature of the basic provider–patient relationship.

THE PROVIDER–PATIENT RELATIONSHIP

As noted at the beginning of this chapter, the provider–patient relationship has historically been the cornerstone of the health-care system. Mergers and acquisitions can often change the relationship of the physician to the organization or to the payers and subsequently alter the nature of the provider–patient relationship. The provider's ability to objectively provide the best-suited care to each patient is a fundamental goal of this relationship, and the quality of this trust often affects the patient's disclosures to the provider and their compliance to maintain optimum health, an aspect of health care that can raise ethical questions. Concerns may arise around a provider's freedom to choose from a broad availability of services to select the best options for a particular patient's care. There could be a perceived (or perhaps actual) encouragement to leverage the merged/acquired vertical or horizontal integrated structures, thereby subtly encouraging the provider to use the services provided by the system such as specialty consults, diagnostic tests, patient-education materials, procedural options, or the availability of clinical trials.

There are also modes of reward that can have an unintended effect of bias in the use of services. Such reward systems may raise the specter or perception of a conflict of interest which raises serious ethical questions. Financial withholds, bonuses, and incentives pose a potential conflict of interest for the providers as they strive to optimize patient care. As an example, a physician noted a potential influencing experience stating, "Last spring I received something completely unexpected: a check for $1200 from a local health maintenance organization along with a letter congratulating me for spending less than predicted on their 100 or so patients under my care during the previous quarter. I got no bonus the next quarter because several of my patients had elective arthroscopies for knee injuries. Nor did I get a bonus from another HMO, because three of their 130 patients under my care had been hospitalized for the previous six months, driving my actual expenditures above those expected for this group" [17]. One can see how this acknowledgement for helping manage costs can raise the subconscious over prioritization of the cost containment focus, if similar acknowledgements are not made for insuring the quality of care. These practices beg the question that notwithstanding the fact that such practices are legal, and that case law has not precluded their application, are these practices ethical and should disclosure by the provider of the existence of potential incentives that could influence care be part of the consent to treat process?

Additionally, mergers and acquisitions can alter the patient's continuity of care with their providers. Acquisitions, closings, health plan changes, and provider panel shifts can all affect the continuity of patient care and frequent changes in plan design and plan costs can cause patients to make choices, intentionally or unintentionally, that also disrupt continuity of care. Are such changes that are undertaken purely for the financial advantage of the provider, and implemented to the detriment of patients ethical?

INFLUENCE ON RESEARCH EFFORTS

Researchers who work within these larger systems may perceive an ever growing distance between themselves and leadership at these complex organizations. Can the dominant priorities of a larger system be a detriment to some of the very "niche focused" scopes of the research enterprise? Might the corporate strategy favor research that is popular in a desired market or is well-funded commercially? Could research with promise of higher financial return be prioritized over alternates likely to meet less profitable but clinically more valuable results? In an organization where interests are managed in accordance with a mindset that financial performance is prominent is a financial scorecard metric rated above clinical impact? These questions remain unanswered but are important in monitoring the impact of health-care mergers and acquisitions on the focus of cancer research and in insuring that care be provided with a high degree of ethical concern.

WHAT IS THE COLLECTIVE ASSESSMENT OF THE ESCALATING MERGER ACTIVITY?

Has the escalation in health-care consolidations increased value for the health-care consumer? What risks and hazards do these changes impose on the cancer caregiver? These questions are difficult to answer precisely and there are, of course, opinions on both sides. Much depends on an individual's perspective. Looking at macroimpacts that can be measured in large high-level analyses may overlook some very serious challenges on the microlevel where the individual patients actually interface with the system of care, and where ethical concerns may be present. Given the difficulty and lag time in measuring the impact of organizational changes in health care, there is reason to foster further debate. It is important to recognize that there are thoughtful arguments on both sides of this issue.

From a patient's perspective, there appears to be confidence in the care afforded by larger integrated systems. A recent marketing research study found that 62% of consumers were more likely to choose a hospital that was part of a system [18]. Reasons cited for this belief include the thought that a larger system may offer better quality and more coordinated care. There may be access to skilled specialists, better, or more modern supportive services, and, possibly, more amenities.

There are reasons to believe the efficiency and value gains of mergers are being seen in some markets. Openness to investor-owned models has enabled cities such as Detroit to prevent the closure of inner city hospitals with high indigent populations through the sale to the for-profit Vanguard Health System [19]. The American Hospital Association asserts that mergers enable hospitals to become more competitive through economies of scale and the elimination of duplicative services and technology [20]. Kenneth Davis MD, president, and CEO of Mount Sinai Health System in New York City, in a Wall Street Journal commentary opined that, "Hospital mergers can reduce unnecessary overlap in regional health-care offerings. For example, after Mount Sinai Health System's merger with Continuum Health Partners last year, the health system ended up with two kidney transplant centers, located only a mile and a half apart. Combining the two centers increased efficiency and eliminated unnecessary costs while ensuring all patients access to world-class transplantation care. It is far more beneficial for patients to have one center that performs many hundreds of specialty procedures each year rather than multiple facilities that each conducts only dozens." Dr. Davis suggests that such mergers also preserve needed care in communities when he states, "Critical services, such as pediatrics, psychiatry and obstetrics, which often rely heavily on Medicaid for payment, can leave hospitals at a financial loss. In a successful merger, the reduction of back-office expenses and elimination of clinical duplication—for example, consolidating three behavioral-health impatient units which are rarely full into two full units—can allow several hospitals in the larger network to continue offering these services. Stand-alone community hospitals may have to eliminate the costly services to survive." [21] These sentiments are echoed by many industry consultants and observational studies that suggest that the merger and acquisition activity has been a contributor to the dropping rate of health-care inflation, has helped curb excess capacity in some areas, improved coverage and access, and accelerated the spread of best practices and enhanced quality.

Not all are convinced that industry consolidations are actually driving health-care systems for the better. There are reasonable concerns that excessive consolidation is resulting in a loss of competition and an increase in costs. David Jones, California's elected insurance commissioner, cautions that consumers, employers, and medical providers could all be harmed by the megamergers such as that of Anthem Inc. with UnitedHealth Group. He states, "Generally speaking further consolidation of the health insurance industry is not a good thing for customers" [22]. David Lanky, chief executive of The Pacific Business Group on Health which represents large employers such as Disney and Boeing commented, "There is a history of large organizations becoming less willing to disclose data on their prices, utilization, or outcomes, and using their market power to command higher prices" [22].

The Synthesis Project from the Robert Wood Johnson Foundation summarized their assessment of the realized outcomes of health-care mergers and acquisitions as related to costs, prices, and quality. Their findings, surprisingly, suggested that hospital consolidations generally resulted in higher prices, and in an already concentrated market the increase could be as much as 20%. It was their conclusion that physician–hospital consolidation led to neither improved quality nor reduced costs. Overall, the findings suggest that growth without true integration of practice does not guarantee value [23]. This conclusion is consistent with earlier studies published in Health Affairs that found no statistical difference in the average spending per day between system and nonsystem hospitals; managed care prices were actually higher in the system hospitals. The decline in competition is considered a likely contributor in failing to realize cost or quality gains in mergers and acquisitions [24].

RESPONDING TO THE ETHICAL CHALLENGES

Despite the often conflicting evidence, the trend toward consolidation and growth continues. While the landscape created by the myriad of mergers and acquisitions presents potential hazards, it is not a foregone conclusion that we succumb to those hazards. Being cognizant of the factors that may cause bias in decision-making gives the health-care professional the opportunity to expose and evaluate these pressures in light of the best interests of the patient. All members of the health-care community should stay abreast of the stated intents of a merger and acquisition and work toward insuring beneficial outcomes are realized in the execution and delivery of ethical patient care as well as ethical research endeavors.

References

[1] Pope C. How the affordable care act fuels health care market consolidation. The Heritage Foundation Backgrounder Available from: ⟨http://www.heritage.org/research/reports/2014/08/how-the-affordable-care-act-fuels-health-care-market-consolidation⟩; 2014.

[2] Powderly H. Cardinal health closes $1.1 billion takeover of Harvard Drug Group. Available from: ⟨http://www.healthcarefinancenews.com/news/cardinal-health-closes-11-billion-takeover-harvard-drug-group⟩; 2015.

[3] Powderly H. Acadia Healthcare buys three behavioral health systems, two overseas healthcare finance Available from: ⟨http://www.healthcarefinancenews.com/news/acadia-healthcare-buys-behavioral-health-systems-two-overseas⟩; 2015.

[4] Morse S. Illinois Health and Science buys radiopharmaceutical manufacturer. IBA Molecular. Available from: ⟨http://www.healthcarefinancenews.com/news/illinois-health-and-science-buys-radiopharmaceutical-manufacturer-iba-molecular⟩; 2015.

[5] Company B. Global healthcare private equity report 2015. Available from: ⟨http://www.bain.com/Images/REPORT_Healthcare_Private_Equity_Report_2015.pdf⟩; 2015.

[6] AHAR Center. Update: number of system-affiliated vs independent community hospitals, 1999–2013. Available from: ⟨https://aharesourcecenter.wordpress.com/2015/04/02/update-number-of-system-affiliated-vs-independent-community-hospitals-1993-2013/⟩; 2015.

[7] Poses RM. Money vs mission—How generic managers vs physicians think. Health Care Renewal. Available from: ⟨http://hcrenewal.blogspot.com/2014/07/money-vs-mission-how-generic-managers.html⟩; 2014.

[8] Kutscher B. Fewer hospitals have positive margins as they face financial squeeze. Available from: ⟨http://www.modernhealthcare.com/article/20140621/MAGAZINE/306219968⟩; 2014.

[9] Brino A. Barnabas Health and Robert Wood Johnson to merge, create New Jersey's biggest system. Available from: ⟨http://www.healthcarefinancenews.com/news/barnabas-health-and-robert-wood-johnson-merge-create-new-jerseys-biggest-system⟩; 2015.

[10] Morse S. Intermountain to take over group purchasing company Amerinet. Available from: ⟨http://www.healthcarefinancenews.com/news/intermountain-take-over-group-purchasing-company-amerinet⟩; 2015.

[11] Brino A. Barnabas Health and Robert Wood Johnson health system sign historic agreement. Available from: ⟨http://www.barnabashealth.org/Press-Center/Barnabas-Health-News/2015/Barnabas-Health-And-Robert-Wood-Johnson-Health-S.aspx⟩; 2015.

[12] Brino A. Aetna buys Humana for $37 billion in largest-ever insurance merger. Available from: ⟨http://www.healthcarefinancenews.com/news/aetna-buys-humana-largest-ever-ma⟩; 2015.

[13] Miliard M. Cerner seals the deal with Siemens. Available from: ⟨http://www.healthcareitnews.com/news/cerner-seals-deal-siemens⟩; 2015.

[14] Brown B. Top 7 healthcare trends and challenges from our financial expert. Available from: ⟨https://www.healthcatalyst.com/top-healthcare-trends-challenges⟩; 2015.

[15] Punke H. 2 Physician groups form mercy health physician partners Becker's Healthcare: Becker's hospital review. Available from: ⟨http://www.beckershospitalreview.com/hospital-physician-relationships/2-physician-groups-form-mercy-health-physician-partners.html⟩; 2013.

[16] Laidlaw K. The influential big three: credit rating agency forecasts. Available from: ⟨http://www.lancasterpollard.com/NewsDetail/tci-fe-credit-rating-agency-forecasts⟩; 2014.

[17] Blumenthal D. Effects of market reforms on doctors and their patients. Health Affairs 1996;15(2):170–84.

[18] Donohue R. Stand-alone hospital or health system: five reasons why consumers choose the system. National Research Corporation; 2014.

[19] Burgdorfer RJS, Jordan Shields J, Brown, TC, Werling KA, Walker BC. Current trends in hospital mergers and acquisitions. Juniper advisory. Available from: ⟨http://www.juniperadvisory.com/current-trends-in-hospital-mergers-and-acquisitions-log-in-required/⟩; 2012.

[20] Association AH Fundamental transformation of the hospital field. Available from: ⟨https://www.google.com/url?sa = t&rct = j&q = &esrc = s&source = web&cd = 1&cad = rja&uact = 8&ved = 0ahUKEwjx5bDXrK_NAhWi64MKHYUODR8QFggcMAA&url = http%3A%2F%2Fwww.aha.org%2Fcontent%2F13%2Ffundamentaltransform.pdf&usg = AFQjCNH1sw7Dn1I3QCOKk0o8mBTmg-Yg_A&sig2 = dlo4PtujhyGnH6aLZXb54A⟩; 2012.

[21] Davis KL. Hospital mergers can lower costs and improve medical care. Wall Street J 2014;. Available from: ⟨https://www.wsj..com/articles/kenneth-l-davis-hospital-mergers-can-lower-costs-and-improve-care-1410823048⟩

[22] Terhune C. Health insurance mergers don't benefit consumers. California regulator warns Los Angeles Times. Available from: ⟨http://www.latimes.com/business/la-fi-health-deals-competition-20150617-story.html⟩; 2015.

[23] Gaynor MT, R. The impact of hospital consolidation—update the synthesis project. Robert Wood Johnson Foundation. Available from: ⟨http://www.rwjf.org/content/dam/farm/reports/issue_briefs/2012/rwjf73261⟩; 2012.

[24] Cuellar AE, Gertler PJ. How the expansion of hospital systems has affected consumers. Health Affairs 2005; 24(1):213–19.

CHAPTER

16

Legal and Compliance

Holly O. Rumbaugh, Krista M. Barnes and Amarjyot S. Purewal

The University of Texas MD Anderson Cancer Center, Houston, TX, United States

INTRODUCTION

All academic medical centers face challenging legal issues due to the highly regulated nature of both health care and research. While not entirely unique to cancer, there are a few legal issues that are more common in the cancer context such as those raised by the prevalence of genetic/genomic information in cancer hospitals, the government's interpretation of cancer as a "sensitive" diagnosis under federal privacy law, certain medical malpractice arguments and challenges, and the prevalence of experimental therapies and how they relate to the pharmaceutical industry partnerships.

HERITABILITY/GENETICS

In cancer hospitals, genetic information is commonplace. Breast cancer patients are routinely tested for BRCA1 and BRCA2 gene mutations, and references to the patient's BRCA status are discussed throughout the patient's medical record. The genetic test results are, in the minds of pathologists, "just another lab." Yet, the law often treats genetic information as "special."

Genetic Information as "Special"

In Texas, for example, genetic information is confidential and requires patient consent to disclose, even for purposes that the Health Insurance Portability and Accountability Act (HIPAA) [1] would allow disclosure without patient authorization, such as treatment [2]. This could mean that a patient's medical record cannot be shared with the patient's other treating physicians without obtaining special written permission from the patient, thereby potentially delaying or stalling the ability to share crucial information quickly.

Ethical Challenges in Oncology.
DOI: http://dx.doi.org/10.1016/B978-0-12-803831-4.00016-6

HIPAA recognizes that health care providers need to be able to share treatment-related information freely and has a broad exception to the authorization requirement for treatment, payment, and healthcare operations [3]. The only treatment-related exception under Texas law is for disclosures to relatives of a deceased person if the disclosure is made to treat the blood relatives of the decedent (more on this later) [4]. As a result, separate operational workflows with respect to consent and release of information may need to be created for genetic information.

Genetic Information Nondiscrimination Act Partial Protections

In 2008, GINA added to existing HIPAA protections by prohibiting use of genetic information to discriminate against individuals in the realms of health insurance and employment [5]. While GINA is a step in the right direction, there are some gaps in the protections it offers. GINA does not:

- Prohibit group health plans or individual insurers from increasing premiums based on manifested diseases or conditions; [6]
- Prohibit health plans or individual insurers from making payment determinations based on genetic information; [7]
- Prohibit discrimination in the realms of life insurance, disability insurance, or long-term care insurance;
- Prohibit small businesses (those with less than 15 employees) from discriminating against individuals in employment decisions.

As a result, if an individual is hesitant to undergo genetic testing to determine whether they may be predisposed to certain cancers, based on fears that they may not be able to get life insurance as a result, the individual's fears are not irrational. Cancer centers should consider developing information sheets about genetic testing and the legal protections that are and are not available to individuals, so patients can take all relevant factors into account when making decisions about which tests to undergo.

HIPAA contains a provision that may help patients with this concern, albeit only those patients with financial resources. If a patient pays out of pocket for a service and requests that a health care provider refrain from telling their health plan about a particular item or service, health care providers are required to comply with that request [8]. For example, if a patient wanted genetic testing and was willing and able to pay out of pocket for the test, the health care provider would be prohibited from disclosing anything about the test to the patient's health plan. Although a thoughtful attempt at protecting individual privacy, in the realm of genetic testing, this regulation protects only those who can afford to pay the high price for the tests. Additionally, even if a patient were to pay out of pocket for a genetic test and successfully keep the results from their health plan if the outcome of the test drove future treatment decisions (e.g., a double mastectomy or a more expensive drug regimen) for which the patient planned to seek insurance coverage, the patient may eventually be forced to disclose the information anyway, in order to prove that the chosen treatment regimen was medically necessary.

Although GINA doesn't protect individuals from discrimination based on manifested disease, patients who already have cancer may still be worried about genetic testing because of the other results that may be generated from the testing which is typically done in panels. These other results, or "incidental findings," pose additional legal and ethical challenges. Similarly, because of the heritable nature of many genetic mutations, there is a push to share genetic information with relatives of cancer patients and to encourage those relatives to undergo testing themselves, both for clinical and research purposes. Performing genetic tests and genetic counseling for relatives, who may live all over the country (or the world) raises even more legal issues such as medical practice constraints.

Incidental Findings and the Duty to Warn

As mentioned above, genetic testing is typically done in panels of tests. As a result, the potential exists for discovering information about other conditions that the patient and her care team were not necessarily looking for. In the case of "actionable" incidental findings (findings that someone can actually do something about), the first question that arises is usually whether the patient must be told.

Although the law varies from state to state for genetic testing performed in a clinical context, physicians probably have a duty to warn/inform patients about actionable incidental findings. Physicians are generally subject to a duty to act in the patient's best interest and to act with reasonable care [9]. The question that would usually be posed in a lawsuit would be: "What would a similarly situated reasonable physician have done in this situation?" If the answer is that the standard of care was to warn the patient about the incidental finding, then a physician who fails to do so may be found to have acted negligently. But this determination depends on how much is known about the condition that was the subject of the finding. This is a constantly moving target in an era where we continually discover more about genetic mutations and their implications.

In the research context, the duty to warn is less clear. If a patient undergoes genetic testing in the context of a research study, whether a duty to warn of incidental findings exists is probably most dependent upon the language of the informed consent document signed by the patient [10]. For example, if the informed consent says that incidental findings may arise and that they will not be returned to the patient, it is far less likely that a researcher would be blamed for failing to return those results. As a result, it is highly advisable to address the potential for incidental findings and how they will be handled in the study's informed consent document and to act in accordance with the document.

The ideal scenario (from an ethical standpoint) would probably be to give participants the choice: "There's potential for incidental findings to results from genetic testing. If we find something unexpected that you may be able to do something about, would you want to know? Yes/No." While ideal, this poses major operational challenges in terms of tracking patients, tracking results, and tracking advances in medical knowledge about all of the tests on the panel over a long period of time. For these reasons, many researchers may choose not to return incidental findings. In the realm of legal liability, making promises may create a duty, and you don't want to make promises that you can't keep.

A related question is whether there's a legal and/or ethical duty to warn relatives of a tested patient about genetic results that may run in families (whether the primary genetic test result or incidental findings). The case law on this topic varies from state to state. In Florida, in 1995, a court found that that a duty to warn relatives about a condition that may run in families may exist but that it is satisfied by telling the patient that the condition may run in families and urging the patient to share information with his relatives [11]. In contrast, a 1996 New Jersey decision found that there may be a duty to warn relatives and that warning only the patient may not be enough [12]. Arguably, the Florida decision (that no duty exists to warn relatives directly) is the better one because it considers the role of privacy law in this analysis (although, to be fair, both cases predated HIPAA).

Privacy law throws a wrench into the traditional common law analysis of duty to warn about genetic findings. Take the following hypothetical example:

> A patient with breast cancer undergoes genetic testing and finds out that she has a BRCA1 mutation. The doctor advises the patient to tell her two adult daughters about the finding so they can be tested as well and discuss preventative measures with their physicians. The patient says she doesn't want them to know and she's not going to tell them. The doctor believes that this is the wrong decision and the daughters have a right to know. Does the doctor have a duty to warn the daughters? Is the doctor even allowed to warn the daughters if she wanted to?

Let's assume this doctor and patient reside in Texas. In Texas, courts have been reluctant to establish a duty to warn relatives [13]. We'll assume that there's no duty to warn the daughters. But the doctor feels strongly that it is her ethical duty to do so. The next question is "can she"? HIPAA allows for treatment-related disclosures of a patient's protected health information (PHI; genetic information or other) without the patient's authorization. This disclosure can be for treatment of the patient herself "or for the treatment of others" [14]. Under HIPAA (federal law), the doctor could disclose the patient's information to the daughters' treating physicians. However, when a state has a more restrictive privacy law than HIPAA, state law applies, and Texas has one. The Texas Occupations Code prohibits disclosure of genetic information without the patient's consent [15]. The only exception is when the patient is deceased and the relative needs to know for the relative's own diagnosis purposes [15]. The doctor finds herself in an ethical dilemma: Because the patient is alive, the patient controls the release of her genetic information and the doctor risks violating state genetic confidentiality law if she discloses the patient's information to the daughters.

RETURN OF GENETIC FINDINGS AFTER DEATH

The unfortunate reality in cancer hospitals is a high mortality rate. It is possible that a patient will die before being informed of his or her genetic results (particularly in the research context if return of results or incidental findings was contemplated). The privacy rules that control disclosure of health information and genetic information after death are complex and vary from state to state.

Under HIPAA, when someone dies, their PHI is still protected for 50 years. This means that PHI (including genetic test results) still cannot be released without the patient's authorization. When the patient dies, the patient's HIPAA rights fall to the patient's "personal representative." The personal representative is the person authorized under state law to make decisions on behalf of the patient. Unfortunately for health care providers, the identity of the "personal representative" may differ depending on whether the patient is alive or deceased. In Texas, for example, a Medical Power of Attorney (if the patient has been deemed incompetent) or a court-appointed guardian may be a personal representative and exercise privacy rights, such as signing HIPAA authorizations on behalf of the patient while the patient is alive. However, once the patient dies, the power shifts to the representative of the patient's estate (who may or may not be the same person who had that power before the person died). If the estate has no representative, the power falls to the deceased patient's next of kin, in a particular order (e.g., surviving spouse, surviving children, etc.).

In addition to creating a lot of operational confusion about who has the right to authorize disclosure of the deceased patient's genetic information, this legal framework gives rise to many practical and ethical dilemmas. Consider the following examples (and assume they are occurring in Texas, where the state genetic confidentiality law discussed above exists):

- A patient did not want to disclose his genetic information to their relatives. He was explicit about it when he was alive. However, now that he has died, the decision falls into the hands of his eldest child, who was designated the representative of his estate. The eldest child wants to know the results of his father's genetic tests and has the right to authorize disclosure of those results to himself or to others.
- Imagine the same scenario as above, but there's no representative of the estate. The spouse now holds rights to disclosure of the patient's information. She wants to respect her husband's wishes and refrain from telling the children about the genetic information. She refuses to sign a HIPAA authorizations allowing for disclosure of the results to her kids. But the children want to know and their attorney tells them about the state law that authorizes disclosure of genetic information to blood relatives of deceased individuals, if needed to treat the relatives. They ask the hospital to disclose their father's genetic information to their physicians. The hospital finds itself caught between the deceased patient's express wishes (and that of his personal representative under HIPAA) and state law which allows for the disclosure [16].

And this is just one state. Highly specialized cancer hospitals may treat patients from around the country (and the world). Trying to manage the various state laws that may apply to a given patient's situation is legally and operationally overwhelming and may have raise huge ethical dilemmas.

RESTRICTIONS ON THE PRACTICE OF MEDICINE AND GENETIC COUNSELING

Although it is becoming more common, cancer genetics is still a highly specialized field. Health care providers with these specialized capabilities want to be able to help individuals across the country (and undoubtedly, the world) obtain access to genetic testing and counseling. However, state laws governing the practice of medicine and genetic counseling often make it difficult for licensed health care professionals to share their expertise with patients or patients' relatives, especially those who live in other jurisdictions. Consider the following hypothetical example:

> Jane, who lives in Texas, has been fighting cancer and recently found out that she has a rare genetic mutation that the mainstream medical community knows little about but is the focus of research at the academic medical center where she's being treated. She has been urging her sister to be tested. Her sister lives on a ranch in rural South Dakota and cannot afford to travel to Texas to see Jane's doctor, but *could* travel to the nearest city to have her sample taken and sent to Jane's doctor in Texas. Jane asks her doctor if she can help her sister by receiving the labs, running the test, and counseling her sister by phone. Jane's doctor says she wants to help but doesn't think she can do it. Why?

There may be a number of factors contributing to Jane's doctor's decision:

- Jane's doctor is not licensed to practice medicine in South Dakota. Unless Jane's sister travels to Texas, Jane's doctor may risk violating South Dakota's state licensure and medical practice rules if she treats Jane's sister.
- This plan may require Jane's doctor to engage in telemedicine, which may not be covered by Jane's sister's health plan.
- Texas Medical Board rules governing telemedicine [17] impose a number of restrictions on the practice of telemedicine, including a face-to-face visit prior to provision of telemedicine services unless the telemedicine services are provided at an "established medical site." Compliance with the rules will require a great deal of planning and may not be possible in this particular situation, depending on whether Jane's sister has access to an established medical site with the right technical capabilities.
- Some States, including South Dakota, require genetic counselors to be licensed. Texas, however, does not. The Texas genetic counselor would risk violating South Dakota law if she were to counsel Jane's sister without being licensed in that state [18].

Unfortunately, the hurdles and uncertainty may prohibit Jane's sister from getting the highly specialized genetic testing that Jane wants her to have.

CANCER AS "SENSITIVE" HEALTH INFORMATION

The HIPAA Privacy Rule regulates the use and disclosure of PHI by "covered entities," which include hospitals.

In order to meet the definition of PHI, information needs to be two things: Health information and identifiable. Information that constitutes PHI is subject to a complex set of regulatory standards governing confidentiality, security, use, disclosure, and patient rights. Therefore, the first step in any HIPAA analysis, including the determination as to whether a particular unauthorized use or disclosure of PHI constitutes a "breach" that must be reported to the affected individual and the federal government, is whether the information at issue constitutes PHI.

Consider the following scenarios:

- An employee at Gotham General Hospital printed out a list of patient names on hospital letterhead (under the heading, "Patient Name"). There was no other information on the sheet of paper. The employee placed the paper in their laptop bag to take home so they could remotely review patient records in preparation for patient visits the next day. On the way home, while the employee ran into the dry cleaner to pick up her dry cleaning, someone stole the laptop bag. The employee immediately called the police and the Compliance Department at the hospital, worried that the theft of the paper may constitute a privacy breach. Does it?
- An employee at Gotham Cancer Hospital printed out a list of patient names on hospital letterhead. There was no other information on the sheet of paper. The employee placed the paper in their laptop bag to take home so they could remotely review patient records in preparation for patient visits the next day. On the way home, while the employee ran into the dry cleaner to pick up her dry cleaning, someone stole the laptop bag. The employee immediately called the police and the Compliance Department at the hospital, worried that the theft of the paper may constitute a privacy breach. Does it? There is only one difference in these scenarios: the name of the hospital. Yet the second scenario is more likely to constitute a breach. Why?

The presence of an individual's name on a hospital letterhead implies that the individual is a patient of that hospital. While some may argue that the fact that a particular patient (let's call him Bruce Wayne) is a patient of Gotham General Hospital is "health information," many would not make this logical step. Arguably, there's nothing specific conveyed about Bruce Wayne's health from the fact that he was at some point seen at Gotham General. He could have gone there for any reason.

In contrast, if Bruce Wayne's name is on the Gotham Cancer Hospital letterhead, people will likely infer that he has cancer. In a 2000 regulatory discussion, the Department of Health and Human Services' Office for Civil Rights grouped cancer with HIV/AIDS, sexually transmitted diseases, and mental health treatment information when discussing the need for increased privacy protections [19]. They stated that privacy is even more important when it comes to cancer than your average diagnosis (e.g., appendicitis), because people fear that the stigma associated with the diagnosis may affect their employability and insurance coverage. Arguably, the inclusion of Bruce Wayne's name on a cancer hospital's letterhead poses a greater risk of compromise to his privacy than the inclusion of his name on the general hospital's letterhead and is, therefore, more likely to be deemed a "breach."

In addition to breach analyses, cancer hospitals face other operational privacy issues that other hospitals may not. Under the HIPAA Privacy Rule, patients have a right to request restrictions on the disclosure of their PHI and hospitals attempt to accommodate such requests when reasonable and feasible. Patients may request that the hospital refrain from sending bills to their homes with the hospital's logo on the envelope because they don't want the town mail carrier or even other members of the household to see that they are receiving a bill from a cancer hospital. Although a request to use a blank envelope to send bills seems feasible and arguably reasonable, in a cancer hospital with a large volume of patients, isolating a particular patient's bill to package it differently than the thousands of other bills sent out each day or week is an operational nightmare. Additionally, patients are less likely to open a blank envelope than one with an identifiable logo which may hinder payment collection.

Similarly, when making phone calls to patients, cancer hospital employees may grapple with whether to use the hospital's name when calling a contact number for a patient or leaving a message. If someone left Bruce Wayne a message saying, "Hi, this is John. I'm calling for Bruce Wayne. Please give me a call back at xxx-xxx-xxxx," Bruce would be far less likely to return the call than if he knew where John was calling from. Failing to state where one is calling from is a terrible business practice. Yet Bruce may also complain if John leaves a message on his machine saying he is calling from Gotham Cancer Center and it turns out that his friend Robin happens to play back the message first and didn't yet know that Bruce had cancer. Because of the potential for inadvertent disclosures like these, legal counsel may advise cancer hospitals to refrain from stating their name when leaving messages or calling patients at work [20]. Operationally speaking, this is problematic.

TERMINALLY ILL CANCER PATIENTS AND END-OF-LIFE LEGAL CONCERNS

Because cancer is often an aggressive and terminal illness, patients may face end-of-life concerns and considerations that are not necessarily presented to patients being treated for curable diseases. For example, medical advanced directives are encouraged to be completed early into the cancer patient's treatment in order to allow the patient to vocalize and formalize their treatment decisions while they are still legally competent to do so. Once a patient is determined to no longer have decision-making capacity, decisions regarding end-of-life care fall on the shoulders of the identified medical power of attorney or the proper person identified under the law to make such decisions. The treatment decisions to be made for terminal cancer patients may involve questions of whether or not to discontinue chemotherapy and to proceed with comfort care, whether or not a do-not-resuscitate (DNR) order should be implemented, or whether removal of life support is desired or is in the patient's best interest. These are extremely hard decisions for surrogate decision makers, and the process can invoke intense and stressful situations for families and care providers. It is important for care providers to timely document into the medical record each discussion that is held with the surrogate decision maker(s) or family member(s) regarding treatment decisions, so that all communications regarding the patient's care are captured, other care team members are kept apprised of the most up-to-date decisions about the patient's care, and such information is available should it be needed at a future date for legal purposes or other reasons.

The type of care provided to a terminally ill cancer patient may appear different from care provided to other patients who are not facing terminal illness. A severely ill cancer patient, whose body may be extremely fragile, feeble, and weak, may reach a point when they no longer benefit from aggressive chemotherapy and/or radiation treatment. In fact, such aggressive treatment could do more harm to the patient. Accordingly, there often comes a time when an aspect of the patient's treatment needs to be curtailed or modified, yet family members who are grieving themselves and are witnessing their relative slowly and painfully succumb to their terminal cancer, desire to continue aggressive treatment in the hopes it will provide benefit or cure; this may occur even when the family member is well informed that the aggressive treatment is not the proper standard of care for the patient. Problems can arise when the treatment appropriately stops but the family member feels as though providers are "giving up" or "not trying hard enough." The family member may later attempt to assert a claim (legal or otherwise) that the provider "failed to treat" the patient or was medically negligent by not providing all possible care. The reality is that the standard care for a terminally ill cancer patient who is actively dying is often quite different from the standard care that would be provided to a similar-appearing patient who does not have end-stage cancer or who is not actively dying. Such modification of care should not imply that a provider was negligent or that he or she failed to properly treat according to acceptable standards of care.

CONCLUSION

In conclusion, Federal and state laws generally gain additional complexity in the context of cancer care and treatment. The special handling of genetic information and cancer diagnoses as sensitive information often requires cancer hospitals to navigate a difficult balance between protecting an individual patient's right to privacy (in life and after death) with the wishes of the patient's family members to know more about their relative's cancer (like its hereditability and whether to seek potentially life-saving early intervention for themselves or their children in turn). Similarly, otherwise routine processes of consent and advance care planning can become fraught with emotion and uncertainty amidst the fight against cancer, making difficult end-of-life discussions and decisions even harder, such as whether to proceed cautiously and preserve the quality of a patient's remaining days or risk shortening a patient's already limited or grim prognosis by aggressively attacking the cancer with a hopeful yet unproven experimental treatment (with results that may not benefit the patient receiving the treatment, but might one day contribute to the general understanding of the disease). Addressing these unique compliance and legal challenges head on is a complex but essential process in which cancer hospitals should engage in order to satisfy the ultimate goal of providing the highest level of care to the patients they serve while staying within the proper legal boundaries.

References

[1] Health Insurance Portability and Accountability Act of 1996 (HIPAA), Pub. L. No. 104-191 (1996).
[2] Tex. Occ. Code §§ 58.102, 58.103.
[3] 45 C.F.R. § 164.506(c).
[4] *Id.* at 58.103(a)(4).
[5] Genetic Information Nondiscrimination Act of 2008, Pub. L. No. 110-233 (2008), § 2.
[6] 74 Fed. Reg. 51,664, 51,667 (Oct. 7, 2009).
[7] *Id.* at 51,668.
[8] 45 C.F.R. § 164.522(a)(1)(vi).

[9] See, e.g., Pate v. Threlkel, 661 So.2d 278, 280 (Fla. 1995); see also Angela Liang, The Argument Against a Physician's Duty to Warn for Genetic Diseases: The Conflicts Created by Safer v. Pack, 1 J. Health Care L. & Pol'y 437, 439 (1998).
[10] Pike, Rothenberg, and Bekman, *Finding Fault? Exploring Legal Duties to Return Incidental Findings in Genomic Research*, 102 Georgetown Law Journal 795 (2014).
[11] Pate v. Threlkel, 661 So.2d 278.
[12] *Safer v. Estate of Pack*, 677 A.2d 1188 (N.J. Sup. Ct. App. DIv. 1996).
[13] See *Thapar v. Zezulka*, 994 S.W.2d 635 (Tex. 1999) (finding no mental health professional duty to warn third parties); *Santa Rosa Health Care Corp. v. Garcia*, 964 S.W.2d 940 (Tex. 1998) (finding no duty to warn a patient's spouse about potential HIV exposure).
[14] 45 C.F.R. §164.506(c)(1)-(2); *see also* HHS OCR FAQ, "Does the HIPAA Privacy Rule permit doctors, nurses, and other health care providers to share patient health information for treatment purposes without the patient's authorization? Answer: Yes. The Privacy Rule allows those doctors, nurses, hospitals, laboratory technicians, and other health care providers that are covered entities to use or disclose protected health information, such as X-rays, laboratory and pathology reports, diagnoses, and other medical information for treatment purposes without the patient's authorization. This includes sharing the information to consult with other providers, including providers who are not covered entities, to treat a different patient, or to refer the patient," available at: http://www.hhs.gov/ocr/privacy/hipaa/faq/disclosures/481.html.
[15] Tex. Occ. Code §§ 58.102, 58.103.
[16] The author contends that although ethically challenging and risky from a litigation standpoint, the law would allow for disclosure of the information to the children's physicians, under both state law and HIPAA (if it is information that is needed to treat the children).
[17] 22 Tex. Admin. Code § 174.
[18] SDCL 36-36-4.
[19] 65 Fed. Reg. 82,462, 82,777 (Dec. 28, 2000).
[20] The HIPAA rules recognize that hospitals need to make certain disclosures of PHI in order to operate efficiently. For example, a hospital may need to call out a patient's name in a waiting area, or post the patient's name on a whiteboard so the care team can quickly see the patient's name and location. These types of disclosures of PHI, like the hospital logo on the envelope or the "I'm calling from Gotham Cancer Center" on the answering machine, are permissible "incidental disclosures of PHI." Nevertheless, incidental disclosures of PHI have the potential to generate plenty of complaints from unhappy patients, and therefore plenty of work for the hospital's lawyers.

CHAPTER

17

Diversity and Health Care*

Harry R. Gibbs and Colleen M. Gallagher

The University of Texas MD Anderson Cancer Center, Houston, TX, United States

INTRODUCTION

Ethics, like morals, principles, and even religious beliefs, are systems of behavior designed to ensure the survival and benefit of a group of people. These systems vary widely across cultural and ethnic lines. Acknowledging the existence of these systems and their impact on daily interactions is an essential step toward successfully functioning in an increasingly integrated global community. For the purposes of this chapter, we will confine ourselves to those guidelines and recommendations pertinent to the successful relationship between a health care provider and a patient or group of patients seeking medical help or advice. Specifically, while there have been a number of articles addressing medical ethics across differences, we will focus on those issues that occur with regularity when dealing with cancer and its management. We will look at the concept of cancer itself across cultures, some critical clinical decisions that must be made during the course of cancer management, and the role cultural differences play in the use of cancer clinical research trials. Finally, we will offer some suggestions for the health care provider that may enhance the patient-provider relationship across cultural and philosophical differences.

WHAT IS CANCER?

For Western medicine, cancer has a biological basis. It is an entity that lends itself to scientific analysis and specific therapies based upon experiments and trials. Even malignancies that may have an environmental or behavioral component (diet, sun exposure, smoking), the emphasis of cancer management revolves around screening procedures,

* The editors would like to thank Dr. Gibbs for his contributions to the understanding of diversity in health care and for sharing his perspectives specifically for this chapter. Unfortunately, Dr. Gibbs passed away before the chapter was finalized. Dr. Gallagher edited and finalized the chapter.

Ethical Challenges in Oncology.
DOI: http://dx.doi.org/10.1016/B978-0-12-803831-4.00017-8

surgery, radiation, and chemotherapy. Malignancy is seen as an individual problem, and the majority of treatment is focused on that individual's biology.

For most of the world, however, cancer and other severe illness are emotionally felt to be intimately associated with lifestyles and choices: What type of person are you, what type of people were your ancestors, how you have treated others, how others have treated you...in this life or even previous ones. It is our relationship with nature and others that determine our future course in life [1]. Karma, fate, and susto are just some of the terms used to explain why one comes down with cancer. In addition, many cultures around the world and even developed countries embrace the concepts of curses, voodoo, etc., which can remotely affect the health of another individual. A simple Google search will result in an abundance of resources for casting spells designed to either cause cancer or cause it to spread. The idea that these causes of malignancy can be treated with pills, X-rays, or transplants is alien to many patients who harbor these belief systems. If one believes that a higher or supernatural power caused the growth of cancerous cells, then he or she will not believe that a man-made medical treatment will be able to destroy them.

The discourse surrounding cancer in our own culture is somewhat unique as well. Discussions on common conditions like diabetes, arthritis, and hypertension center on the concepts of *controlling* or *managing* them. The focus is on maintaining these conditions as opposed to eradicating them. When it comes to discussions of cancer, we resolve to cure it as if it were a short-term event such as pneumonia or another fleeting infection. The concept of a short-term curative process may be compatible with some of the behavior-based ideas of cancer cure ("right the wrong," find the one who cursed you, etc.). However, the actual management of one's cancer is usually a more prolonged process, which frequently can make the patient feel temporarily worse, suffer temporary or permanent disfigurement, or not produce any significant benefit for some period of time. In some situations, the cancer may reoccur.

Realizing that not everyone has the same view of cancer (from its origin to its purpose) is an important first step in establishing a trusting and understanding relationship between the patient, provider, and other stakeholders in the process.

CANCER DECISIONS

Western medicine has changed dramatically over the years. Initially, Western medicine functioned based on a "doctor knows best" philosophy where such things as placebo medicine were widely used. The philosophy has shifted to become more of a patient-centered model where the patient is kept well-informed (by the provider or by their own initiatives via social media and the Internet) and is made a part of every significant decision regarding their medical management. Interestingly, while this relatively new approach has improved the patient experience, there is not much data to show that it has significantly improved survival or cure [2]. This patient-empowering approach, however, is not common in many cultures. When patients from other cultures present to a Western-practicing physician, communication conflicts can occur at the very beginning of the patient-provider relationship and last throughout the management process, including end of life decision-making.

Initial Conversations

A recent study of Western medicine practicing internists from both the United States and China by Mitchell Feldman et al. [3] demonstrated that Chinese physicians, and those of many Asian countries, are more concerned than US physicians over the impact that being informed of a cancer diagnosis may have on the family of the patient, rather than on the patient himself/herself. The study concluded that this difference is rooted in differing perspectives. The US physicians were focused on the patient's right to be informed whereas the Chinese physicians were focused on the wants and needs of the patient's entire family unit. This reluctance to tell the patient a diagnosis is frequently shared by the family who feels that it will be of no benefit and might even cause harm. In their study, only 2% of US physicians would comply with a wife's request to withhold cancer information from her husband versus 80% of Chinese physicians who would do as relative requested. To a US physician, the thought of withholding such information would generally be an anathema to both medical as well as legal practitioners.

The subject of informed consent is one that is difficult for all patients. It can pose additional unique challenges for patients from other cultures. The first obvious challenge is one of language of origin. Informed means informed. Both written and oral descriptions of treatment plans, options, etc., must be transmitted in such a way that the patient and other stakeholders understand them to be able to make the best possible decisions. How the information is delivered may be more important than what information is delivered.

For example, in many Arab countries, verbal consent is as binding as written consent. Taking the extra (and American standard) step to obtain written consent is considered an implication of mistrust. Also, the method for obtaining verbal consent is crucial. Arab-Americans do not jump directly into engaging in critical conversations. They would rather take the time to gather more information about the topics and the individuals involved first. Moreover, family members are typically used to discuss medical information before making decisions. This process can be a source of conflict for Western providers who feel they do not have the time to get into long conversations with extended family members. For example, during an overbooked clinic or who are trying to get consent for treatment in the middle of a moment of crisis or just prior to a procedure, a provider's need for speed and a family member's need for a personal connection will be at odds.

Hospital Stay

The management of cancer involves a significant invasion of one's privacy, habits, and life decisions. Multiple medications, invasive or unusual procedures, strange environments, and isolation are the hallmarks of cancer management. Each of these can cause conflict with one's usual way of being. Certain patients have restrictions on diets, blood products, organically derived medication, etc. Muslim patients practice wudu [4], which requires them to cleanse themselves multiple times per day prior to prayer. The use of IV's, pumps, and other medical devices can seriously interfere with these practices. Other cultures may associate certain colored sheets and clothing with death that would otherwise have been benign to Western eyes. Religious icons and other practices (i.e., incense, saging, and frequent prayers facing certain directions) are all important to these patients, as is the use of alternative medications, ointments, stones, coins, and other religious talismans.

Even medical facilities as they are currently designed pose potential issues as we become a multicultural society. Most of the world does not use the western-style toilet on a regular basis. This fact should lead to questions about how we should design future restrooms. The Western practice of limiting examination and patient rooms to a certain number of visitors may conflict with those who view their "immediate" family as a much larger group beyond parents and siblings or spouse and children. As we continue to try and make patients' hospital experiences and clinic stays more comfortable, we have to be cognizant of what stations are available on the televisions as well as what magazines are placed in the clinic waiting rooms. Paradoxically, the more we focus on improving the patient's experience, the more we have to realize that patients are unique entities and not just parts of one large group.

End of Life

All physicians, regardless of their specialty, must deal with end of life issues with one patient or another; however, the nature of cancer makes end of life issues more prevalent than the norm. Historically, in-depth discussions of end of life options were few and far between. Death occurred as a result of acute infection, trauma, and "old age." In many Eastern and native American cultures, death is simply another part of the cycle of life. The thought of expending large amounts of financial and emotional capital to sustain someone "beyond his time" is not part of that paradigm. The concept of such things as palliative care, do not resuscitate ("DNR") orders, and end of life decisions is a result of modern, predominantly western, technology, as well as the belief that an individual's life is precious and should be sustained. Moreover, many physicians in this country feel that assisting in terminating one's life for any reason is contrary to the Hippocratic Oath. Subscribers of this theory maintain a "do everything possible" attitude throughout the patient's course of disease. African-Americans are much more likely to want aggressive medical care at end of life and are less likely to have DNR orders than European American patients [5]. This is partly due to a lack of trust in the American medical system and concern that the physician is not honest during end of life discussions. Korean and Mexican American patients are more likely than European American patients to believe that their families should make these decisions rather than the patients themselves. For many cultures, the process of dying is an intimate one. Hospice utilization, for example, is very low among minority communities especially Asian ones. Members of those communities may prefer to die in the privacy of their home with their loved ones.

Another important aspect of this discussion is the fact that traditional Western-trained physicians view a patient's death as a sign of professional failure. A prominent surgeon once told one of the authors of this chapter that if he were not perfect, his patients would die. This surgeon conceded that it is a matter of course that everyone dies, but he felt that once patients were under his care, he was both medically and morally responsible for their survival.

CLINICAL CANCER RESEARCH

Although clinical research studies are a major part of every branch of medicine, they are particularly associated with the prevention and management of cancer. A significant number of cancer patients are currently participating in clinical and prevention trials or have been in the past. The majority of Western physicians believe that participation in these trials has a significant benefit in both survival outcomes as well as quality of life. Unfortunately, these benefits are not seen across all ethnic groups [6]. It is well-known that women, ethnic minorities, and those patients from cultures outside the United States have a lower participation rate than European Americans. A recent review noted that less than 5% of National Institute of Health ("NIH") sponsored trial participants are nonwhite, and less than 2% of clinical cancer studies focus on nonwhite racial or ethnic groups [7]. For example, although African-Americans have the highest incidence of cancer in this country, they have, along with Hispanic Americans, the lowest rates of participation in cancer trials. Interestingly, this is not the case for pediatric trials, which suggests a possible generational component in addition to an ethnic one. Reasons for these disparities are numerous and include such things as the accuracy of reporting data, both by researchers as well as individuals. A colleague who immigrated to the United States from South America noted that she did not realize she was "Hispanic" until she arrived in the United States. In addition, the constant alteration of reporting standards designed to portray a more complete picture of the ethnic background of patients ("two or more races," "white Hispanic" versus "white Nonhispanic") will, in the near future, complicate matters even further.

There are also significant ethical challenges present when discussing clinical trial participation with patients in a multicultural environment. Many of the pitfalls fall into three distinct categories: Mistrust, misunderstanding, and miscommunication.

Mistrust

As noted above, African-Americans have a consistently low participation rate in cancer clinical trials. One of the primary factors in this refusal to participate is mistrust of the American medical system. Historical evidence of medical abuse of African-Americans, including unauthorized gynecological and sterilization procedures as well as the infamous Tuskegee study [8], have caused many patients and stakeholders to hold the view that the medical profession does not have their best interests at heart. Many of these cases involved the patients being lied to about the nature of the procedures and intentionally harmed. Some years ago, the Office of Institutional Diversity at The University of Texas MD Anderson Cancer Center conducted focus groups with African-Americans in the Houston area. The participants who did not currently have cancer but were in one of the many higher than normal risk categories for developing a malignancy. Those individuals over the age of 40 indicated that they would not go to an academic cancer center or participate in a clinical trial. Many of them cited the Tuskegee study as one of the reasons for their refusal. Interestingly, this view was not shared by those under the age of 40. This distrust in the American medical system held by older African-Americans pervades the entire provider-patient interaction, especially if the patient sees someone who does not look like him/her in any position of authority or decision-making capability.

Another unfortunate legacy of clinical research comes from a well-intentioned effort by researchers to conduct studies in various multicultural communities. Sadly, this process frequently led to patients from the community being used as subjects in studies with no clear benefit being demonstrated to the community. Such a pattern breeds mistrust. This "study them and then leave" philosophy leads one to wonder about the ethics of developing a treatment using patients from lower socioeconomic communities. Members of those communities typically cannot afford or obtain access to the fruits of the clinical research. Thankfully, this philosophy is less common today.

Misunderstanding

As previously noted, patients from different worldviews have a perception of cancer, its effects, and outcomes. Trying to understand the concepts of biological treatment for a lifestyle, moral, religious, or ethical issue can be a daunting task. In many cultures, the health care provider is the subject matter authority for what ails you, and your role is to follow advice. The idea of a randomized trial, for example, is quite confusing to many as it is felt that the physician should already know what the best treatment is. The possibility of being placed in a group of patients not getting the best treatment is both frightening and confusing for a potential study participant.

Failure to adequately explain treatment plans and patient expectations is a frequent issue, as many Western physicians assume that all patients have a similar idea of what medical treatment entails. However, even our sense of time is culturally determined. In many countries, punctuality is defined in terms of an event rather than the clock. Unfortunately, this can lead to patients being labeled as noncompliant because of late or missed clinical appointments. To avoid a misunderstanding, the importance of timely and consistent attendance at appointments must be communicated clearly at the inception of their participation in a study.

The use of alternative medicine is very common throughout the world. In the United States, we ingest multivitamins, vitamin-enriched water, probiotics, and similar products for no particularly good medical reason, and at great cost both financially and injury related to the agent involved or impurity of what is given. Patients from other cultures have similar practices that are not necessarily familiar to Western physicians. A lack of understanding of the importance of these practices and, even worse, a dismissal of their efficacy can widen the gap between patients and providers. This is true, especially, in those cases where the alternative treatment may interfere with standard care.

A clear area of misunderstanding can arise when communicating expectations of the participants from clinical trials. Being told that you may go through some strenuous treatment that is expensive, time-consuming, and may possibly be the cause of a worsening of symptoms is difficult, at best, to hear. The difficulty is compounded when a patient is told there may be a relatively small chance of remission. Many patients, and particularly those who are not intimately familiar with scientific experimentation, do not understand the goals of "tumor reduction," and the purpose of Phase 1, 2, or 3 trials. Moreover, describing outcomes in terms of percent remission is incomprehensible to many. As mentioned before, many patients view death as part of the cycle of life and have a much more global and long-term outlook for survival. Expensive treatments for the individual with no guaranteed outcome seem to be more harmful to the family and future of the "clan."

Another area of mistrust must be laid at the feet of the provider. Bias and prejudice can cause researchers not to actively pursue women or minorities as potential participants in clinical trials. Studies have shown that unconscious as well as conscious bias has been the cause of low minority participation [9]. These biases manifest themselves in preconceived attitudes about patient compliance, ability to pay, difficulty in dealing with language and cultural barriers, and providers own feelings about the person's, race, gender, sexual orientation, or religious beliefs.

Miscommunication

Many articles have been written about addressing language barriers in effective communication across cultures [10]. In the arena of clinical trials, however, additional ethical questions arise over how much information to convey, how to convey it, and to whom it should be conveyed. The standard Western model of written informed consent (i.e., one on one interviews, clinical care conferences) does not fit those who come from different world views. Our tendency to fragment care information into multiple conversations with physicians, nurses, mid-level providers, pharmacists, social workers, translators, etc., can be extremely confusing and frightening to others. Even the addition of "someone from that culture" into the conversation mix can be viewed as an intrusion and possibly even patronizing. Also, our own views of what constitutes a good faith effort to reduce cultural communication barriers are sometimes inadequate. It is the practice at many hospitals and clinics to ask patients if they have any cultural issues that needs to be addressed at the beginning of treatment or trial. This may be useful for those patients who have had experience with Western practices that conflict with their own (diet, clothing preferences, gender of health care professional, etc.). However, it is a meaningless question for someone who is unaware that their long-held beliefs and lifestyle are somehow different from what they are going to be experiencing.

RECOMMENDATIONS

Since the basis of all intercultural communications involves conversations which are individual, contextual, fluid, and organic, it is somewhat artificial to give a list of "how to's" that would encompass all encounters. Nonetheless, we will try to present a set of broad guidelines and checkpoints that hopefully will aid the health care provider to navigate these critical waters.

The center concept of all patient-provider interactions is trust. Physicians are not law enforcement officers, and patients cannot be forced to comply with recommendations, protocols, treatment plans, and directions. The patient (and all stakeholders) must honestly believe that the provider understands their needs and is truly acting in their best interests.

Similarly, the provider must have faith in the stakeholders and must believe that they understand what the future holds and what role they will play in the management of this disease. Without this level of mutual understanding, there is no coordinated action.

CANCER DECISIONS

Initial Conversations

Understand That Your View of Cancer is Your View of Cancer, Not the Only View of Cancer

Physicians trained in Western medicine view cancer with a scientific, individualistic, and sometimes reductionist lens. All patients do not universally subscribe to this concept. Similarly, physicians who come to the West from other cultures harbor different views as well. Providers must have a clear understanding of not only how they view cancer, but also how they view their patients through their own personal, medical, ethical, and religious lens. A law student who feels that she cannot ever morally defend a guilty client can choose not to go into trial law. Similarly, if a physician feels that she would have difficulty dealing with a particular type of patient, she should look at that view early in her career, rather than wait until a critical encounter occurs [11,12].

The Initial Conversation Should be Between the Primary Health Provider and the Key Stakeholders

As noted above, trust begins at the beginning. A provider should not leave the initial encounter to surrogates. Once the relationship is established, then it is permissible to introduce other members of the health care team. It would also be a good time to explain in detail what role they will play, and to assure them that you, the primary provider, will be intimately involved in all major health care decisions. Patients come to your institution because of the physicians' reputations, not the physician assistants or advance practice nurses. Be sure to find out who the patient wants included in these discussions. In some case, there may have to be some negotiations as to the size of the group, but holding patients to the extreme "patient only" limitation is counterproductive to establishing the requisite trust.

Be Humble and Inquisitive [13]

If this is the first time that you have interacted with someone from a different culture, state that up front and point out that this will be a learning situation for both sides. Frequently, there are visual clues that a patient may be from a different world view (accent, physical appearance, clothing, etc.), but be aware that just because someone has an accent or epicanthal folds does not mean they are a foreigner. Also, be aware that someone who looks like you does not necessarily mean that they share your belief systems. For example patients from East Texas are not the same as patients from West Texas. Constantly check for levels of understanding. One way to do this is to ask the patient to repeat critical instructions to be sure that they understand. In the initial conversation, it is probably more critical to listen than it is to talk. Treating cancer is a marathon and not a sprint. You will have plenty of time after the initial conversation to share what you need to share.

Be Respectful

Although some beliefs that you hear may sound odd and even outlandish, a patient may sincerely hold these beliefs, so you have to adjust to their understanding and sensitivity. Ask more about the beliefs rather than trying to change them. The key in the initial conversation is to find a common belief system from which the provider can operate, not to attempt to convert one side to the other.

HAVE A RESOURCE

Widen Your Support System to Include Individuals From Various Cultural Backgrounds

They may provide guidance and wisdom about certain beliefs that may be foreign to you. Use them frequently but do not drag them into the conference with the patient. Bringing in a complete stranger into the room to discuss private feelings and situations often erodes trust. Utilize these resources. *Be content with small victories.*

Modern hospital and clinic guidelines frequently put time limits on the length of initial visits. This puts providers under strain to "get all of the business done" in a certain amount of time. The more cultural differences there are, the more difficult this becomes. However, in the long run, it is better to be patient with a patient who is not totally in synch with the Western way of academic medicine. Patience is key. Rushing someone into signing protocols, arranging donor databases, or similar transactions will result in critical treatment snafus down the line. Extra time during the initial visit or arranging additional visits may be necessary to ensure an effective patient-provider relationship throughout the clinical course.

HOSPITAL STAY

Revisit, Revisit, Revisit

Many academic centers have a process whereby the initial provider is not assigned to the inpatient service at the time, and the patient is "transferred" to another physician during the course of stay. This can be very disturbing to someone who has felt that the provider bond has been made with one individual and now a stranger is taking over his care. It is imperative that this transition is made clear and is made seamlessly. We recommend that the primary provider continue to visit the patient while he is in the hospital and constantly provide reassurance that he is being continually informed of his clinical status. Assure him that his health care team is keeping him involved in the major decisions. While this requires a little bit more work on the part of the primary physician, it pays off in great dividends down the line. Similarly, the inpatient attending should constantly reassure the patient that she is in contact with the primary provider.

Speak With One Voice

During a hospital stay the patient is seen by a host of hospital employees including nurses, mid-level providers, dieticians, social workers, chaplains, volunteers, housekeeping, case workers, physical therapists, and laboratory technicians as well as their physicians. It is important that the patient and stakeholders understand who these people are, and what roles they play in the patient's care. The daily schedule and its purpose needs to be explained to each patient.

CULTURAL NEEDS ASSESSMENT

Many hospitals are making attempts to understand the cultural needs of their patients, including dietary preference, use of icons and symbols, religious practices, and the like. Unfortunately, this information frequently reduces itself to a questionnaire or a simple inquiry such as "do you have any cultural concerns?" It should be noted that all patients have preferences, and they do not view them as barriers; for them it is their way of life. Conversations with floor personnel will yield more valuable information than simple questions or a written questionnaire. In addition, frequent check-ins during the course of the stay will uncover problems before they become critical. This does not mean that the hospital must acquiesce to all demands, but compromises can usually be made.

END OF LIFE DISCUSSIONS

Lay the Groundwork Early

Understandably, providers do not like these discussions. The fact that they are not usually trained in the techniques of giving bad news while in medical school compounds the issue. Therefore, there is a tendency to wait until death is imminent before broaching this topic. Waiting until the crisis occurs is never a good practice. For most cancers, death is a real possibility, and the patient and stakeholders should be aware of this possibility as early in the treatment course as possible. This allows ample time for the patient and stakeholders to process the possibility of death, to ask questions and to prepare.

The Discussion Should be Between the Primary Provider and the Patient/Stakeholders

Recently, a relative of one of the authors (HG) was admitted with end-stage lung disease. During a visit, he was approached by the research nurse and told that there was nothing more that could be done, and the team was considering sending the patient to a hospice environment. She then asked him if he had had "the end of life talk with her yet." The author pointed out that he was there in the capacity of a relative and not as a member of the care team. He made it clear to the care team that it was not his responsibility to have "that talk" with the patient. This example illustrates the common reluctance that providers have when approaching conversations about end of life decisions with the patient. These talks should include both the patient and the stakeholders. They should be held with the primary physician, not the research nurse, mid-level provider, shift nurse, or even the inpatient attending (if he/she is not the primary physician of record).

Determine the Decision-maker

Western medicine views the patient as the primary decision-maker for critical issues, yet this is not the case in many cultures. Be sure that the care team knows who the primary decision-maker is. This should be established early in the relationship. If the patient declares that major decisions will be made by someone other than himself/herself, respect that. Be sure to document it in the patient's file and communicate it to the rest of the care team. Of course, if the situation calls for legal documentation of a delegation of decision-making authority, make sure that exists. It is also a good idea to periodically check in with the patient to be sure that this decision still holds.

If Possible, For Large Family Groups, Try to Designate a Spokesperson

Despite meticulous planning, emergencies occur at any time. For those patients who have a large group of stakeholders, it is sometimes difficult to gather them all at once to give out critical information. It becomes especially difficult if some of them do not reside locally. The phrase "the cousin from California" is frequently used in Texas to describe a relative who comes out of nowhere during a critical moment and demands to be brought up to speed before any major decision is made. The team should know who the primary stakeholders are and should request that they designate one or a few individuals to contact in case of an emergency. The designated emergency contacts can then fill in the rest of the family at a later time. Again, documentation of this communication plan and notification of the care team is essential.

Use Available Resources

Most hospitals have an ethics team. In some cases they are on call 24 hours a day to help providers navigate critical medical decisions with ethical implications that occur. Consider utilizing these services in preparation for having difficult conversations as it may result in better patient/provider relationships than first declaring to the patient what you have planned and then meeting resistance. Certainly utilize this resource when your first encounter resistance. Again, having the awareness of the patient's and family's feelings is extremely important. As noted above, many African-Americans and those from Middle Eastern countries view verbal messages to be more important than any written document. Just because someone has signed a DNR form or living will does not mean that, in the heat of the moment, they will attest that that is their wish.

CLINICAL CANCER RESEARCH

Include the Community From Start to Finish and Beyond

All too often a researcher enters a community with a carefully packaged detailed protocol that may look good to the NIH but may be impractical to implement in a specific community because of factors such as personal expense, time constraints, or misunderstanding about outcomes. Researchers should involve community members at the very beginning of the design stages of the protocol, and be sure to give them appropriate credit as well. Many savvy researchers have community representatives or advisory groups intimately involved in all protocols [14]. Start by making connections with trusted sources like churches/houses of faith, community organizations, community leaders, community health care providers, sororities/fraternities [15]. These people and organizations speak the potential participants' language (literally and figuratively) and are enmeshed in the culture. They can be an invaluable asset to you in trying to get the listening ear of potential participants. To avoid the previously mentioned "study them and then leave" phenomenon, maintain a presence in the community beyond the sunset of the trial. Continue to offer screenings, lectures, and similar services.

Make the Study Accessible For the Participants

One of the commonly cited barriers to minority participation in clinical trials is accessibility whether it is dealing with work obligations, childcare, or eldercare [15]. Hold study orientation sessions to allow potential participants to make informed decisions and to poll them on their daily schedules and responsibilities. Schedule study-related appointments outside of working hours. Make onsite childcare or eldercare available or schedule study-related appointments around the participants that need to provide childcare or eldercare. In the event that potential participants do not have their own transportation or do not have access to public transportation, either provide transportation to and from appointments or conduct home visits if your budget allows.

Diversify your Clinical Research Personnel

Members of the community want to see people like them at both ends of the study, and especially as part of the design and implementation teams. Whenever possible, use a diverse research team.

Partner

In many cases, a medical institution can assist the community in ways that may not be directly related to the specific study (i.e., screening, prevention lectures, blood drives). Ask the community what it needs and provide it if possible. By agreeing to participate in your study, the participants are giving you something of value—data. Create a quid pro quo relationship where you provide them with something of value to them and their community. Share the benefits of the partnership by sharing some of the financial profit and giving appropriate credit for contributions to the research. In community-based research this is called "community engagement" and is now customary. These actions could help preserve the goodwill built up and, hopefully, leave a positive impact that could trickle down to future generations.

CONCLUSION

Diversity is not just "us" dealing with "them." Everyone has a different view about their bodies, their health, and their future. As we become a more global society, these differences are coming to the forefront. Even within Western society the advent of social media has exposed multiple variant points of view from patients who would historically have been viewed as "typical" patients. In the 21st century, diversity really means customized patient care. The DNA is surrounded by a cell that is surrounded by an organ that is surrounded by a person, who is surrounded by a community. All of these factors need to be considered when addressing issues of cancer and its management. This requires a set of skills that may not have been covered during primary training, including the ability to self-reflect, empathize, accept variant viewpoints, and communicate in a variety of ways to make the relationship effective.

"Diversity ain't nothin' but a conversation" [16].

References

[1] Tromprnaars F, Hampden-Turner C. Riding the waves of culture. New York: McGraw–Hill; 1988.

[2] Rathert C, Wyrwich M. Boren S. Patient-centered care and outcomes: a systematic review of the literature. Med Care Res Rev 2012;70(4):351–79.

[3] Feldman M, Zhang J. Cummings S. Chinese and U.S. internists adhere to different ethical standards. J Gen Intern Med 1999;14(8):469–73.

[4] Kopec D, Han L. Islam and the healthcare environment: designing patient rooms. HERD: Health Environ Res Design J 2008;1(4):111–21.

[5] Kagawa-Singer M, Blackhall L. Negotiating cross-cultural issues at the end of life. JAMA 2001;286(23): 2992–3001.

[6] Albain K, Unger J, Crowley J, Coltman C, Hershman D. Racial disparities in cancer survival among randomized clinical trials patients of the southwest oncology group. J Natl Cancer I 2009;101(14):984–92.

[7] Chen M. Minority clinical trials participation and analysis still lag 20 years after federal mandate. UCDAVIS Health System News, March 18, 2014.

[8] Corbie-Smith G, Thomas SB, Williams MV, Moody-Ayers S. Attitudes and beliefs of African Americans toward participation in medical research. J Gen Intern Med 1999;14(9):537–46.

[9] Van Ryan M, Saha S. Exploring unconscious bias in disparities research and medical education. JAMA 2011;306(9):995–6.

[10] Kermani F, Virk KP. Language and culture in global clinical trials. Applied Clinical Trials Online. Jun 01, 2011.

[11] Diekema DS. Cross-cultural issues and diverse belief March Ethics in Medicine. University of Washington School of Medicine; 2011

[12] Phillip A. Pediatrician refuses to treat baby with lesbian parents and there's nothing illegal about it. Washington Post. 19 02 2015. Cited 13 September 2015.

[13] Kagawa-Singer M, Kassim-Lakha S. A strategy to reduce cross-cultural miscommunication and increase the likelihood of improving health outcomes. Acad Med 2003;78(6):577–87.

[14] Emanuel E, Wendler D, Killen J, Grady C. What makes clinical research in developing countries ethical? The benchmarks of ethical research. J Infect Dis 2004;189(5):930–7.
[15] Hansen PJ, Drost M, Rivera RM, Paula-Lopes FF, et al. Adverse impact of heat stress on embryo production: causes and strategies for mitigation. Theriogenology 2001;55(1):91–103, (Janauary 1).
[16] Gibbs H. Diversity ain't nothin' but a conversation presentation. The University of Texas MD Anderson Cancer Center, April 9, 2012.

CHAPTER

18

An Ethical Framework for Disclosing Individual Genetic Findings to Patients, Research Participants, and Relatives

Jeffrey S. Farroni, Jessica A. Moore and Colleen M. Gallagher

The University of Texas MD Anderson Cancer Center, Houston, TX, United States

INTRODUCTION

Even at the dawn of the genetics revolution, experts of the day advocated a prudent approach to utilize these technologies in an ethical fashion [1,2]. That is not to say that scientific progress need be impaired, but merely that we should do our best to anticipate, and mitigate, the risks inherent to recombinant methodologies. Today we are contemplating an embodiment of that technology, whereby we are able to inform patients and relatives of current or potential disease risk based upon prognostic markers. Screening for genetic markers of disease holds the potential to not only better treat and care for our patients but also exponentially impact the lives of those that may be at risk for cancer. The increased pace and power of our understanding of cancer screening, diagnosis, and treatment also instills within us a greater responsibility in the use of that knowledge. The practice of medicine and conduct of research are value-laden, moral enterprises, and, as such, we should execute that moral agency with due diligence.

Part of the challenge that arises in disclosing individual genetic findings is the implication the data has as a "future diary" for not only the patient but the patient's relatives [3]. What emerges is a dilemma as to what we should do with information that has broader impact. Do new obligations arise to people for whom we have information that may impact their health? How far do these obligations extend? We must not forget that our primary duty is to care for the patient in front of us over a hypothetical patient we do not have. We must respect the values and honor the dignity of the patient at our bedside

Ethical Challenges in Oncology.
DOI: http://dx.doi.org/10.1016/B978-0-12-803831-4.00018-X

when deciding how best to utilize important clinical and research findings. We would not *always* defer to the patient in every case, but merely that much deference must be given to the patients for which we have an established fiduciary relationship. There are many careful considerations when developing a policy for the disclosure of genetic or genomic information. Fundamentally, we must discern: (1) What information must we disclose? May disclose? Or may not disclose? (2) To whom will this information be disclosed and how do we identify the recipients? (3) Who decides what information is disclosed? (4) What is the disclosure process? (5) How is the disclosure process administrated?

For the sake of brevity, some terms will be used in a generalized fashion. For example, when speaking in terms of genetic or genomic information, the analysis and framework could easily be applicable to other individually unique data substrates like proteomics. Other terms are used in a more limited way for pragmatic reasons, such as "relatives." The use of "relatives" is restricted to first-degree family members. It becomes untenable to develop a disclosure policy if a duty of care is created for multigeneration family structures. Please see Table 18.1 for definitions of key terms used in this chapter.

A careful balance must be struck between providing important, clinically relevant information while mitigating over-reporting information that may confound health care decisions. The information disclosed should be limited, actionable with appropriate safeguards, and administrative oversight to protect the patient's privacy. Also, currently existing services and infrastructure should be leveraged to reduce cost, decrease resource utilization, and lower administrative barriers.

TABLE 18.1 Definitions of Key Terms and Types of Genetic Findings

Genetic/Genomic—Although this chapter primary focuses on genetic or genomic findings, many of the comments, recommendations and ethical issues are also applicable to other data sets and platforms, e.g., proteomic, metabolomics.

Individual genetic finding—Is a finding concerning an individual patient or research participant that has potential health or reproductive importance and is discovered in the course of clinical care or conducting research that is beyond either the scope of the current the aims of the study [4].

For the purposes here, individual genetic findings includes the follow subgroups [5]:

1. *Anticipatable incidental finding*	a finding that is known to be associated with a test or procedure.
2. *Unanticipatable incidental finding*	a finding that could not have been anticipated given the current state of scientific knowledge.
3. *Secondary finding*	a finding that is actively sought by a practitioner that is not the primary target.
4. *Discovery finding*	the results of a broad or wide-ranging test that was intended to reveal anything of interest.

Actionable—A genetic individual genetic finding where there are established therapeutic or preventive interventions or other available actions that have the potential to change the clinical course of the disease [6].

Relatives—Refers to first-degree family members, e.g., mother, father, sister, brother, daughter, and son.

ETHICAL CONSIDERATIONS

The decision of whether or not to disclose individual genetic findings is a difficult one, fraught with many ethical and practical landmines. Part of the difficulty lies in the embodiment of the practice of medicine itself. As professions, medicine and clinical research aim to improve human health, whether individually, at the patient's bedside, or societally, through the development of better diagnostics, treatments, and preventative measures. We are at the emergence of an unprecedented time, where technology and specialization are ever increasing our understanding of disease. Adopting policies to disclose genetic individual genetic findings holds the potential to exponentially impact the health of many people. Although the prospect is alluring, particularly within the "knowledge is progress" narrative, very careful reflection must take place to ensure the protection and honoring the professional/ethical obligations to our current patients. We must balance the potential good of genetic risk factor disclosure to the potential harms such information could wield.

Many commentators articulate the important distinction between treatment and research [7]. We proscribe different duties and obligations depending upon the context: Fiduciary duty to patients [8] and a protective duty to research participants [9]. The doctor–patient relationship imbues a duty to promote the benefit of the patient and respect their choices and autonomy. The researcher–participant relationship incorporates societal trust and benefit with an appropriate balance of risks. The emergence of genetic/genomic technologies and information is challenging the normative boundaries of these relationships [10]. We must make a distinction between the technological imperative (the existence of test/information is justification to use it) and the ethical imperative (moral justification arises from the individual benefit) in the utilization of genetic individual genetic findings [10]. As stated in the introduction, there is a pervasive notion of genetic exceptionalism. The idea that there is something inherently special about our genes and the information it contains [3]. The inherent nature of our genetics is that the information contained therein is relevant to not only ourselves but to our ancestors and progeny as well. The strong value of individual autonomy faces an existential challenge, as genetic information arguably does not solely "belong" to that individual. Rather than compromise the right of autonomy, it must evolve with emerging technology and shifting values, placing a greater burden of protection by the custodians of that technology and information [10]. Not everyone agrees with the notion of genetic exceptionalism, whether it is for philosophical or pragmatic reasons [11,12].

Some have argued that clinicians and researchers have a duty of "ancillary care," which extends beyond what is required to safely carry out the research [13]. This would include an obligation to provide participants with limited individual genetic findings due to the fact that they have entrusted the researchers with their personal information [14,15]. An ethical rationale for returning genetic individual genetic findings exists based upon honoring a patient/research participant's autonomy and potential clinical benefits [4]. Generally, many patients and research participants see a benefit from, and want, individual genetic findings [16,17]. Benefits of disclosing actionable information include an awareness of the disease, reducing risk factors for disease, seeking treatment, and even the positive psychosocial impact of having that knowledge. Many individuals are particularly interested in findings that may have serious consequences [18–23]. Motivations vary ranging from a strong desire to know all they can [24] to the view that it is earned through participating in research [25,26]. Some feel that returning individual genetic findings is a way of honoring patient autonomy and their medical interests [27]. It should be noted that disclosure should be limited to actionable findings tailored to health and reproductive interests [28].

Less certain is whether or not information disclosure to a patient/research participant's relatives is appropriate. There are cases where disclosure is permissible, e.g., the guardianship responsibility over a child [29]. Our primary obligations rest with our patients and research participants who are bearing disease and treatment burden and research risks. We should defer to their wishes, were possible, through informed consent as to the appropriateness of disclosing individual genetic findings to relatives.

Risks of disclosure include violating the autonomy or liberty interest of not only the patient/research participant but also the individual receiving the information. It is difficult to justify the unilateral imposition of genetic individual genetic findings on someone who may not want that information. There are cases of very real psychological and cultural harms imposed by the unwanted disclosure of genetic information [30,31]. Recipients can be overwhelmed with the information which necessitates a need for appropriate counseling and processes for returning results [17].

Another concern is the exclusion of information that may have great health impact and clinical significance for someone. Does honoring a patient/research participant's autonomy to not disclose information to a relative outweigh the interest of that recipient from having the information? Is it just that potentially willful harm may come to someone who does not have pertinent information? What duty, if any, do we have to individuals for which we have clinically relevant information but who are not patients? This last question is critical in that we have a commitment to serving societal goods and engendering public trust through clinical care and research; yet, where do we draw the reasonable, practical boundaries of intervention? How do we balance conflicts between patient needs and society's demands?

It will be imperative to develop evidence-based approaches to information disclosure that clearly recognize the "continuous spectrum of causality, possible clinical utility, and psychosocial impact" represented by such powerful tools as whole genome sequencing [17]. Current and emerging technologies offer tremendous hope for actionable findings but new models for returning results will be challenging [17]. It is recommended that a scientific review committee conduct an evidence-based assessment of actionable findings and a separate administrative service implement the disclosure program. The task will be daunting as there is no consensus on what information should be disclosed [4,32–38]. Data/information could be "binned" based upon the level of evidence showing association with disease risk and potential interventions [39]. For example, one mechanism for returning individual genetic findings includes establishing disclosure criteria, analyze in concordance with those criteria, and identify and contact the individuals that should receive the results [28]. Regardless of the reporting mechanism, an ethical framework is predicated on ensuring informed consent and honoring an individual's autonomy in deciding to participate in this enterprise.

Practical Considerations

Although an ethical argument can be made for the limited release of some individual genetic findings, many practical considerations become important to reflect upon when implementing a disclosure program. These considerations include: Who will receive this information, what information will be released, who will disclose this information, and what actions will take place after disclosure? The following are questions for inquiry:

Results Returned to Whom?

- Should results be conveyed to the patient or participant directly or to that individual's primary health care providers?
- Is it possible or appropriate to keep genetic results, or at least incidental genetic findings, out of the individual's medical record?
- When, if ever, does a third party's interest in these data trump a patient's?
- Must all known family members be notified of clinically relevant genetic findings? What if the mechanism chosen for notification excludes potential family members, e.g. people with unlisted phone numbers, limited access to the internet, do not use social media, etc.?
- Will an investigation need to be conducted for relevant family members or do we take patient at their word?
- Should there just be a fixed policy, whereby only children, siblings, and parents are notified?
- How will we know that a search has been thorough enough?
- What if the patient expressly indicates that they do not want their genetic information shared with a particular family member?
- What if the family is maligned, e.g., patient makes claim of physical/sexual abuse of close family member, or perhaps there is tangible proof of abuse?
- Does the reason for exclusion by the patient really matter? (Do they have to have a good enough reason, e.g., "I just don't want to!" versus "He/she abused me, I have documentation to prove it!")

Another debate exists regarding the permission or obligation to disclose genetic information to family members who may be at risk from a heritable condition. The 2013 American College of Medical Genetics policy statement, supported by a recent statement by the Pediatric Committee on Bioethics, argues that clinicians have an obligation to report significant individual genetic findings to the parents of children who have been tested about the personal health implications for both the child and the parent or other family members of these results [40]. Despite a few court cases and some debate based on discomfort with withholding important information from family members, especially if that information is "actionable," a consensus remains that physicians should not report genetic data to relatives without the patient's consent [41]. Chan and colleagues outline in a target article in the American Journal of Bioethics the plan that they defined and followed regarding disclosure of individual research results from whole-exome sequencing to deceased participants' relatives in the ClinSeq study. In this report, they stipulate that they will only disclose results of potential clinical benefit that have been confirmed in a laboratory that has been certified for providing clinical test results to relatives who actively seek the information, through the executor to the patient's estate. The authors enumerate three common reasons why disclosure of individual genetic results is generally not made to relatives: If there is reason to believe the decedent would have objected to the release of information, if the wishes of the potential recipient are unknown, or if researchers lack the analytic or clinical resources to responsibly disclose the findings [42].

Many ethicists would argue for a more stringent standard, where the participant or patient must specifically consent to disclosure to relatives or other named individuals during the informed consent process or in a subsequent documented conversation. Additionally, it might also be recommended that the family members seeking the information sign a request to receive information or a release to be contacted.

It would be logistically and economically impractical to "genealogically investigate" each patient/research participant. We have to resign ourselves that family dynamics can be complicated. Despite the justice concern that all relevant family members should be informed, it should be the patient who decides which relatives receive information. The exception would be if the patient/research participant is deceased, and there exists a strong rationale to disclose information.

How Much Information to Disclose?

Will all of the findings be returned or just actionable findings? This identifies a need for an independent scientific review committee to determine actionable findings due to evolving nature of genetic/genomic technology and understanding. Part of the analysis in determining whether or not to disclose information is predicated upon the relative clinical impact and benefit. Table 18.2 presents a summary of a possible analytical framework based upon the potential for benefit and actionability of the information.

Results Returned by Whom?

Many scientists and ethicists, who agree that there exists an obligation to report individual genetic findings to participants and patients who freely choose to receive them, are still concerned with who is the appropriate person to provide this information. In both, the clinical and the research, settings, the one seeking the primary information for which the genetic test is needed may not be qualified to interpret the individual genetic finding. Furthermore, this person, although well-intended, may not be the appropriate person to disclose this information in a scientifically informed and psychologically sensitive manner. In other words, it is likely beyond the scope of practice of the researcher or clinician. The concern that naturally follows is whether the research study or diagnostic clinic is equipped practically or financially with the recommended personnel such as genetic counselors or medical geneticists who would have the appropriate expertise within their scope of practice to adequately inform patients and participants of the implications of the discovered individual genetic findings [43–46]. Alternatively, the medical professionals and research investigators seeking primary genetic tests that may, or likely will, return secondary and individual genetic findings may choose to seek additional information and training relevant to informing their patients and participants in a responsible manner that is informed and sensitive [47,48]. The ACMG is currently developing a set of clinical decision support tools for this purpose [40].

TABLE 18.2 Classification of Individual Genetic Findings

Category	Relevant Individual Genetic Finding
Strong Net Benefit **Disclose** to research participant as an individual genetic finding, unless s/he elected not to know.	• Information revealing a **condition likely to be life-threatening.** • Information revealing a **condition likely to be grave that can be avoided or ameliorated.** • Genetic information revealing significant risk of a condition likely to be life-threatening. • Genetic information that can be used to avoid or ameliorate a condition likely to be grave. • Genetic information that can be used in reproductive decision-making: (1) To avoid significant risk for offspring of a condition likely to be life-threatening or grave or (2) to ameliorate a condition likely to be life-threatening or grave.
Possible Net Benefit **May disclose** to research participant as an individual genetic findings, unless s/he elected not to know.	• Information revealing a nonfatal condition that is likely to be grave or serious but that cannot be avoided or ameliorated, when a research participant is likely to deem that information important. • Genetic information revealing significant risk of a condition likely to be grave or serious, when that risk cannot be modified but a research participant is likely to deem that information important. • Genetic information that is likely to be deemed important by a research participant and can be used in reproductive decision-making: (1) to avoid significant risk for offspring of a condition likely to be serious or (2) to ameliorate a condition likely to be serious.
Unlikely Net Benefit **Do not disclose** to research participant as an individual genetic finding.	• Information revealing a condition that is not likely to be of serious health or reproductive importance. • Information whose likely health or reproductive importance cannot be ascertained.

Source: Wolf SM, et al. Managing incidental findings in human subjects research: analysis and recommendations. J Law Med Ethics 2008;36 (2):219–48, 211.

What Happens After Disclosure of Genetic Findings?

- What if the recipient of the genetic findings cannot afford follow up testing and/or care?
- Has the institution established a fiduciary relationship with the recipient?
- Does the institution now have an obligation to provide follow up care/services to the recipient?

INFORMED CONSENT

Informed consent is an integral part of medical decision-making and decisions to participate in research whether that decision is a consideration of optional procedures or the primary purpose of the test and, therefore, required for discovery of the answers sought. To many of us in science, medicine, and ethics, it seems this is an issue with a simple answer. Of course, the practitioner or investigator has an obligation to seek informed consent regarding anticipated or sought secondary findings and/or foreseen or unforeseen potential individual genetic findings. But some debate still exists regarding both whether consent for the test beforehand and disclosure of the result afterwards must be sought and respected [28,34,40,49–52].

Most notable is the 2013 American College of Medical Genetics Policy Statement that argues that soliciting and honoring patient and/or parent preferences for return of significant incidental results or permitting one to opt-out of receiving reports of these results would be impractical and unreasonably onerous on the laboratory and the clinician and, therefore, should not be practiced [40]. The only option for the patient who does not want to receive individual genetic findings would be to opt-out of sequencing altogether, thereby not having access to the primary sought-after answers for which the sequencing or specific genetic test was intended [52]. There has been a significant response to disregard this recommendation, as well as a call to investigate further all of the ramifications of both sides of the debate [52,53]. The ACMG recommendations do clearly support pre-test and post-test counseling for the patient or legally authorized representative and the referral to appropriate professionals for further testing and or interpretation of results. The statement also affirms the principle of respect for patient autonomy in the provision of information about the implications and limitations of the tests, as well as the risks, benefits, and alternatives [40,51].

The right to informed consent in the clinical and research setting before a procedure is performed [54–58] and the right to refuse treatment or information [59,60] are long-standing ethical standards, supported by legal precedents in many cases. Even the ACMG affirms the right to refuse unwanted individual genetic findings in clinical genomic sequencing in a 2012 statement [40,51,52]. Similarly, the National Heart, Lung, and Blood Institute working group in their 2010 updated *Ethical and Practical Guidelines for Reporting Genetic Research Results to Study Participants* affirmed that genetic results should be offered to study participants if the finding has important health implications for the participant with established and significant risks, the finding is actionable, the test is analytically valid and the study participant has opted to receive his/her individual genetic results as part of the informed consent process. The report states that researcher should outline plans for return of results in consent forms and processes, giving the participant the opportunity to opt-in or opt-out. Although they acknowledge some exceptions may apply where the researcher's IRB should be consulted for guidance, the overarching recommendation is to respect the wishes of the participant [43,50,61].

On December 12, 2013, the Presidential Commission for the Study of Bioethics Issues released its report, *Anticipate and Communicate: Ethical Management of individual genetic findings in the Clinical, Research and Direct-to-Consumer Contexts.* The aim of this report was to complement and broaden the scope of the recommendations made in the October 2012 report, *Privacy and Progress in Whole Genome Sequencing* in regard to the practical, legal and ethical implications of incidental and secondary findings in multiple contexts related to health and well-being [47]. In preparing the October 2012 Commission report, "the Commission's goal was to find the most feasible ways to reconcile the enormous medical potential of whole genome sequencing with the pressing privacy and data issues raised by the rapid emergence of low-cost whole genome sequencing." The Commission was concerned about the variation in protections afforded individuals by states regarding collection and use of genetic data with a focus on both ethical and legal obligations to both patients and research participants, supported by both HIPAA and Common Rule protections and the gaps that exist within and beyond these regulations [47].

The December 2013 report provides the scientific community with five overarching recommendations that are relevant to multiple contexts, focusing on clinical, research, and direct-to-consumer settings as well as nine context-specific recommendations that built upon the more general ones. Informed consent factors prominently between both the general and the specific recommendations and is primary in each listing of context-specific recommendations. These recommendations focus on incidental and secondary findings arising from large-scale genetic sequencing, tests of biological specimens, and imaging. The five broad and overarching recommendations state that:

1. Informed consent and open communication between providers and potential recipients is essential, which necessitates that recipients in any setting be informed about the possibility and disclosure plans for incidental or secondary findings;
2. development of guidelines that categorize likely findings and outline best practices for management by professional groups;
3. funding should be provided for research regarding evolving types and frequencies of findings, potential costs, benefits and harms, and recipient and practitioner preferences about incidental and secondary findings;
4. preparation of materials and enhanced education of practitioners, IRB members, and potential recipients about ethical, practical, and legal considerations raised by these findings is needed; and
5. provision of access to information and guidance needed to make informed decisions about what tests to undergo, what kind of information to seek, and what to do with information once it is received due to the concern that affordable access to care and quality information about incidental and secondary findings, before and after testing, can be potentially lifesaving is essential [47].

Related to informed consent specifically in the clinical and research settings, the focus of this section of our guidance, the commission recommended that the practitioner respect both the right to know about the possibility of individual genetic findings and the choice, whether to receive those results.

Clinical Context: Clinicians should make patients aware that incidental and secondary findings are a possible, or likely, result of the tests or procedures being conducted. Clinicians should engage in shared decision-making with patients about the scope of findings that will be communicated and the steps to be taken upon discovery of individual genetic findings. Clinicians should respect a patient's preference not to know about incidental or secondary findings to the extent consistent with the clinician's fiduciary duty.

Research Context: During the informed consent process, researchers should convey to participants the scope of potential incidental or secondary findings, whether such findings will be disclosed, the process for disclosing these findings, and whether and how participants might opt out of receiving certain types of findings [47].

Preferences of Patients, Participants, and Providers

A number of studies have been undertaken to ascertain the preferences of patients, participants, and providers about returning incidental genetic findings discovered in the course of clinical testing or research endeavors [17,61–64]. Many have found a consistent indication of interest in knowing about individual genetic findings, which has lead, a growing number of federal agencies, expert panels, and authors to recommend that at least some genomic individual genetic findings be made available to individuals [17,62]. This continually growing consensus begs the question of how to address the concern about how best to obtain informed consent for return of individual genetic findings. Expert guidelines, federal regulations, and emerging legal arguments agree that preferences should be obtained, disclosure of foreseeable risks and benefits is required, and other aspects of informed consent are expected regarding return of individual genetic findings. Dissenting positions among experts cite feasibility and cost issues, as well as time constraints with these expectations [62]. Lohn and colleagues report that similar studies reveal that lay persons "emphasized the importance of patient autonomy and empowerment, while the genetics professionals believed that the return of individual genetic findings should be limited due to available resources and the potential to burden patients" [64]. This concern for the burden to patients and participants was echoed in several studies [17,61,62,64].

The Lohn study focused on which factors would influence a genetic professional's (genetic counselors and medical geneticists) decision to disclose individual genetic findings discovered in the course of clinical investigations. In addition to the most common factors: Condition-specific factors, accuracy of the test, and evidence of pathogenicity, a significant number of respondents indicated that pre-test counseling to establish the patient's preferences for disclosure would influence their decision [64]. Similar to other studies, the Lohn survey also asked respondents about which results they felt were most important to disclose. Appelbaum and colleagues have attempted to ascertain the preferences of participants and researchers regarding whether individual genetic finding should be reported, which ones, how those decisions are made and the consequences of doing so, as well as their views on the informed consent process and how it should be conducted. They report that a majority of researchers believe that participants should be able to choose which results they would want to receive from among a list of options. With some variation in degree, researchers and research participants agreed that potential benefits and risks, impact on family, return of results to family upon impairment or death, potential information from subsequent studies of banked tissue or information, data security, and placement of information in their medical record should all be disclosed in an informed consent document [62]. Simon and colleagues survey of IRB chairs revealed similar concerns. This study asked IRB chairs if seven categories of information should be included in a consent document. The categories are listed here in descending order of reported importance: Prospect of and study disclosure policy on genetic individual genetic findings; genetic incidental finding (GIF) management plans; potential clinical significance of GIFs; potential risks of GIF disclosure; elicit GIF disclosure preferences; definition of GIFs; and duration of duty to disclose. A number of recommendations were offered on these and additional topics by the authors [61].

Types of Consent

The different types of consent documents provide a means of communicating information about the study or procedure with varying degrees of specificity. In research, narrow consent documents provide information about the benefits and risks of a specific study, while broad consents provide the opportunity to agree to use of data or samples in future studies with greater flexibility. Research participants may vary in their desire to be contacted for re-consent for each study. While investigators may argue that re-contact is expensive, time-consuming and cumbersome and additionally may create an increased risk of breach of privacy [65].

A number of studies suggest or demonstrate a plan to re-contact the patient or participant if potentially significant findings are discovered to ask the individual at that time if they would like the results confirmed and disclosed, as an alternative to asking up front [42,66], while other advocate strongly for upfront and specific consent to receive results and or disclose results to relatives or physicians [34,67]. Regardless of whether permission is sought up-front or the patient/participant is re-contacted, the consensus practice is to obtain specific consent from the patient/participant for testing, return of results, and potential disclosure to other parties. See Table 18.3 for a summary comparison of informed consent models.

There are a number of suggested informed consent templates tailored to genomic research available both in the literature on the topic and on the websites of relevant professional institutes and tasked working groups. These include guidance from the National Human Genome Research Institute (NHGRI) and the Electronic Medical Records and Genomics (eMERGE) Network Consent and Community Workgroup Informed Consent Task Force and publications from NIH funded studies.

TABLE 18.3 Comparison of Informed Consent Models

	How Informed Consent Is Obtained	Advantages	Disadvantages
General/broad/ blanket Consent	• Permission for all future studies of provided sample given by patient/participant at time of sample collection.	• Minimizes burden to investigators. • Facilitates extensive use of specimens.	• Offers participants very little control over the use of their samples.
Tiered Consent	• Patients/participants are presented with a variety of options from which to grant permission (blanket consent, permission for uses related to original intent/study, specify re-contact all uses).	• Strike a balance between respecting participants and not unduly limiting scientific advancement.	• Allows participant to define level of control with which they are comfortable.
Specific/explicit Consent	• Participants must be re-contacted and asked to give consent for each new use of their specimens.	• Gives participants more control.	• Limits scientific use due to cost and practicality of re-contact.

Consent templates: *[19,41,61,68].*

Source: Mello MM, Wolf LE. The Havasupai Indian tribe case—lessons for research involving stored biologic samples. N Engl J Med 2010;363(3):204–7.

The informed consent elements tailored to genomics research offered by the NHGRI include the expected required elements: Purpose of research project; description of the research procedures; financial compensation, costs and commercialization; potential benefits of participating in the project; potential risks of participating in the project; confidentiality; returning results to research participants; withdrawal; alternatives to participating in the project; voluntary participation; and contact information. In the returning results to participants section of the expanded recommendations, the authors refer to a consensus statement relevant to our purpose and supportive of our conclusions.

Two recommendations address the return of results:

1. Personal genome projects should have an established process approved by a research ethics committee for evaluating whether findings (incidental or otherwise) meet criteria for offering results to individual participants. This process should be highlighted in the initial consent and should acknowledge the participants' rights not to know certain results.
2. The process of identifying and disclosing research results should involve professionals with the appropriate expertise required to provide the participant with sufficient interpretive information. In general, the results offered should be scientifically valid, confirmed, and should have significant implications for the subject's health and well-being. Plans to return other forms of data-such as significant nonhealth-related data should be built into the study design and governance structure

As in many other guidance documents, the authors caution the investigator to consider return of results carefully due to the implications to the recipient's health and well-beings, both positive and negative; the potential to increase stress and anxiety; the harms of incorrect information and subsequent inappropriate testing or medical treatment; and the time and expertise required to avoid these harms [68].

The authors also offer suggestions regarding particular risks that should be addressed that include: broad data sharing, privacy protections, unwanted results, shared risks with family or identifiable populations, uncertainty of meaning related to findings, confidentiality, psychological and social risks, and physical risks of the tests. There is also a suggestion to include a statement regarding GINA protections [68]. This document and others are helpful resources to help the investigator or clinician consider the important elements of informed consent for genetic studies as well as a useful resource for sample language [19,68].

FRAMEWORK FOR DISCLOSING INDIVIDUAL GENETIC INFORMATION

A disclosure framework needs to balance many conflicting interests. This framework strives for practical policies that offer the best care of our patients, maximize clinical and societal benefit, all while being feasible and sustainable. Table 18.4 summarizes overarching key elements for an ethical disclosure framework. These elements include adequate informed consent; adherence to applicable laws/regulations; a centralized, streamlined process; and evidence-based oversight and genetic counseling.

TABLE 18.4 Key Points of an Ethical Genetic Disclosure Framework

1. Patients and study participants must provide informed consent for the release of their genetic information.
2. Patients and study participants must provide valid HIPAA authorization to disclose protected health information.[a]
3. Signed release forms from patient/research participant's relatives indicating that they wish to receive genetic information.
4. Centralized notification service is created that will set disclosure policies and implement the disclosure process. Existing institutional structures/procedures/personnel may be incorporated into this entity.
5. An in-house scientific committee will create, maintain, and regularly update a List of actionable findings [40,69].
6. Genetic counseling services or referral to local services will be offered to individuals who are provided actionable findings.

[a]*See 45 CFR 164.508(c) for further guidance.*

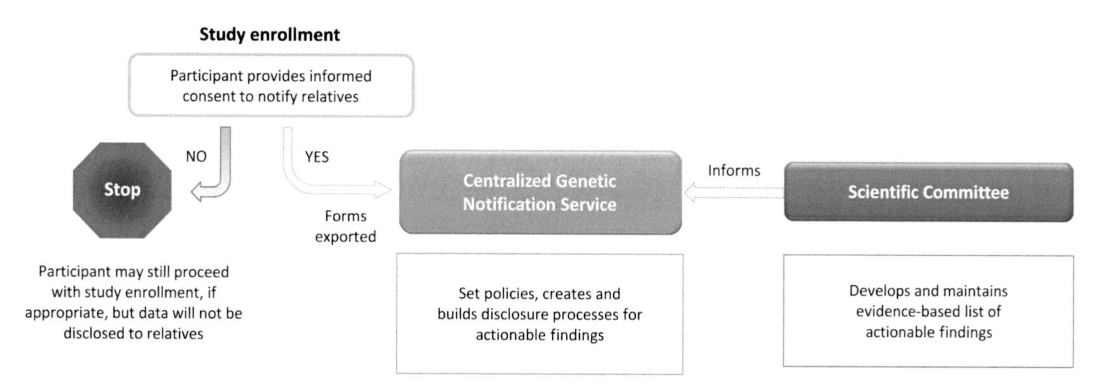

FIGURE 18.1 **Initiation of disclosure pathway for prospective research participants during study enrollment.** The presumptive research participant will provide proper informed consent for disclosure individual genetic information to first order relatives. Opting out of disclosure will not exclude a potential research participant. Documentation of participant preferences will be submitted to a centralized notification service. The notification service develops and implements appropriate disclosure and counseling procedures. An in-house, multidisciplinary scientific committee will determine the actionable findings based upon existing guidelines, standards of practice, and evidence.

Disclosure of Individual Genetic Findings—Research

Potential research participants should be approached during study enrollment with an option to have individual genetic information disclosed to them and their relatives (Fig. 18.1).

If they choose not to receive or disclose results, then they shall proceed with study enrollment as appropriate. Deciding not to participate in genetic information disclosure should not be an ancillary exclusion criterion for participation in a research protocol.

Research results may be offered to participants when they have been informed and have consented to both participate in research and receive such findings (Fig. 18.2).

Participants should be aware that these findings are investigational in nature and that results will be validated in a CLIA-certified lab prior to disclosure. The notification service will receive information regarding the findings and evaluate whether or not they are

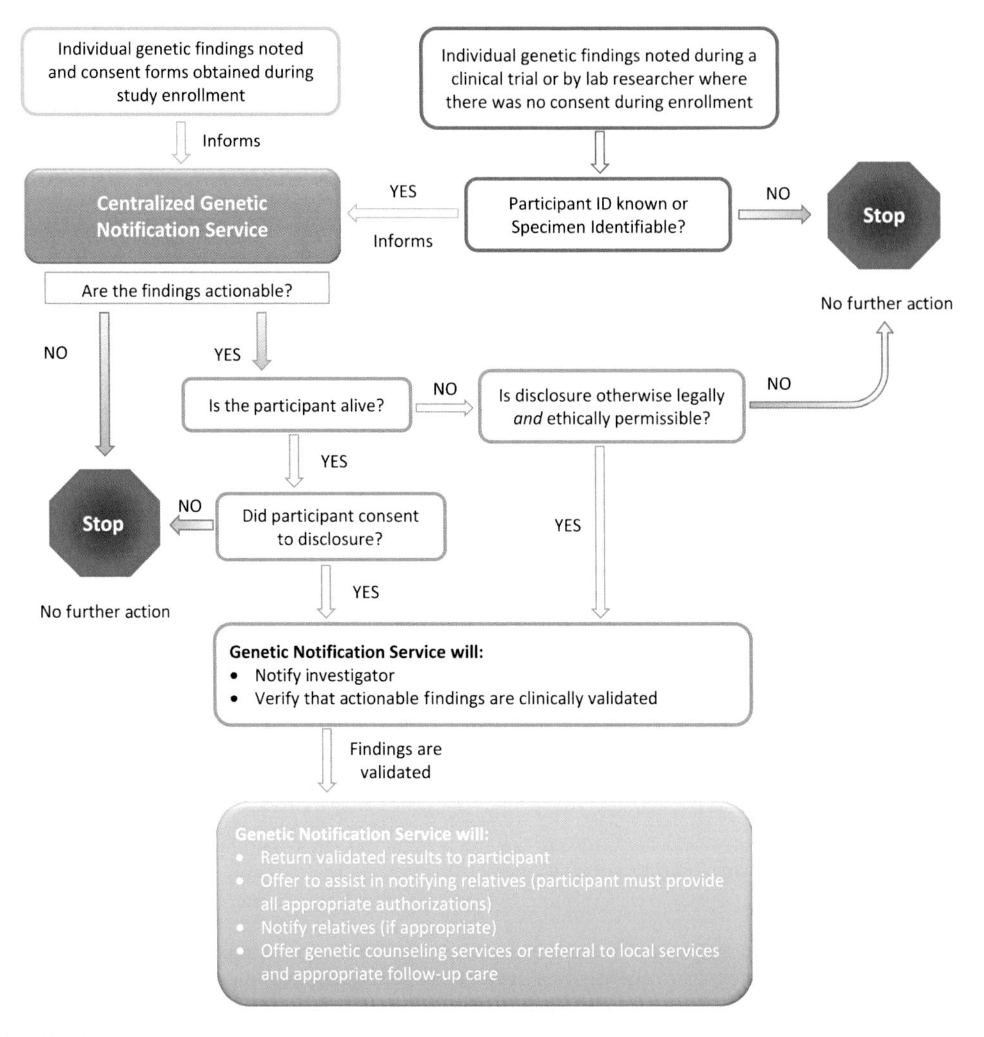

FIGURE 18.2 **Disclosure framework during the conduct of research.** Individual genetic findings, unrelated to study aims, will be sent to the Genetic Notification Service (GNS). The GNS will assess whether or not the findings are actionable. If so and if the participant provided informed consent to receive those results, the GNS will notify the investigator and verify that the results have been clinically validated. If the participant is deceased, it may still be legally and ethically permissible to disclose actionable findings. The GNS will develop policies for those circumstances. Upon validation testing, the participant or personal representative may opt to allow the notification service to disclose findings to relatives. The service will offer or provide referrals to genetic counseling services for individuals that have received actionable findings.

actionable based upon the List created by the Scientific Committee. If the findings are actionable and disclosure is authorized, the notification service may continue the disclosure process.

Only clinically validated results may be returned to a participant and their relatives. Upon receipt of actionable results, the participant shall be offered genetic counseling services (or referral). If the patient would like assistance in notifying first-degree relatives, they must complete a valid HIPAA authorization. The participant shall indicate which first-degree relatives are to be contacted and have those individuals complete a release to receive actionable findings. This ensures that actionable finding are disclosed per the participants wishes and recipients indeed wish to acquire that information for their health needs. The Genetic Notification Service (GNS) will collect and store these forms. Both the participant and relatives will be offered either in-house genetic counseling or referral to local services.

It is possible that individual genetic findings may arise from the conduct of clinical trials or other research on banked specimens where consent and authorization were not acquired. Information may be sent to the GNS if the individual participant is known or samples are able to be de-identified. If the participant is living, the GNS may notify him/her in an appropriate fashion that findings are available. The participant will then have an opportunity to decide whether or not to receive such information and/or provide authorization and release documents.

If the participant is deceased, then information may be disclosed where legally *and* ethically appropriate, e.g., 45 CFR 164.502(f)-(g), 45 CFR 164.510(b). Interpretations of laws, such as HIPAA, GINA, and ACA, may inform policies to permit disclosure of a decedent's individual genetic information, but they shall be interpreted through a lens of prudent, ethical reflection. *Legal permissibility should not be equated to ethical permissibility.* Further ethical analysis may be conducted at such time as these policies are developed. If the GNS determines that disclosure exceptions are deemed to apply to a decedent's individual genetic information, then the process continues as described above.

Disclosure of Individual Genetic Findings—Clinical

Individual genetic findings may be disclosed to patients and their relatives. The GNS may facilitate the disclosure of individual genetic information during the course of clinical care and treatment (Fig. 18.3).

In the typical context, a patient will provide informed consent for genetic testing when that procedure is appropriate. Upon receipt of genetic test results, the physician may notify the GNS who will in turn, inform the physician as to which findings are actionable. The physician may choose to either discuss these findings with the patient or allow the GNS to contact the patient. Although the GNS is empowered to discuss individual genetic findings with the patient, the Service will defer to the physician's preference so as to not interfere with the doctor–patient relationship.

Regardless of who will discuss the findings with the patient, the first inquiry will be whether or not the patient wishes to have this information. It is possible that this conversation can be held during the informed consent process. If the patient does not wish to be notified of the findings, then there is no further action needed in this regard. If the patient does want to be informed of the actionable findings, then they will be disclosed along with an offer of in-house genetic counseling or referral to local services.

The patient may then be asked whether or not he/she would like to disclose findings to relatives. The patient will be encouraged to discuss findings with the appropriate relatives.

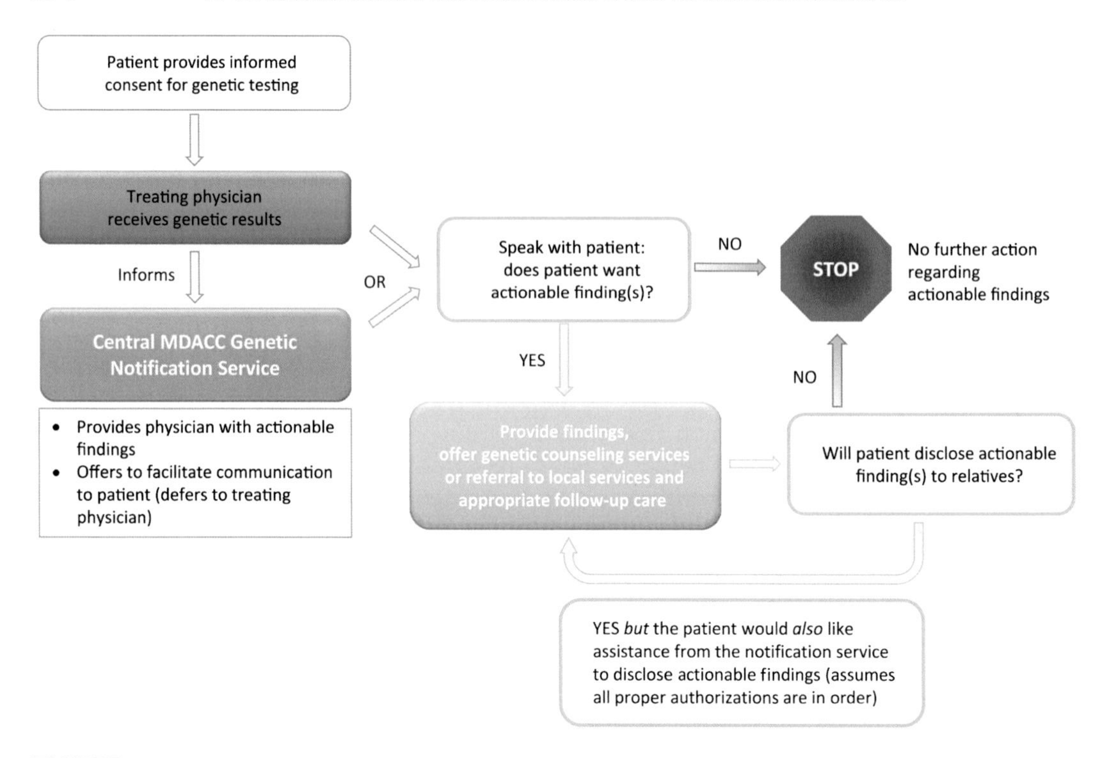

FIGURE 18.3 **Disclosure framework for clinical treatment.** Patients will provide informed consent for genetic testing. The treating physician may inform the Genetic Notification Service (GNS) genetic diagnostic results. The GNS will provide a list of actionable findings to the physician. The physician may then discuss and offer the findings to the patient, as appropriate. The patient shall receive all appropriate follow up care and genetic counseling services. The GNS may also offer to notify the patient giving all due deference to the treating physician's preferences. A patient may then be offered the option of disclosing actionable findings to relatives. The GNS may offer to disclose individual genetic results to relatives if the patient wishes that information disclosed and does not wish to inform their relatives personally. The patient must provide all appropriate authorization and documentation to the GNS in order for relatives to be contacted.

The GNS may notify the appropriate relatives if the patient him/herself both wants the information disclosed but does not wish to personally discuss the findings. The patient shall complete all proper authorizations, including release documents signed by the family members who wish to receive the individual genetic findings. Regardless of who notifies the relatives, we will offer of in-house genetic counseling or referral to local services.

ORGANIZATIONAL ETHICS AND SYSTEMS CONSIDERATIONS FOR IMPLEMENTING A GENETIC INFORMATION DISCLOSURE PROGRAM

There is a need for a plan and implementation process. This plan should include staff with expertise and ability to actually conduct the notifications. This is likely to be genetic counselors, even for initial conversation when people call with questions about whether they should be tested as well as after testing if provided in-house. Other practical concerns include financial resources, personnel to run tests, personnel to appropriately return results, time to run additional tests to confirm findings and patient access to info, and additional tests or specialists. One group, representative of clinical and research expertise, is needed to determine which findings are "actionable." This group should not only determine the list of actionable findings but also periodically review and update said list to keep current with our understanding of genetic determinants of disease. Another, independent group is needed to administrate the process of dealing with the genetic information and maintain a list of what notifications waiting to be made, have been made, and all other administrative functions. This group should cover both clinical and research notifications so that there is consistency and to prevent a duplicative structure.

It will be important to consider the most appropriate means to transmit results. Possible mechanism may include a letter via mail, a phone call from a clinician or genetic counselor, or even in-person (with or without referral). The measure of appropriateness should include practical, economic and psychosocial considerations. These can come from existing resources of the institution and be functioned through existing departments or services. There will be expense related to development of the infrastructure to cover these new elements of the process. The legal and compliance impact of a notification service should also be evaluated. Such considerations include protections from discrimination and privacy concerns, e.g., GINA, HIPAA, and ACA [41,65].

CONCLUSION

In 1965, an engineer by the name of Gordon E. Moore predicted that computational power, as measured by transistor density, would increase in a linear fashion; doubling every two years [70]. This speaks to the dramatic increase in the pace of technological development we have observed over the last half-century. A similar trend has emerged in not only growth of biotechnology but also access to vast amounts of information. As we race toward the $1000 genome, we are only beginning to capture a glimpse of the clinical utility contained in these data. The imperative to leverage this information for the benefit of not only our patients but society in general will only grow. We must exert the same measure of forethought as those who were involved in the early days of the genomic revolution to anticipate the consequences of this technology as we move forward in its implementation [1,2]. Table 18.5 outlines a detailed summary of practice points for consideration in three key areas:

1. Organization/institutional elements and commitments necessary to develop an efficient and effective disclosure policies and process.
2. Disclosure recommendations are provided for several contexts, including returning research results, clinical results and results to relatives.
3. Informed consent requirements to ensure patient autonomy is honored, recipients are provided appropriate information if desired, and informational risks are mitigated.

TABLE 18.5 A Summary of Practice Points for Developing an Ethical Disclosure Framework of Individual Genetic Information to a Patient/Research Participant or Family

Organizational Elements

The institution recognizes and makes a commitment to a duty to provide genetic counseling resources for recipients of individual genetic findings, in the form of offering direct services or referral to local resources.

Create an in-house scientific advisory committee should be established charged with *determining actionable findings*. The role of the committee includes:

1. Conduct evidence-based review of genetic risk factors.
2. Determine actionable individual genetic findings.
3. Regularly review and update actionable finding list.
4. Provide guidance to clinicians, investigators, research institutions, and IRBs regarding when a genetic result is well enough understood and has sufficiently serious clinical implications to justify returning genetic results to patients and study participants.
5. Committee membership should be made up of a multidisciplinary constituency with the following expertise: Clinical genetics/genomics, Research genetics/genomics, Epidemiology, Genetic counseling, Ethics.

Create a centralized administrative service charged with developing the *process* and *policies* of individual genetic finding disclosure. The duties of the service include:

1. Determine which actionable findings may be disclosed to the patient/research participant and/or relatives.
2. Determine which actionable findings may *not* be disclosed to the patient/research participant and/or relatives.
3. Establish a mechanism to determine who will receive information.
4. Organize how findings will be disclosed to the appropriate recipients.
5. Develop policies for quality improvement and oversight.
6. Coordinate the infrastructure and personnel to execute disclosure process.
7. Administer the disclosure program.
8. Develop an appeals process for exceptions to disclosure decisions.
9. Leadership of the service should include a multidisciplinary team consisting of the following expertise: Clinical genetics/genomics, Research genetics/genomics, Pathology, Behavioral Health, Epidemiology, Genetic counseling, Ethics, Legal, Regulatory, Compliance or Risk, Institutional Administration, Patient/Community members.

Engage institutional leadership to obtain their support for the development, design and implementation of the administrative notification plan, including:

1. Finalizing disclosure policies.
2. Conduct needs assessment for implementing disclosure program.
3. Commitment to Personnel.
4. Developing the needed infrastructure.
5. Program scope and cost (direct and indirect).
6. Create implementation plan (location, timeframe, oversight).
7. Create plan for program sustainability.

Disclosure Recommendations

Only *actionable* individual genetic findings should be returned to the patient/research participant and relatives thereof where appropriate. An actionable genetic individual genetic finding is one where there are established therapeutic or preventive interventions or other available actions that have the potential *to change the clinical course of the disease* [6].

(*Continued*)

TABLE 18.5 (Continued)

Individual genetic results should be offered to *study participants* in a timely manner if they meet all of the following criteria [71]:

1. During the informed consent process or subsequently, the patient/study participant has opted to receive his/her individual genetic results.
2. The genetic finding is actionable, that is, there are established therapeutic or preventive interventions or other available actions that have the potential to change the clinical course of the disease.
3. The test is analytically valid and the disclosure plan complies with all applicable laws and standards of practice, e.g., CLIA-certified laboratory.

Individual genetic results may be offered to *patient/study participants' family* in a timely manner if they meet all of the following criteria [71]:

1. During the informed consent process or subsequently, the patient/study participant has opted to disclose his/her individual genetic results to family members *and* the family member has provided a release to be contacted. Disclosure will be limited to first-degree relatives and determined by the patient/research participant.
2. The genetic finding is actionable, that is, there are established therapeutic or preventive interventions or other available actions that have the potential to change the clinical course of the disease.
3. The test is analytically valid and the disclosure plan complies with all applicable laws and standards of practice, e.g., CLIA-certified laboratory.

Investigators may return individual genetic results to *study participants* if the criteria for an obligation to return results are not satisfied (see Recommendation 4) but all of the following apply:
1. The investigator's Institutional Review Board (IRB) has approved the disclosure plan.
2. The test is analytically valid and the disclosure plan complies with all applicable laws.
3. During the informed consent process or subsequently, the study participant has opted to receive his/her individual genetic results.

Informed Consent Recommendations

Clinicians and Investigators shall provide for informed consent for *both* the primary investigation and the possibility of individual genetic findings. The consent should also provide for an opportunity to *opt-in* for receipt of individual genetic findings.

The consent language regarding individual genetic findings should include the following information:

1. Nature of the testing that will create the possibility or likelihood of individual genetic findings.
2. Risks, benefits, and alternatives to this testing.
3. Risks and benefits of the return of results that include individual genetic findings (physical, psychosocial, emotional, and privacy).
4. Implications for the individual and his/her health and well-being as well as the implications for first-degree family members (medical, psychological, financial—employment, insurance).
5. Type of findings that will be disclosed.
6. How the information will be disclosed, and if there is an opportunity to release information to relatives, pre or post mortem or expectation that information will be released to the individual's physician.
7. Choosing not to disclose individual genetic findings will not otherwise negatively impact access to clinical treatment or study participation.

References

[1] Berg P, et al. Asilomar conference on recombinant DNA molecules. Science 1975;188(4192):991–4.
[2] Berg P, et al. Summary statement of the Asilomar conference on recombinant DNA molecules. Proc Natl Acad Sci USA 1975;72(6):1981–4.
[3] Annas GJ. Privacy rules for DNA databanks. Protecting coded 'future diaries'. JAMA 1993;270(19):2346–50.
[4] Wolf SM, et al. Managing incidental findings in human subjects research: analysis and recommendations. J Law Med Ethics 2008;36(2):219–48, 211.
[5] D.o.H.a.H. Services, Editor. Anticipate and communicate: ethical management of incidental and secondary findings in the clinical, research, and direct-to-consumer contexts. Presidential Commission for the Study of Bioethical Issues: Washington, DC; 2013.
[6] Fabsitz RR, et al. Ethical and practical guidelines for reporting genetic research results to study participants: updated guidelines from a National Heart, Lung, and Blood Institute working group. Circ Cardiovasc Genet 2010;3(6):574–80.
[7] Freedman B, Fuks A, Weijer C. Demarcating research and treatment: a systematic approach for the analysis of the ethics of clinical research. Clin Res 1992;40(4):653–60.
[8] Oberman M. Mothers and doctors' orders: unmasking the doctor's fiduciary role in maternal–fetal conflicts. Nw UL Rev 2000;94(2):451–501.
[9] Morreim EH. The clinical investigator as fiduciary: discarding a misguided idea. J Law Med Ethics 2005;33 (3):586–98.
[10] Carter MA. Ethical aspects of genetic testing. Biol Res Nurs 2001;3(1):24–32.
[11] Green MJ, Botkin JR. "Genetic exceptionalism" in medicine: clarifying the differences between genetic and nongenetic tests. Ann Intern Med 2003;138(7):571–5.
[12] Bork RH. The challenges of biology for law. Tex Rev Law Polit 1999;4(1):6.
[13] Richardson HS, Belsky L. The ancillary-care responsibilities of medical researchers: an ethical framework for thinking about the clinical care that researchers owe their subjects. Hastings Center Report 2004;34(1):25–33.
[14] Richardson HS. Incidental findings and ancillary-care obligations. J Law Med Ethics 2008;36(2):256–70, 211.
[15] Miller FG, Mello MM, Joffe S. Incidental findings in human subjects research: what do investigators owe research participants? J Law Med Ethics 2008;36(2):271–9, 211.
[16] Ruiz-Canela M, Valle-Mansilla JI, Sulmasy DP. What research participants want to know about genetic research results: the impact of "genetic exceptionalism". J Empir Res Hum Res Ethics 2011;6(3):39–46.
[17] Biesecker LG. Opportunities and challenges for the integration of massively parallel genomic sequencing into clinical practice: lessons from the ClinSeq project. Genet Med 2012;14(4):393–8.
[18] Beskow LM, Dean E. Informed consent for biorepositories: assessing prospective participants' understanding and opinions. Cancer Epidemiol Biomarkers Prev 2008;17(6):1440–51.
[19] Beskow LM, Smolek SJ. Prospective biorepository participants' perspectives on access to research results. J Empir Res Hum Res Ethics 2009;4(3):99–111.
[20] Kaufman D, et al. Ethical implications of including children in a large biobank for genetic-epidemiologic research: a qualitative study of public opinion. Am J Med Genet C Sem Med Genet 2008;148C(1):31–9.
[21] Murphy J, et al. Public expectations for return of results from large-cohort genetic research. Am J Bioeth 2008;8(11):36–43.
[22] Partridge AH, Winer EP. Informing clinical trial participants about study results. JAMA 2002;288(3):363–5.
[23] Wendler D, Emanuel E. The debate over research on stored biological samples: what do sources think? Arch Intern Med 2002;162(13):1457–62.
[24] Facio FM, et al. Motivators for participation in a whole-genome sequencing study: implications for translational genomics research. Eur J Hum Genet 2011;19(12):1213–17.
[25] Murphy J, et al. Public perspectives on informed consent for biobanking. Am J Public Health 2009; 99(12):2128–34.
[26] Kohane IS, et al. Medicine. Reestablishing the researcher–patient compact. Science 2007;316(5826):836–7.
[27] Illes J, et al. Ethics. Incidental findings in brain imaging research. Science 2006;311(5762):783–4.
[28] Wolf SM, et al. Managing incidental findings and research results in genomic research involving biobanks and archived data sets. Genet Med 2012;14(4):361–84.
[29] Fernandez CV, et al. Recommendations for the return of research results to study participants and guardians: a report from the Children's Oncology Group. J Clin Oncol 2012;30(36):4573–9.

[30] Mello MM, Wolf LE. The Havasupai Indian tribe case—lessons for research involving stored biologic samples. N Engl J Med 2010;363(3):204–7.
[31] Sterling RL. Genetic research among the Havasupai—a cautionary tale. Virtual Mentor 2011;13(2):113–17.
[32] Affleck P. Is it ethical to deny genetic research participants individualised results? J Med Ethics 2009;35 (4):209–13.
[33] Beskow LM, Burke W. Offering individual genetic research results: context matters. Sci Transl Med 2010;2 (38) 38cm20.
[34] Bredenoord AL, et al. Disclosure of individual genetic data to research participants: the debate reconsidered. Trends Genet 2011;27(2):41–7.
[35] Dressler LG. Disclosure of research results from cancer genomic studies: state of the science. Clin Cancer Res 2009;15(13):4270–6.
[36] Fernandez C. Public expectations for return of results—time to stop being paternalistic? Am J Bioeth 2008;8 (11):46–8.
[37] Fernandez CV, Kodish E, Weijer C. Informing study participants of research results: an ethical imperative. IRB 2003;25(3):12–19.
[38] Shalowitz DI, Miller FG. Disclosing individual results of clinical research: implications of respect for participants.. JAMA 2005;294(6):737–40.
[39] Evans JP, Rothschild BB. Return of results: not that complicated? Genet Med 2012;14(4):358–60.
[40] Watson M. Incidental findings in clinical genomics: a clarification. Genet Med 2013;15(8):664–6.
[41] Lolkema MP, et al. Ethical, legal, and counseling challenges surrounding the return of genetic results in oncology. J Clin Oncol 2013;31(15):1842–8.
[42] Chan B, et al. Genomic inheritances: disclosing individual research results from whole-exome sequencing to deceased participants' relatives. Am J Bioeth 2012;12(10):1–8.
[43] Abdul-Karim R, et al. Disclosure of incidental findings from next-generation sequencing in pediatric genomic research. Pediatrics 2013;131(3):564–71.
[44] Costain G, Bassett AS. Incomplete knowledge of the clinical context as a barrier to interpreting incidental genetic research findings. Am J Bioeth 2013;13(2):58–60.
[45] Garrett JR. Reframing the ethical debate regarding incidental findings in genetic research. Am J Bioeth 2013;13(2):44–6.
[46] Ross KM, Reiff M. A perspective from clinical providers and patients: researchers' duty to actively look for genetic incidental findings. Am J Bioeth 2013;13(2):56–8.
[47] Committee on Bioethics. Ethical and policy issues in genetic testing and screening of children. Pediatrics 2013;131(3):620–2.
[48] Borgelt E, Anderson JA, Illes J. Managing incidental findings: lessons from neuroimaging. Am J Bioeth 2013;13(2):46–7.
[49] Bombard Y, Robson M, Offit K. Revealing the incidentalome when targeting the tumor genome. JAMA 2013;310(8):795–6.
[50] The Electronic Medical Records and Genomics (eMERGE) Network Consent and Community Consultation Workgroup Informed Consent Task Force. Consent form examples and model consent language. Issues in genetics: informed consent for genomics research, <http://www.genome.gov/Pages/PolicyEthics/InformedConsent/eMERGEModelLanguage2009-12-15.pdf>; 2009 [accessed 26.9.13 and 15.10.13].
[51] McGuire AL, et al. Ethics and genomic incidental findings. Science 2013;340(6136):1047–8.
[52] Wolf SM, Annas GJ, Elias S. Patient autonomy and incidental findings in clinical genomics. Science 2013;340 (6136):1049–50.
[53] Klitzman R, Appelbaum PS, Chung W. Return of secondary genomic findings vs patient autonomy: implications for medical care. JAMA 2013;310(4):369–70.
[54] Pratt v. Davis. In: Ill App. Illinois Appellate Court; 1905. p. 161.
[55] Schloendorff v. Society of New York Hospital, New York. New York Court of Appeals; 1914. p. 125.
[56] Salgo v. Leland Stanford Jr. University Board of Trustees. In: Pacific reporter. Court of Appeals California; 1957. p. 170.
[57] Nathanson v Kline. In: Pacific reporter. Kansas Supreme Court; 1960. p. 1093.
[58] Office of the Secretary. Protection of human subjects; Belmont Report: notice of report for public comment. Fed Regist 1979;44(76):23191–7.

[59] In re Quinlan. In: Atlantic reporter. Supreme court of New Jersey; 1976. p. 647.
[60] UNESCO Editor. Universal declaration on the human genome and human rights. United Nations; 1997.
[61] Simon CM, et al. Informed consent and genomic incidental findings: IRB chair perspectives. J Empir Res Hum Res Ethics 2011;6(4):53–67.
[62] Appelbaum PS, et al. Informed consent for return of incidental findings in genomic research. Genet Med 2013;16(5):367–73.
[63] Bennette CS, et al. Return of incidental findings in genomic medicine: measuring what patients value—development of an instrument to measure preferences for information from next-generation testing (IMPRINT). Genet Med 2013;15(11):873–81.
[64] Lohn Z, et al. Genetics professionals' perspectives on reporting incidental findings from clinical genome-wide sequencing. Am J Med Genet A 2013;161(3):542–9.
[65] Hudson KL. Genomics, health care, and society. N Engl J Med 2011;365(11):1033–41.
[66] Beskow LM, et al. Offering aggregate results to participants in genomic research: opportunities and challenges. Genet Med 2012;14(4):490–6.
[67] Rothstein MA. Disclosing decedents' research results to relatives violates the HIPAA privacy rule. Am J Bioeth 2012;12(10):16–17.
[68] National Human Genome Research Institute. Informed consent elements tailored to genomic research. Issues in genetics: informed consent for genomics research, <http://www.genome.gov/27026589>; 2012 [accessed 19.05.12 and 15.12.13].
[69] National Human Genome Research Institute. Informed consent. Issues in genetics: health issues in genetics, <http://www.genome.gov/10002332>; 2012 [accessed 28.2.12 and 1.8.13].
[70] Moore GE. Cramming more components onto integrated circuits. Proceedings of the IEEE 1998;86(1):82–5.
[71] Fabsitz RR, et al. Ethical and practical guidelines for reporting genetic research results to study participants: updated guidelines from a National Heart, Lung, and Blood Institute working group. Circ Cardiovasc Genet 2010;3(6):574–80.

CHAPTER

19

Ethical Issues of Cancer Center Administrators

Frank R. Tortorella

The University of Texas MD Anderson Cancer Center, Houston, TX, United States

INTRODUCTION

"Ethics is about doing what's right in the right way with equity and justice" [1]. Ethical decisions are not always time-sensitive dramas with life-threatening consequences; many ethical decisions are made by health care administrators in their daily routine, ensuring that certain processes are in place for staff to provide their services and that budgets are sufficient to provide the appropriate amount and type of resources. The Institute of Medicine's (IOM) 1999 report titled "To Err is Human: Building a Safer Health System" documented the need for improvement in quality and safety and provided specific recommendations referred to as the six aims [2]. The goal of health care administrators is to make the right ethical decisions and adhere to the six aims when designing, implementing, and improving systems and processes. The six aims are for health care to be safe, effective, patient-centered, timely, efficient, and equitable [2]. The IOM report also gave further guidance to health care administrators by including 10 rules for care delivery redesign [3]. If an administrator faces a conflict between quality and safety versus productivity, quality and safety should always be paramount when making the best ethical and business decision.

Health care administrators address ethical challenges each day resulting from conflicts in "competing values, such as personal, organizational, professional, and societal values" [4]. The top two ethical issues for senior health care executives from a 2010 survey conducted by the American College of Healthcare Executives (ACHE) were addressing medical and management issues and creating an environment where ethical issues are openly addressed [5]. The Ethical Policy Statement of ACHE states that the health care administrator should refer to the institution's ethical policies, ethics committee, and ethics consultation service when making ethical decisions [4].

A significant ethical challenge for all health care administrators is to create a culture of safety and quality so that the health care team is effective and responsive to patients and

Ethical Challenges in Oncology.
DOI: http://dx.doi.org/10.1016/B978-0-12-803831-4.00019-1

their family members' needs. The criticality for the health care team in a cancer center to listen to patients and family members is illustrated in the story of Betsy Lehman, a medical reporter, who died in 1994 at a world-renowned cancer center due to continued overdose of cancer medication [1]. The patient and family told the staff that something was not right but no action was taken; the medical review of this case found that the patient died because the health care team did not respond to the patient's and family member's concerns [1]. Creating a patient- and family-centered culture is just one of the myriad of ethical challenges health care administrators face.

This chapter will discuss some of the major ethical responsibilities that all health care administrators have to address. Then the focus will shift to some of the ethical challenges specific to the role of the cancer center administrator. The ethical issues include the need to ensure the appropriate utilization of resources while providing essential services for cancer patients and the availability of sufficient and capable staff to provide services. Additionally, the special psychosocial care and support services important for the cancer patient are discussed. Finally, the priority for cancer center administrators to create a patient- and family-centered approach to health care will be emphasized.

ETHICAL RESPONSIBILITIES OF ALL HEALTH CARE ADMINISTRATORS

All health care administrators have the challenge of balancing ethical issues with running an efficient organization. The concept of "no margin, no mission" is the idea that a hospital must have a healthy financial margin to sustain itself as an organization to deliver its mission [6]. The ACHE Ethical Policy Statement outlines its position to "Promote decision-making that results in the appropriate use of power while balancing individual, organizational, and societal issues" [7]. All health care administrators should "model ethical behavior and instill a culture in which unethical behavior is not tolerated" [8]. In fact, the ACHE has published "The ACHE Ethics Self-Assessment" tool for administrators to identify opportunities for improvement in adhering to ethical principles [9]. The administrator has to contend with the realities of limited resources when trying to fulfill the needs of the patient and institution without compromise. Health care administrators must manage the ethical responsibilities in all aspects of their business in defining their leadership approach; outlining the institution's mission and vision; setting the strategic plan and deploying throughout the organization; managing all the elements of human resources to ensure appropriate staffing levels and capabilities; identifying and meeting customer requirements; utilizing a performance improvement model to drive quality, safety, and efficiency; using data to manage the complex health care operations; and complying with regulatory standards while recognizing the necessity of achieving business results [10]. In all of these components of the health care administrators' responsibilities, ethical issues can arise.

ETHICAL ISSUES FACED BY CANCER CENTER ADMINISTRATORS

Cancer centers are specialty hospitals with a number of unique characteristics due to the type of patients and the services required to obtain this status. In 1971, the first 15 cancer centers were designated by the National Cancer Institute as part of the National Cancer Act [11]. In 2015, there were 68 designated cancer hospitals in the United States [11]. Cancer centers who receive this designation must conduct "clinical research, training, and demonstration of advanced diagnostic and treatment methods relating to cancer" [11]. The administrator of the cancer center has to budget for the cost of the additional services for cancer patients while running an efficient and effective organization to sustain the institution's mission.

Cancer patients have benefited from the Patient Protection and Affordable Care Act (PPACA) passed in 2010, preventing denial of health insurance based on pre-existing conditions [12].

Health care organizations, including cancer centers, can now receive reimbursement from these patients' health insurance, which provides these facilities the funds to pay for the expense associated with the care for these previously uninsured patients. The ethical implications are that the PPACA allows patients with a pre-existing condition to obtain health insurance and, thus, improves the public health issue of access to care for cancer patients. Other ethical imperatives include ensuring the appropriate number of and level of training of staff and cultivating patient- and family-centered care in cancer centers. This chapter illustrates key ethical issues and some necessary components to address these ethical issues (Table 19.1).

TABLE 19.1 Key Ethical Issues and Necessary Components

Ethical Issues	Some Necessary Components to Address the Ethical Issue
Appropriate Utilization of Resources	• Process for case managers to monitor acute care criteria for appropriate bed utilization. • Process to review and determine appropriate level of care transitions for patients who are uninsured, under-insured, or have exhausted their health insurance benefits. • Process for physicians and others to review cases that do not meet medical necessity for care. • Process for clinical team to address medical futility. • Focus services to areas of expertise and refer certain services to external experts.
Appropriate Resources Available	• Psychosocial care needed for social work counseling, advance care planning, spiritual care services, etc. • Room service to promote optimal nutritional intake for cancer inpatients. • Healthy food options in the cancer center cafeterias to promote healthy eating for outpatients and staff. • Language assistance to ensure accurate communication between the limited English proficient patient and the clinical team. • Electronic health record to document and provide easy access to medical and psychosocial care information with policies to address "copy and paste" issues.
Appropriate Staffing	• Low turnover and retention of staff to accumulate the experience to care for cancer patients. • Carefully determine appropriateness of temporary labor for certain jobs with regards to skill level and education on privacy issues. • Productivity adjustments working with certain types of cancer patients.
Appropriate Focus on Patient- and Family-Centered Care	• Patient Family Advisory Council to listen and solicit input from patients and family members. • Institutional self-assessment tool to identify opportunities to build patient- and family-focused culture.

UTILIZATION OF RESOURCES

The cancer center administrator has the ethical obligation to efficiently use resources for patient care to sustain the organization. To systematically monitor resources and achieve efficiency, the administrator must develop and implement a robust utilization management program. Key components of utilization management include a Case Management department with sufficient staffing and software to monitor oncology-specific criteria to ensure patients are in the right level of care at the right time during their hospital stay; a medical director for Case Management to provide clinical leadership; and a Utilization Management Committee to monitor use of institutional resources and take appropriate action. The cancer center administrator must have processes in place to monitor inpatient stays and identify if there are nonmedical reasons preventing a patient's discharge from the hospital and to address these issues in a timely manner. Also, the cancer center administrator must have processes in place to obtain initial authorizations, monitor continued inpatient stay from insurance companies (as required), and alert when reaching lifetime limits on insurance payment. There are cancer patients who have an extended inpatient stay, and the documentation in their electronic health record (EHR) must meet insurance companies' inpatient acute care criteria to ensure reimbursement while the patient remains in the inpatient setting. Sometimes, the medical director of Case Management must review a patient's case with the involved physicians to determine if the patient's acuity requires the patient to be transitioned to a lower level of care, discharged, or to remain in the hospital. The primary function of the institution's Utilization Management Committee is to assure appropriate allocation of institutional resources for the delivery of care to patients. The administrator must provide the necessary resources to mine the appropriate utilization data for this committee to analyze and determine if the extended lengths of stay for particular cancer patients is warranted. The administrator must work with the multidisciplinary team to establish a policy delineating the appropriate membership and functions of this committee [13].

The case manager has to meet patient and physician demands in real-time. There are numerous details to comply with, such as the patients' insurance, and managing the medical needs for the safe movement of patients across the continuum of care. Case managers must also partner with physicians and patients to keep them up-to-date and knowledgeable about the patient's insurance benefits as well as limitations; however, the ethical concern must not be lost in the maze of insurance requirements. As Darlene M. Stromstad, FACHE, president and CEO of Waterbury Hospital, Connecticut stated as a participant on an ACHE ethics panel: "The one ethical thing we must do is to make a commitment to our patients that we will keep them at the center of everything we do" [8]. One key duty of the case manager is to monitor daily the condition of the cancer patients to make sure that they still meet acute care criteria for inpatient hospitalization as well as meet the requirements of the patient's health insurance requirements. The clinical review is to ensure the safe and appropriate care of the patient. The health insurance review is necessary so that the patient's insurance does not deny payment for the services due to technical or administrative reasons. When the patient's attending physician decides to discharge the patient, the case manager must follow-up on numerous details to transition the patient home or to the appropriate medical setting. The cancer center administrator must ensure that case management has sufficient staff to facilitate the safe transfer of patients to the right level of care.

Ethical dilemmas can arise when a physician or patient desires limitless care, as the cancer center does not have endless resources. The cancer center administrator is ultimately responsible to ensure that the cancer center stays fiscally sound to continue to serve the community while providing the appropriate services for the individual cancer patient [14]. Issues of medical futility may arise in these cases when discussing whether "an intervention is futile if it will not achieve its intended goal" [15]. Another way to view medical futility is a "meaningful cutoff point for when, in the course of a progressive illness, continued use of resources and therapeutic interventions (in contrast to management of symptoms) are no longer reasonable and appropriate" [16]. An ethics consultation may help to appreciate perspectives when additional care is sought that is deemed to be futile or when a patient is medically ready to go home but refuses to be discharged. As part of an ethics consultation, the ethicist will communicate with the patient's physician, case manager, patient, nurse, family members, and any other appropriate individual to get all of their perspectives and then render a recommendation. The ethicist supports the need for the inpatient bed to be used in a medically appropriately manner, recognizing that the cancer center must be reserved for those patients with an acute care need. The cancer center must be managed efficiently so that only patients with a cancer diagnosis, who meet acute care criteria, occupy the bed. Oftentimes, the cancer center's occupancy is full; the cancer center administrator must ensure that if a patient no longer requires inpatient care, the process is in place to safely discharge the patient to the medically appropriate setting. With the clinical oversight of the Utilization Management Committee and the Case Management department, the cancer center administrator ensures that the beds are utilized in a medically and fiscally responsible manner.

The demand for certain types of services in a cancer center can be significant, and the administrator has the responsibility to oversee the appropriate use of all resources. For example, the right type of patient must be admitted to the right bed at the right time. If a leukemia patient is not on a designated leukemia nursing unit, the patient will not be able to benefit from the expertise of the nurses trained specifically in leukemia care. Furthermore, the administrator must ensure resources are properly allocated, such as staff, diagnostic services, and other outpatient services. A cancer patient's continuity of care can be impacted if the patient has to wait a significant time period for an inpatient bed, outpatient appointment, or reporting of test results.

Another ethical responsibility of the cancer center administrator is to make sure that the institution provides services in which they have expertise and not duplicate services that are better delivered by other available health care providers. The cancer center administrator has the ethical duty to work with the multidisciplinary team to assess which services should be provided at the cancer center due to the level of specialty care needed and which services are appropriate to be referred to other health care organizations. For example, there is a continuous demand for rehabilitation services to build cancer patients' strength for treatment and perform activities of daily living. The administrator must work with the multidisciplinary team to determine which patients require the expertise of the physical and occupational therapists with many years of experience in oncologic care at the cancer center and which patients are able to receive their physical and occupational therapy in the community. Another example can be found in social work, where staff provides counseling to cancer patients that is focused on removing barriers to care and facilitating coping with a variety of factors such as diagnosis, treatment, and end-of-life. Given this role, social work counseling tends to be provided on a short-term basis, and patients are generally referred to an appropriate community provider for any long-term needs. Referring cancer patients to other health care organizations can be challenging because oftentimes they have developed a trusting relationship with the cancer center and may want to receive all of their care at the cancer center.

APPROPRIATE RESOURCES AVAILABLE

The cancer patient's needs extend beyond medical or surgical intervention and will also include their psychosocial needs. A cancer diagnosis, the treatment process, and all the hardships patients face as a result of their experience can cause significant distress. Therefore, psychosocial care is an important and necessary component of the cancer patient's care and must be made available. The cancer center administrator has to budget the appropriate resources for the psychosocial care of the cancer patient. Table 19.2 lists examples of psychosocial programs and departments at The University of Texas MD Anderson Cancer Center [17].

A cancer patient may receive a diagnosis that will result in significant life changes and require psychosocial services. The administrator has an ethical responsibility to have a process in place to identify patients who require psychosocial services and to refer them to the appropriate service. For example, many cancer patients desire to have spiritual needs addressed in their care plan [18]. Spiritual care includes religion but is also

TABLE 19.2 Examples of Psychosocial Programs and Departments at The University of Texas MD Anderson Cancer Center

Department or Program	Functions
myCancerConnection	One-on-one support connects cancer patients and their caregivers with others who have been there.
Children's Lives Include Moments of Bravery (CLIMB^R)	Support program that uses art and conversation to help children and teens identify and appropriately express complex feelings related to having a parent or grandparent with cancer.
Ethics Consultation Service	Provides assistance with patient care decisions for patients, families, and health professionals.
Kids Inquire, We Inform (KIWI)	Comprehensive program aimed at helping children and teens cope with a loved one's cancer.
Social Work	Provides help to patients, their families and caregivers cope with cancer and assists in dealing with psychological or social barriers to their treatment.
Spiritual Care and Education	Provides spiritual care to patients and their families.

broader and encompasses "finding a sense of ultimate meaning and purpose, connecting with others or with a transcendent, finding hope in the midst of despair, and being able to reconcile or forgive" [18]. Many general acute care hospitals rely on faith groups in the community to provide spiritual care services; however, board certified hospital chaplains, who are trained to help cancer patients with life-threatening illnesses, should be available in the cancer center. Additionally, hospital chaplain students and specialty-trained volunteers can also perform certain services under the appropriate supervision of a certified chaplain.

Positive health benefits have been shown to result from meditation and prayer, demonstrating the benefits of spiritual care for cancer patients when dealing with the stress of their condition, so cancer center administrators have an ethical duty to provide these services [18]. The administrator has an ethical responsibility, while keeping within the context of state and federal law, to provide spiritual space for patients and their families, such as interdenominational chapels and prayer rooms. Often, chaplains provide spiritual care to cancer patients at the end-of-life, which has been shown to positively affect the patient's end-of- life experience [18]. Hospital chaplaincy services should provide 24-hour coverage in a cancer center to be available at the time of the death of a patient, and for their family members and staff. Additionally, chaplains need to visit cancer patients in the outpatient setting, where oftentimes, the patient first learns about their cancer diagnosis and is likely to experience higher levels of distress. The cancer center administrator should optimally provide spiritual care in both the outpatient and inpatient settings. The Commission on Quality in Pastoral Services of the Association of Professional Chaplains (APC) endorses staffing a spiritual care department based on the role and goals of spiritual care in the organization [19].

As mentioned above, the role of social workers in cancer centers can be different than the general acute care hospital. Usually, a key function of hospital social work in the general acute care hospital is discharge planning. However, a social worker in a cancer center should be licensed to provide counseling services for distress and life-changing news as this is a necessary component of care for the cancer patient. The cancer center administrator has a duty to budget these counseling services. The social worker is often challenged with ethically charged, value laden issues when navigating patient and/or family wishes that conflict with the recommendations by the multidisciplinary team or vice versa. Additionally, ethical conflicts can arise when the patient's wishes are not consistent with family preferences, certain economic realities, or appropriate bed utilization. In these types of situations, the social worker may request an ethics consultation so that the patient, family members, and appropriate members of the clinical team involved in these issues come to a resolution.

Another important component of psychosocial care is advance care planning. The cancer center administrator has a duty to implement an effective process for advance care planning, which includes conversations between the patient and members of the health care team to identify the patient's wishes and appropriately document those wishes. The primary advance directive documents include the Living Will, which documents a patient's wishes for their care in the future, and the Medical Power of Attorney for Health Care, which designates an agent to make a patient's health care decisions if the patient becomes incapacitated or incompetent [20].

The concept of advance care planning has a much broader scope than end-of-life planning documents. Advance care planning includes a process for health care providers to have conversations with their patients regarding the patient's wishes, and documenting those wishes in the appropriate place in the EHR so that all caregivers have access to this information to ensure these wishes are met. The Joint Commission requires hospitals to provide written information on advance directives and to inform the patient the extent "to which the hospital is able, unable, or unwilling to honor advance directives" [21]. The regulatory requirement outlined in the 1991 Patient Self-Determination Act (PSDA) requires hospitals receiving Medicare reimbursement to ask all patients upon admission if they completed an advance directive and to provide advance directive information and forms [22]. The cancer center administrator must adhere to these regulatory requirements as part of a robust advance care planning program. For example, physicians and other health care providers should be comfortable with having advance care planning discussions with patients at the appropriate time. Usually, these conversations can be more thoughtful and productive prior to admissions. Many patients do not want to discuss advance care planning during admission or during their inpatient stay. Ideally, advance care planning conversations with members of the multidisciplinary care team should begin early in the care of the cancer patient in the outpatient setting. However, members of the multidisciplinary team should timely discuss advanced care planning if life expectancy is likely to be less than one year. The administrator must ensure that members of the multidisciplinary team are provided educational sessions on when to approach these important discussions and how to conduct these conversations.

Sensitivity is required when broaching the subject of advance care planning with patients whose cultural background is inconsistent with this type of conversation. Advance care planning discussions should focus on the patient's wishes which may change over time and should be documented in the EHR in a designated area to ensure access when needed urgently. The administrator must ensure that this advance directive documentation is accessible for all members of the clinical team to easily locate in the EHR.

In addition to the necessary clinical and psychosocial services the cancer patient needs, the administrator has an ethical responsibility to provide appropriate support services. The administrator must budget adequate funds to supplement certain support services to meet the special needs of cancer patients. The administrator has the challenge to balance the cost of these services with the organization's budget, insuring that care remains clinically effective and fully ethical. For example, nutrition plays an important role in cancer care beyond the inpatient bed [23]. Malnutrition is common in cancer patients due to the disease and/or treatment. Timely screening and referral of patients for nutrition assessment and intervention is essential to minimize the adverse effects of malnutrition, such as delayed wound healing, increased risk of infection, and longer length of stay [24]. However, the availability of nutrition services may be very limited in outpatient treatment settings. For example, the Commission on Cancer (CoC) is a voluntary accreditation program for cancer centers and requires a policy to access nutrition services but does not mandate a nutrition professional onsite in the outpatient setting [25]. As more patients receive treatment in the outpatient setting, the cancer center administrator should provide clinical nutrition services for outpatients. The Association of Community Cancer Centers (ACCC) is an organization committed to advancing access to quality cancer care for all patients. ACCC recommends a nutrition professional provide care for patients and their families, especially those at risk of nutritional problems, as well as work with the patient, family, and multidisciplinary team to provide nutritional assessment and intervention across the continuum of care [26]. However, these guidelines were developed to reflect the optimal components for a cancer program.

Many general acute care hospitals have implemented some form of room service as part of their marketing campaigns as an amenity to attract patients. However, room service in the cancer center is much more than a patient amenity. The cancer center administrator recognizes that nutrition plays a key role in the care of a cancer patient. Providing patients with a broader variety of appropriate food options to choose from allows patients to have a sense of normalcy and autonomy during a time when they may feel a significant loss of control. Administrators have to budget for extended hours to keep room service open and develop processes to ensure accuracy of meal orders, deliver meals within 45 minutes, and keep food at the correct temperature. Food safety is especially important for infection-prone, high-risk cancer patients. This type of room service requires staff with hotel-style room service experience whose salaries are higher than a typical food service worker in a hospital setting. Hotel-style room service also requires an executive chef whose salary is above a cook to prepare options that are healthy, adhere to dietary needs, and yet remain appetizing both visually and olfactory. Additionally, the cancer center needs to invest in an information technology infrastructure that supports a call center for patients to order meals, monitor compliance with special dietary restrictions, and track orders and delivery time. Hotel-style room service contributes significantly to the quality of care of cancer patients by allowing patients to order the food they want to eat and when they want to eat it. Therefore, this service contributes to the patient getting adequate nutrition, which is often a challenge for many cancer patients. Thus, the cancer center administrator has the obligation to budget the expense for hotel-style room service so that optimal nutritional care is provided to the cancer patient.

In addition to having healthy food options on the hotel-style room service menu, the cancer center administrator has the ethical responsibility to offer healthy foods and post nutritional information in the cancer center's cafeterias for their outpatient and employee patrons. At The University of Texas MD Anderson Cancer Center, the American Institute for Cancer Research guidelines are used to create a healthy food environment [27]. The administrator is sending a message about the importance of nutrition and supporting an environment for wellness in the cafeterias by providing healthy food options and point-of-purchase nutritional information on calories, fat, sodium, etc. While nutritional information is important for everyone, the cancer patient can be particularly interested in this information as an important part of their overall care. Additionally, food choices in the cancer center cafeterias should provide healthy options for those involved in caring for the patient, including caregivers and cancer center employees.

Another ethical issue at a cancer center relates to limited English proficient patients having appropriate access to interpretive services. Timely and accurate interpretation is critical for the cancer patient to understand the medical information given to them from the multidisciplinary team. Cancer center administrators must adhere to regulations that require live face-to-face interpretation [28]. For many cancer center patients, their primary language may not be English. Patients may ask a family member to interpret, but there is a risk that the family member may incorrectly interpret certain information, raising both ethical and legal concerns and placing the patient at potential risk. Also, the patient may not disclose certain information that he or she does not want the family member to know but is key information needed by the physician. Oftentimes, the medical information that the physician or nurse must relay to the cancer patient is medically complex and requires a professional interpreter who is trained in medical terminology. There are often language, ethical, and cultural issues that exist interpreting to non-English-speaking cancer patients. In some instances, the hospital's Language Assistance department may not have an employed staff interpreter for a certain language. When this situation occurs, the administrator can purchase interpreter services which are available by telephone, on-demand video, or a contract face-to-face interpreter service. Decreased wait time is a benefit of telephone and on-demand video interpreter services. However, the multidisciplinary team may need live face-to-face interpreter services, depending upon the type of medical information to be interpreted. When hiring contract interpreters to work in a cancer center, the cancer center administrator must delineate the specific educational requirements and experience to get the appropriate interpreters needed. The cancer center administrator has the ethical responsibility to provide professional language assistance services with interpreters who are not only trained in medical terminology but also have a cross-cultural understanding to aid in the interpretation [28].

There are several ethical issues for the cancer center administrator regarding the EHR. The cancer center should have an EHR to fulfill the organization's duty to accurately capture all the important health information to treat the cancer patient; support coordination of care; comply with complex reimbursement models for appropriate and fair claims payment; and ensure research priorities are addressed. The EHR should be designed so that clinical staff can locate and document key medical and psychosocial information quickly and easily. Additionally, the administrator must implement a clinical documentation improvement program with clinical documentation specialists with either a nursing or coding background to partner with physicians to ensure that the documentation in the EHR accurately records the patient's illness, treatment, and ongoing status. With the implementation of the International Classification of Diseases, Tenth Revision (ICD-10) in October 2015, the number of codes to document a patient's condition increased from approximately 13,000 to approximately 68,000 [29]. This increase in codes will give far greater specificity in the description of the patient's condition, which should improve the accuracy of documentation for patient care and reimbursement as well as aid in reporting quality metrics and conducting research.

Incomplete or inaccurate health record documentation has far reaching implications. A key ethical consideration with the EHR is the "copy and paste" function, oftentimes referred to as "copy forward" or "cloning." For busy clinicians with time constraints, there is a strong temptation to use the "copy and paste" function. While the "copy and paste" function may serve an appropriate function in the EHR, when used incorrectly this function can create an incomplete picture of the patient's status and the treatment delivered. The cancer center administrator has the responsibility to implement an appropriate policy to govern the use of the "copy and paste" function, as well as conduct medical record review to monitor the appropriate use of this function to avoid unethical and potentially dangerous patient errors.

Patients and their representatives are becoming more consumer driven and desire access to health information in the EHR. At The University of Texas MD Anderson Cancer Center, an online patient portal is available for patients to view their health information. The patient portal creates ethical issues such as what information is to be included and who has access. Another ethical consideration for the administrator is when a patient identifies an error in the EHR that they want removed, not just amended which is the usual action; or, if the patient wants certain information that was told to their physician in confidence extracted from the EHR. An ethical concern that cancer center administrators have with an EHR is that providers can lose significant patient-provider interaction due to real-time documentation with providers facing the computer screen versus the patient. These situations are potential ethical dilemmas for the cancer center administrator in complying with the organization's regulatory responsibilities governing patient health information.

APPROPRIATE STAFFING LEVELS

The health care industry has staff shortages in certain positions, and the cancer center administrator must contend with these staffing limitations and yet provide the necessary services to patients. Additionally, the cancer center administrator has the challenge that some employees even after expensive recruiting and on-boarding processes find that they are not interested in working in a cancer center environment. First-year turnover can be high in a cancer center because treating cancer patients is rewarding but yet challenging. This turnover is costly for the cancer center and can affect the quality of patient care when employees are not retained and staff lack the sufficient years of valuable cancer care experience. Thus, the administrator has to monitor costly first-year turnover and find innovative solutions to make sure the right candidates are recruited and hired. The goal is to recruit the right person who will remain in the job and gather years of experience, which will improve the quality of care for the cancer patient. For example, one such tactic to reduce turnover and improve the recruitment process was implemented in the Rehabilitation Services department at The University of Texas MD Anderson Cancer Center which had experienced high turnover in the past. Nationally, physical and occupational therapy departments have staffing shortages exacerbating the problem to remain fully staffed [30,31]. One of the main reasons for the high turnover at The University of Texas MD Anderson Cancer Center was that physical and occupational therapists would be hired and complete a lengthy and costly orientation to only find that some therapists did not want to work with cancer patients. Two different strategies to address turnover issues which achieved significant results are shown in Table 19.3.

Another tactic for staff shortages that is sometimes appropriate, is hiring contract labor. The administrator of the cancer center must carefully assess the pros and cons of this option. One consideration is the specialized skills that the clinical staff must have for caring for cancer patients that a contract workforce may lack. Additionally, the administrator must assess the amount of desired control over staff education for their workforce. For example, ethics and patient privacy training are very important for the workforce in a cancer center to understand and follow. The administrator should assess whether to have

TABLE 19.3 Rehabilitation Services Recruitment Strategies and Results to Reduce Turnover at The University of Texas MD Anderson Cancer Center

Strategy #1:	Result:
• Candidates must pass a rigorous interview to identify motivation and ability to collaborate and be flexible: New approach recognizes that technical skills can be taught as needed.	• Favorable 10% turnover compared to reported industry averages ranging from 13.5% for physical therapists to 20.7% for occupational therapists [32,33].
Strategy #2:	Results:
• Offer 6–13 weeks student internships having 40–50 students each year work in Rehabilitation Service: New approach exposes students to the care of the cancer patient and provides the student insight into determining their preferred patient population to treat.	• Approximately 8%–10% of the students are hired to fill vacancies. • Advantages include knowing the students to screen out poor candidates and a significantly lower orientation training period. • Rehabilitation Services department staff exposed to the newest techniques from students and offer teaching opportunities for interested staff.

their staff employed by the cancer center to ensure certain levels of training and skills are met rather than to fill certain positions with contract workers.

Another ethical issue with staffing is adjusting staffing levels and productivity targets due to the special circumstances in the cancer center. The administrator must balance meeting productivity standards while providing quality care for cancer patients. For example, the cancer center administrator has to factor into the productivity standard for physical and occupational therapists the significant amount of time to motivate cancer patients who are weak from their treatments and are required to perform the necessary therapies to build strength for their next treatment or to live at home. In fact, these types of staffing level and productivity issues must be considered in all the clinical and support departments. The administrator has the ethical responsibility that productivity metrics, such as staff to workload ratios, incorporate these special issues for treating cancer patients and to monitor these metrics as part of their fiduciary duty to the institution.

PATIENT- AND FAMILY-CENTERED CARE

The story of Betsy Lehman is compelling and illustrates the importance for multidisciplinary care teams and administrators in cancer centers to partner with patients and their families and adopt the practice of patient- and family-centered care. This approach to care recognizes that patients and family members have unique and important information to share about the quality and safety of care and that not including them in the patient's care planning would be unethical [1]. Questions that the administrator might ask are: "Is there patient involvement in developing the strategic plan; where in the strategic plan process is the inclusion of the patients' voice; and are patient voices included in the various committees and projects of the cancer center?" All members of the health care team should solicit information from the patient and family members to improve patient safety and the quality of care. Additionally, best practice recommends implementing a Patient Family

Advisory Council with both patients and staff members to obtain feedback from patients on designing operational systems, educational processes, and facility design. For example at The University of Texas MD Anderson Cancer Center, there is a Patient Family Advisory Council with approximately 27 patient and family members and 10 staff members. Patients give feedback on multiple issues, including but limited to advance care planning tools and processes, inpatient room design, the patient needs assessment tool, patient education materials, way-finding, and the visitation policy. Patient members also sit on other committees, including the Safety Committee, Nursing Congress, and the Psychosocial Council. In the Safety Committee, to emphasize the importance of patient input on safety matters, there is a special portion of every meeting dedicated for patient feedback.

Cancer centers should conduct an assessment to identify opportunities for improvement to strengthen the approach to patient- and family-centered care. The American Hospital Association has a self-assessment tool with a set of key indicators in multiple areas, including leadership, mission, definitions of quality, patient and family advisors, patterns of care, educational materials, charting and documentation, patient and family support, quality improvement, personnel, and environment and design [34]. After completing an assessment, action plans should be created and implemented to further build patient- and family-centeredness approaches.

CONCLUSION

The cancer center administrator confronts ethical challenges each day in managing the cancer center. A number of these ethical issues relate to utilization management, medical futility, psychosocial care, staffing levels and training, and patient- and family-centered care. In addressing these challenges, he or she must always ask first "what is best for the patient?" Then the administrator must identify the necessary resources and develop the appropriate systematic processes to implement the services most efficiently and effectively. Finally, the cancer center administrator must provide for the special needs of the cancer patient, while assuring the appropriate utilization of resources to ensure the institution remains viable to serve those in need of cancer care.

Some of the major ethical dilemmas confronting cancer center administrators that will continue to expand in the future relate to decreasing operating margins, increasing value-based care payment methodologies, and increasing consumerism [35]. Constraints on resources from reduced operating margins may present further ethical questions to cancer center administrators on how to reduce costs while providing cancer patients the necessary services and ensuring quality and safety. However, the shift in payment to value-based care should promote quality and safety and reward the cancer center administrator's efforts in these areas. Additionally, the trend towards increased consumerism emphasizes the need for cancer center administrators to advance patient- and family-centered care so systems and processes are designed and improved based on feedback from patients and their families. The cancer center administrator is ultimately responsible for the cancer center to continue to serve the community. When presented with challenges and opportunities that may have ethical implications, the cancer center administrator should always ask and do "what is best for the patient."

References

[1] Piper LE. The ethical leadership challenge: creating a culture of patient- and family-centered care in the hospital setting. Health Care Manager 2011;30(2):125–32.

[2] Improving the 21st-century health care system. Institute of medicine crossing the quality chasm: a new health system for the 21st Century. Washington, DC: National Academy Press; 2004. p. 39–60.

[3] Formulating new rules to redesign and improve care. Institute of medicine crossing the quality chasm: a new health system for the 21st Century. Washington, DC: National Academy Press; 2004. p. 61–88.

[4] Anonymous. Ethical decision making for healthcare executives. Healthcare Executive 2012;27(5):102–3.

[5] Bowen DJ, Weil PA. Examining the code. ACHE's code of ethics highlights challenges faced by healthcare leaders. Healthcare Executive 2011;26(4):39–42.

[6] Sedgwick D. Leading long-term care, <http://leadinglongtermcare.com/tag/non-profit/>; 2010.

[7] Ethical decision making for healthcare executives. American college of healthcare executives (ACHE). 2016. Available from: <https://www.ache.org/policy/decision.cfm>.

[8] Buell JM. Ethical leadership in uncertain times. Healthcare Executive 2015;30(3):31–4.

[9] Squazzo JD. Ethical challenges and responsibilities of leaders. Health Care Executive 2012;33–8.

[10] 2013–14 Health care criteria for performance excellence. Baldridge performance excellence program. National Institute of Standards and Technology, United States Department of Commerce; 2014.

[11] History of the NCI Cancer Centers Program, Available from: <www.cancer.gov/research/nci-role/cancer-centers/history>; 2012.

[12] Patient Protection and Affordable Care Act USA 2010 [cited 219–212]. Available from: <http://www.gpo.gov/fdsys/granule/PLAW-111publ148/PLAW-111publ148/content-detail.html>.

[13] Michelman MS, Mass S, Ukanowicz D. The role of physician advisors. Optimizing the physician advisor in case management. Marblehead, MA: HCPro, Inc; 2008. p. 23–45.

[14] Terra SM, Powell SK. Is a determination of medical futility ethical? Professional Case Management 2012;17(3):103–6.

[15] Nelson WA, Macauley RC. Balancing issues of medical futility. Health Care Executive 2015;30(2):48–51.

[16] Ewer MS. The definition of medical futility: are we trying to define the wrong term? Heart Lung 2001;30(1):3–4.

[17] The University of Texas MD Anderson Cancer Center. Patient and family support, Available from: <http://www.mdanderson.org/patient-and-cancer-information/guide-to-md-anderson/patient-and-family-support/index.html>; 2015.

[18] Puchalski CM. Addressing the spiritual needs of patients. In: Angelos P, editor. Ethical issues in cancer patient care. Cancer Treatment and Research. 2nd New York: Springer; 2008. p. 79–91.

[19] Staffing for quality chaplaincy care services. A position paper of the APC Commission on Quality in Pastoral Services. 2009. Available from: <http://http://www.professionalchaplains.org/files/resources/reading_room/chaplain_to_patient_ratios_staffing_for_quality.pdf>.

[20] Detering K, Silveira MJ. Advance care planning and advance directives 2015: Wolters Kluwer, <http://www.uptodate.com/contents/advance-care-planning-and-advance-directives>; 2015.

[21] The Joint Commission, Requirements related to the provision of culturally competent patient-centered care hospital accreditation program; 2009 [HAP. Sect. Standard RI.01.05.01].

[22] Angelos P, Kapadia MR. Physicians and cancer patients: communication and advance directives. In: Angelos P, editor. Ethical issues in cancer patient care. Cancer treatment and research. 2nd New York: Springer; 2008. p. 13–28.

[23] National Cancer Institute. Nutrition in cancer care (PDQ^{R}), <http://www.cancer.gov/about-cancer/treatment/side-effects/appetite-loss/nutrition-pdq#section/_164>; 2015.

[24] Tappenden KA, Quatrara B, Parkhurst ML, Malone AM, Fanjiang G, Zeigler TR. Critical role of nutrition in improving quality care: an interdisciplinary call to action to address adult hospital malnutrition. J Parenter Enteral Nutr 2013;37:482–97.

[25] Commission on Cancer and the American Cancer Society. Collaborative action plan guide. Cancer program standards 2012: ensuring patient-centered care ACS-CoC collaborative action plan guide, Available from: <https://www.facs.org/~/media/files/quality%20programs/cancer/clp/collaborative%20action%20plan%20guide.ashx>; 2013.

[26] Association of Community Cancer Centers. Cancer program guidelines, Available from: <https://www.accc-cancer.org/publications/CancerProgramGuidelines-4.asp>; 2012.

[27] American Institute for Cancer Research. AICR's guidelines for cancer survivors, Available from: <http://www.aicr.org/patients-survivors/aicrs-guidelines-for-cancer.html>.

[28] Guidance to federal financial assistance to recipients regarding Title VI prohibition against national origin discrimination affecting limited English proficient persons. Department of Health and Human Services; 2003. p. 47311–23.

[29] Kuehn L. Preparing for ICD-10-CM in physician practices. J AHIMA 2009;80(8):26–9.

[30] Lin V, Zhang X, Dixon P. Occupational therapy workforce in the United States: forecasting nationwide shortages. PM&R 2015;7(9):946–54.

[31] Zimbelman JL, Juraschek SP, Zhang X, Lin VW-H. Physical therapy workforce in the United States: forecasting nationwide shortages. PM&R 2010;2:1021–9.

[32] American Physical Therapy Association. Today's physical therapist: a comprehensive review of a 21st-century health care profession, Available from: <http://www.apta.org/uploadedFiles/APTAorg/Practice_and_Patient_Care/PR_and_Marketing/Market_to_Professionals/TodaysPhysicalTherapist.pdf>; 2011.

[33] American Occupational Therapy Association. Why and how do OT practitioners leave, Available from: <http://www.aota.org/education-careers/advance-career/salary-workforce-survey/why-ot-ota-leave-jobs.aspx>; 2015.

[34] Strategies for leadership. Patient- and family-centered care. A hospital self-assessment inventory. Institute for Family Centered Care. American Hospital Association; 2010.

[35] Lens into the future. Health System CEO Interviews. Deloitte; 2015.

CHAPTER

20

Ethics and Information Technology

Leslie A. Kian and Scott D. Eastman

The University of Texas MD Anderson Cancer Center, Houston, TX, United States

INTRODUCTION

Information technology in health care has enabled both healthcare providers and patients to experience a host of benefits and, at the same time, many opportunities for harm. Ethicists play a significant role in managing healthcare information because they support autonomy, beneficence, nonmaleficence, and justice. Information technology presents challenges to these very same principles. Since technology is so pervasive in our everyday lives including health care, there are an overwhelming number of examples to cover in a single chapter. However, here are a few major subject areas that introduce some thought-provoking ethical conversations.

ELECTRONIC MEDICAL RECORDS

The transition from a paper medical record to an electronic medical record has presented a great deal of benefit to both patients and providers. Accessibility has no longer been restricted to having a physical folder in hand but instead anyone can access a patient's medical record from virtually anywhere and at any time. In fact, medical documentation can be viewed on many platforms and devices that include computers, laptops, tablets, cell phones, and more. A doctor who gets a call at night about a patient can review the patient's latest information before making any clinical decisions without leaving home. Alternatively, a patient can view their information at any time where healthcare organizations have provided web portals for this purpose, also on multiple platforms and devices. Patients can not only look up their medical information but also can check on appointments, communicate with their provider, and have a convenient place for information resources already vetted by their healthcare organization.

A fully functional electronic medical record includes results reporting, structured clinical documentation, and electronic order entry. Results reporting provides the ability to view laboratory or pathological results, diagnostic images, and reports from all

Ethical Challenges in Oncology.
DOI: http://dx.doi.org/10.1016/B978-0-12-803831-4.00030-0

providers. Structured clinical documentation is the ability to record health information in discrete data fields. Often forms are predesigned for each discipline to record data that are stored in a manner that allows us to submit queries to database(s) that we previously were unable to perform with transcribed documents. Electronic order entry is also a function of recording discrete data fields but with clinical orders that represent treatments, procedures, prescriptions, etc. Having discrete data for both clinical documentation and order entry not only allows the reporting of these data but also the ability to analyze them to explore trends and correlations between patients and their encounters. Whole populations can be reviewed for these trends to track how clinical practice affect outcomes. For example, an informatics analyst could use business intelligence tools to study how many lung cancer patients received an opioid for pain levels recorded above a five. Previously, medical records needed to be reviewed to abstract that information from verbose, transcribed reports after identifying which patients were diagnosed with lung cancer.

Additional benefits of a fully functional electronic medical record include the ability to deploy clinical rules and establish best practice pathways supported by evidence-based medicine. Imagine a scenario where a medication is being prescribed using the electronic order entry function and allergies were recorded using the structured clinical documentation function. A clinical rule could provide a timely message to the ordering clinician that a particular medication is contraindicated due to a recorded allergy at the time the clinician attempted to place the order. The use of discrete data fields can also identify patients who are eligible for clinical pathways that represent best practices and present these options to the clinician in a timely fashion. These examples represent an untold number of benefits that can be available.

Despite the many benefits of an electronic medical record, there are also challenges that present an ethical dilemma. Accessibility may be a beneficial feature but it is also a security concern. A patient's privacy has always been an important issue for any healthcare organization. A paper medical record is inherently limited in terms of accessibility due to the physical nature of the medium. Electronic medical records, however, could potentially be accessed from anywhere in the world. Even from within the confines of the healthcare organization caring for a patient, all those who are accessing a medical record require scrutiny and oversight.

COMMUNICATIONS

As technology has advanced, communication is one of the aspects of our daily lives that has seen the greatest change in recent years. Paper mail, referred to informally and somewhat derogatorily as "snail mail" has been replaced with e-mail, texting, and communication via social media. Telephone landlines and answering machines have been replaced with mobile phones, voicemail, and the ability to directly connect with almost anyone from most any place at any time. E-commerce via the Internet has made it possible to purchase almost any good or service around the clock and customer service from these organizations is generally available at any hour via telephone or the Internet.

As has been the case with most new technology, these changes bring benefits in addition to new risks that must be addressed. The benefits of easier, more frequent communication must be weighed against concerns about privacy, safety, security, and many other issues that were not risks to the same degree prior to the advent of current communication technology. The convenience of being able to reach others at all hours often is accompanied by the expectation that we are available when needed, regardless of the day or hour. The added safety of having a telephone available in the event of an emergency while traveling must be balanced against the dangers of distracted driving caused whenever the operator of a vehicle is unable to resist the temptation to view an electronic device. Information about almost any conceivable subject is available instantly but the validity of what is available via the Internet is often suspect [1]. These are just a few examples of the many benefits and challenges that have come about due to advances in communications technology.

These changes in technology, and to our society in general, affect the medical field and specifically influence communication between patients, physicians, and medical care teams. Expectations about the speed and ease of communication in other aspects of daily life influence what is expected in healthcare settings. Many who are able to chat via a computer or smartphone with a representative of the cable company, the bank, and most of the other organizations that they interact with on a regular basis are likely to expect similar availability from doctors, hospitals, and other medical providers. Organizations must evolve in order to meet the changing expectations of patients and their families. As a part of these efforts, processes designed with a paper medical record in mind often must be modified in order to adapt to our current electronic world. Successful organizations will incorporate communications technology into clinical operations in ways that will better meet the needs of the patients. Advanced clinical technology, electronic health records, and other information systems are extremely important and systems that have great potential to improve the patient's experience cannot be overlooked. Healthcare providers should use technology to facilitate more timely communication with those that they care for and leverage these same channels of information to help patients develop better knowledge and understanding of their health status and any recommended interventions. Changes of this sort have great potential to improve the patient's experience with the health care received and may in fact result in superior outcomes when better information results in improved compliance with a physician's instructions. It is critical to implement any changes in the methods of communication between patients and the care team in such a way as to protect the patient's privacy. A breach that compromises sensitive patient information will completely overshadow any perceived benefits of improved communication.

A significant technology that affected communications is the mobile telephone. As of 2015, an estimated 92% of American adults owned a cellular telephone. Additionally, approximately 43% of adults live in a cellular-only household without a landline telephone [2]. What was once a luxury item or a novelty is now viewed as a completely ordinary aspect of daily life, and in fact, many see the cellular phone as an absolute necessity. Mobile phones are so vital to daily life that the US government implemented a program to provide them at little or no cost to those who are unable to purchase a phone without assistance. In addition to voice communication, today's cellular phones are useful for sending and receiving text messages, often referred to as simply texting. This technology first developed in 1992 has grown exponentially since it was introduced to consumers.

Recent data indicates that more text messages are sent than mobile phone calls are placed each year [3]. Texting is now a routine medium of communication, used not only to communicate with family and friends, but as a method of identity verification and fraud prevention by banks and other financial institutions. This demonstrates the level of integration of this technology into everyday life. The use of texting in medicine from providers to patients has been a topic of research interest. Studies have found that patients and in the case of children and adolescents, the parents of patients are open to health communication via text messaging [4]. This same research indicates that this form of communication is ideal for distributing general information to patients and may not be appropriate for detailed patient-specific health information. In oncology, there are significant opportunities to leverage this technology for the benefit of patients. We have achieved improvements in cancer survival rates through advances on many fronts. One of theses has been improved early detection through better screening. Text message reminders as well as smartphone applications are and will continue to be employed to remind patients when screening is due and can be used to provide coaching and support for patients who have agreed to participate in potentially beneficial programs such as smoking cessation [5]. The early achievements of these programs point to the need to develop additional initiatives that leverage mobile technology to improve the healthcare services delivered to patients. As discussed previously, the success and continued existence of any electronic notification program is dependent on meeting the ethical obligation to protect the privacy of the participants.

The rise in the use of electronic mail or e-mail is another technological advance that has had a significant impact on medicine with ethical overtones. E-mail has been described as combining elements of both letter writing and a spoken conversation [6]. An exchange of e-mails moves at a much faster pace than trading handwritten letters and in many cases is more spontaneous as well. Many users view e-mail as less formal, so for better or worse, the reduction in time required to compose a message also increases the speed of communication. Unlike an oral discussion, there is a written record of the thoughts expressed by each of the parties. Depending on the e-mail system, a record of all e-mails sent and received may be maintained even after the user deletes messages. The creation of a written record is perhaps the most beneficial aspect of the use of e-mail in medicine. A patient who has questions for a physician or the care team has an opportunity to carefully formulate the question and ask multiple questions in a single e-mail message. The physician or other care provider can be more confident that any messages sent to patients are more likely to be received correctly when delivered in writing as opposed to orally. Because of the increased clarity, an e-mail exchange may take less time for all involved that a discussion of the same material in person or over the telephone. Also, because the parties in an e-mail exchange are not required to be in the same place as with a face-to-face conversation or available at the same time as with a telephone call, the e-mail conversation may be more convenient for one or both parties than either of the traditional alternatives. As an additional benefit, either party has the opportunity to go back and review the question asked or an answer provided at any point in the future if there is a need to do so. This can be extremely beneficial for the patient, because many medical situations, especially in oncology, are stressful and even with the assistance of a family member or other caregiver, the patient may have difficulty recalling everything that he or she was told. In these instances, a written record may improve understanding and increase compliance with any instructions that were provided by the medical team. For all these reasons, e-mail has the potential for improving the relationship between a physician and a patient, as long as it is viewed as an enhancement to and not a replacement for face-to-face communication [7]. Studies of the attitudes of patients toward e-mail have found that the majority of patients who were not yet communicating with their physician via e-mail were interested in doing so in the future [8]. Given the potential benefits of e-mail use to both the provider and the patient and the general acceptance of this method of communication, healthcare providers should implement programs that encourage the responsible use of e-mail communication between patients and providers. Despite the widespread use of electronic mail, any programs should still be elective on the part of the patient to allow for those who may have reservations about using this mode of communication. In addition, we must take careful precautions to keep the information contained in any e-mails confidential. In recent years, there have been numerous examples of the consequences that result when private information contained on e-mails servers becomes public.

Social media are another form of electronic communication that are primarily web sites that feature content contributed by users. This content often includes thoughts, ideas, photographs, videos, and other media. The mechanics of the various platforms vary but in general, users have control over who is able to view the posted material and have a mechanism such as "following" or "friending" others so that it is possible to create connections with people, businesses, or other organizations that the user selects. Social media is an evolving phenomenon as new web sites are launched from time to time and others decline in popularity as they lose users to one of the newer alternatives.

Facebook is currently considered the most popular social media network worldwide. Data released by the company in the second quarter of 2016 indicates that Facebook has more than 1.7 billion active monthly users, defined as those who had logged in to the site at least once during the prior 30 days [9]. Facebook centers on friendships established among users. Once a friendship is declared (an act known as "friending"), the friend has access to the comments, photos, and other content posted to the site by one's friends. The subject of a physician and patients entering into a Facebook relationship has been the subject of considerable discussion and debate [10]. The chief concern is that this action may imply or perhaps, in fact, create a relationship that is beyond the expected businesslike arrangement that should exist between the parties [11]. Additionally, depending on what a patient or a physician reveals via Facebook, other dilemmas may arise. For example, if a patient reveals recreational drug use or symptoms that were previously unreported to the physician, what obligation does the physician have to act on this information or to include such information in the electronic medical record? If the physician elects to mention such an issue at a future clinic visit, what reaction will he or she receive from the patient? In the event that a physician shares personal information or describes conduct that could be viewed as unprofessional, what impact might that have on the relationships with his or her other patients? Is the physician jeopardizing his or her safety by sharing personal information or creating a legal or ethical risk by discussing his or her conduct outside of working hours [10]? Individual opinions and organizational polices vary, but these are issues that must be given careful consideration as the use of social media in medicine becomes more widespread.

One of the major challenges of the Internet in general and social media in particular is the proliferation of information that is either unproven or has been absolutely demonstrated to be false [12,13]. Various conspiracy theories and medical myths seem to have a found a strange sort of immortality via the Internet. For example, there are vocal proponents online of the idea that the cure for all cancer was discovered in the year 1934 but has been kept secret for financial reasons, underarm deodorant leads to breast cancer [14], and the use of a cellular phone may cause one to develop a brain tumor. This sort of speculation and myth clearly predates the advent of the World Wide Web, but the Internet provides a much more powerful voice to those who wish to advance suspect ideas and gives anyone, regardless of their qualifications to speak on a particular topic, the ability to reach exponentially more people than was possible in the past. Because of the power of social media to reach so many at little or no cost, healthcare organizations should look for opportunities to disseminate truthful, scientifically proven, and ethical information to the public. We can be hopeful that information coming from credible sources will counter that which is coming from those with few credentials and suspect motivation [15]. With respect to individual physicians, because of the concerns discussed previously, they must proceed with caution as they incorporate social media into their practices, but it is clear that there are opportunities to strengthen doctor–patient relationships through social media. These improved relationships can lead to greater patient satisfaction and have the potential to move the healthcare system in a more patient-centered direction [16].

The adoption of these communication technologies has ethical implications and should be considered in conjunction with generally accepted ethical principles. The relevant principles for this discussion are autonomy, beneficence/nonmaleficence, and privacy/confidentiality [17,18]. The first principle is that of autonomy, which in this context means that a patient should be allowed to make choices regarding his or her health care and decisions as to how medical information will be used. With respect to communication technology, the principle of autonomy dictates that patients should be able to make choices about their communication with healthcare providers. In the opinion of many ethicists, patients should be allowed to "opt-in" to various modes of communication depending on their personal comfort with various technologies. For example, a patient should decide if he or she would like to receive health reminders via text message and have an option, but not a mandate, of communicating with a physician via e-mail.

The second principle of beneficence/nonmaleficence mandates that we "do good" for and avoid "doing harm" to the patient. Improved communication has great potential to achieve gains in both of these areas. If a patient has access to better information about his or her own health, improved knowledge of the treatments or other care recommendations, and is regularly reminded when a health screening is due, the patient is likely to derive benefit and potentially better outcomes from the healthcare system. In much the same way, a better-informed patient may avoid the harm associated with inaccurate or incomplete health information, poor understanding of a treatment plan or other recommendations, or other gaps in knowledge that can lead to poor choices and adverse outcomes.

The third principle is that of privacy and confidentiality. Medical information is among the most sensitive and confidential data that exists for most people. The risk of a security breach and the exposure of personal information increases dramatically when this data exists on the Internet or is transmitted using mobile devices. A perfect system with no security vulnerability does not yet exist, but every effort must be made to secure this information, and the additional risks that are created as new modes of communication are added must be carefully studied and mitigated using the best methods available.

Advances in technology have changed and will continue to influence the communication between healthcare providers and their patients. The relationships that exist between patient and care providers are influenced either positively or negatively by the nature of this communication. When implemented correctly and in accordance with ethical principles, communication technologies have the potential to improve the experiences of all the parties involved in a healthcare interaction. If these same ethical principles are compromised, then the damage to an organization's reputation and finances can be severe. Under the current system, the healthcare market is extremely competitive, and in all but the smallest towns, patients have choices as to where they seek care. Following a breach of sensitive data, potential patients are very likely to avoid the provider that compromised private information. In addition to lost revenue, these providers may face civil and criminal penalties, depending on the nature of the violation and the applicable laws that were broken.

PATIENT ACCESS

An aspect of a patient's right to autonomy includes access to their medical record information as well as communication with their care provider. An electronic medical record facilitates the availability of medical information to the patient by the ability to send this information electronically (e.g., e-mail) or by providing a web portal by which they can view information that is posted. As long as a protected method is provided by managing usernames, passwords, and a secured web site, their privacy can hopefully be assured. Today, information security requires consideration of the many platforms that exists beyond the standard desktop computer workstation [19]. Access can be acquired on cell phones, tablets, watches, etc., which all have variations in security protocols. Over time, security improved to prevent hacking with newer and more advanced operating systems. Support for older platforms and operating systems can be quite challenging and often it has been best to maintain only the more current systems [20]. This, however, presents barriers to patients who cannot afford new devices.

One issue encountered regarding the efficiency inherent with electronic data that should be considered involves results reporting. It is not difficult to post laboratory or pathology results automatically from the medical record or the source system to the patient portal. However, the presentation of that information should be timed so that the provider can review it and perhaps has the option to discuss the results prior to posting the information without explanation. Otherwise, unnecessary concerns or misinterpretation could result on the part of the patient.

Patient portals are also used to offer reliable resources for information that can be tailored to the patient's needs (i.e., information about their illness or disease, healthful tips, available services, and sources for assistance). These resources can be vetted by the healthcare organization or provider for reliability rather than a patient surfing the web and having to sort through unproven sources of information. An additional feature now used with patient portals is the enabling of communication between patients and their providers. Patients can conveniently post inquiries online at any time and receive responses sometimes faster than by making phone calls and leaving messages or having to wait for their next appointment.

RESEARCH

Information technology has played a most important role in advancing oncology research from one perspective: The ability to process exponentially increasing amounts of data and to do so faster [21]. Sequencing the first human genome took approximately 13 years and cost $1 billion. Today, sequencing a human genome takes one to two days and costs less than $10,000 [22]. Our growing computational abilities are not the only factor that contributed to this research. Knowledge was openly shared among scientists internationally over the course of those 13 years. Since that time, many important discoveries using this knowledge has enhanced our understanding. For example, a major breakthrough in oncology research has been experienced through genomics. It is difficult, however, to observe the same level of collaboration, internationally or nationally, when it comes to oncology research. Although there are many oncology research projects benefiting from genomic research, many of them are conducted with data that are managed by separate entities. In addition, access to all available therapies are limited. This raises ethical concerns where justice, a fourth important ethical principle, is weighed. An attempt to resolve these concerns has recently been presented in an initiative announced by President Barack Obama called the Cancer Moonshot [23] during his State of the Union address on January 12, 2016. In this address, he placed Vice President Joe Biden in charge of "mission control"[24,25]. Although this is not first time an initiative focused on eliminating cancer was announced, today's information technology is in a unique position to facilitate the many programs that will be launched to address both the access to cancer care and the sharing of research information. For example, the National Cancer Institute (NCI) will partner with the White House Presidential Innovation Fellows to redesign how patients and oncologists access information about clinical trials. Cancer clinical data hosted on the website www.cancer.gov will be made available through direct access using an application programming interface (API), which will enable third-party entities to develop applications that can integrate and search through its database making this information more available to patients, physicians, and caregivers. In addition, there are new federal incentive for cancer care where the Centers of Medicare and Medicaid Services (CMS) will enroll almost 200 physician practices that include more than 3200 oncologists in the Oncology Care Model. This program will provide approximately $6 billion in cancer patient care for an estimated 155,000 beneficiaries each year. An example of facilitating collaboration efforts as part of the Cancer Moonshot is the Department of Energy (DOE) partnering with the NCI to launch three pilot projects that will harness almost 100 cancer researchers, providers, computer scientists, and engineers to analyze data from preclinical cancer models, molecular interaction data for ras (mutant retrovirus-associated DNA sequences have been identified in several different cancers) [26], and cancer surveillance data from four DOE National Laboratories in conjunction with the NCI Frederick National Laboratory for Cancer Research. Another example is a program that includes the open access of sharing cancer data through the Genomic Data Commons (GDC). The Foundation Medicine doubling the number of patients represented in the NCI's GDC to over 32,000 accumulated in the last month. GDC has already share more than five petabytes of genomic data on almost 30 tumor types that included additional clinical data. The GDC database is built and overseen by the University of Chicago Center for Data Intensive Science and the Ontario Institute for Cancer Research. A significant point announced by Vice President Biden at the 2016 annual American Society of Clinical Oncology (ASCO) meeting in Chicago was that, "All information from trials funded by NCI from this point on will have to be submitted to the database" [27].

As the use of electronic medical records becomes mainstream, the importance of sharing data is no longer restricted to research. Another example where the sharing of data is advantageous also exists with the clinical data that are gathered in the process of providing care to patients. The more discrete data are collected in the patient care process through electronic medical record applications, the more information becomes available to collectively monitor and measure healthcare outcomes. Instead of traditionally collecting or extracting data specific to a research hypothesis, the routine gathering of discrete data has been occurring on a regular basis via the structured clinical documentation functionality found in today's electronic medical records applications. These routinely gathered data can be leveraged to ascertain the most recent trends in outcomes that could influence clinical decisions being made while treating a patient [28]. The more data available, the more statistically significant the results.

In large healthcare systems that share a common electronic medical records (or a healthcare organization with a large volume of patients), the more data are available to dynamically support (or alter) clinical decisions moving forward. Knowledge is advanced as more data are accumulated within that system. This phenomenon essentially creates a continuously learning healthcare system [29]. Current practices influenced by regulation, however, restrict the widespread sharing of patient data to deidentified information. Despite the current restrictions, some entities have found ways to share their data and maintain current regulatory privacy rules. For example, Vizient's University Hospital Consortium (UHC) now has over 400 members sharing data for comparative analysis and subsequent quality improvement projects [30].

Health Care Systems Research Network (formerly known as HMO Research Network) is a consortium of 20 healthcare organizations that share patient data that includes cancer, pharmacy, vital signs, laboratory results, demographics, and more [31]. In these two examples, members pay membership and maintenance fees to support the administrative and data infrastructure. The members who have the resources to leverage these data sets enjoy the ability to fund projects that help them improve the efficiency of their operations and efficacy of their patient care. Unfortunately, there are significantly more healthcare organizations that are unable to afford these benefits. Moreover, the learning that takes place in these improvement initiatives are primarily the result of retrospective studies. The Institute of Medicine (IOM) studies propose that a more robust sharing of patient data includes the use of all clinical data as a source that is used for the public good and a centralized resource for advancing clinical knowledge [32]. This concept takes us beyond exclusive membership organizations that share data but also challenges the current regulations surrounding patient privacy. Moreover, the IOM is encouraging us to leverage these data for more dynamic applications where clinical decision support provides ongoing evidence that is applied on a more timely basis during the patient care process. In order to accomplish these goals that benefit all patients, several major changes will have to take place. First, current privacy regulations will require change before greater sharing of routine clinical data are encouraged. Second, electronic medical records would require a common language beyond the healthcare standards found today (commonly described as "interoperability"). Lastly and more importantly, we will need to apply the appropriate balance between the ethical principles that maintain a patient's right to privacy while serving our patient population's needs for timely clinical decision support that leads to improved efficacy in our care.

SUMMARY

Advances in information technology continues to evolve and will continue to change our abilities in electronic medical record functionality, communications in different forms, how we access health information, and how we manage data in both research and everyday patient care processes. As a result, we can expect these changes will have enormous ethical implications. The topics presented here are only a few examples of where we have gained an advantage and, at the same time, have experienced ethical challenges. Careful thought must be given to how we apply the technology so that we protect the patient and safeguard healthcare data. This in turn can have a positive impact on the patient experience and increase the quality, safety, efficiency, and effectiveness in the treatment and care provided to our patients.

References

[1] McKearney TC, Mckearney RM. The quality and accuracy of internet information on the subject of ear tubes. Int J Pediatr Otorhinolaryngol 2013;77:894–7.

[2] Rainie L, Zickuhr K. Americans' views on mobile etiquette. Pew Research Center 2015.

[3] Goldfarb J, Kayssi A, Devon K, Rossos PG, Cil TD. Smartphones and patient care: exploring the use of text-based messaging for patient-related communication. Surg Innov 2016;23(3):305–8.

[4] Rand CM, Blumkin A, Vincelli P, Katsetos V, Szilagyi PG. Parent preferences for communicating with their adolescent's provider using new technologies. J Adolesc Health 2015;57:299–304.

[5] Abroms LC, Boal AL, Simmens SJ, Mendel JA, Windsor RA. A randomized trial of Text2Quit. Am J Prev Med 2014;47(3):242–50.

[6] Kane B, Sands D. Guidelines for the clinical use of electronic mail with patients. The AMIA Internet Working Group, Task Force on Guidelines for the Use of Clinic-Patient Electronic Mail. J Am Med Inform Assoc 1998;5:101–11.

[7] Katzen C, Solan MJ, Dicker AP. E-mail and oncology: a survey of radiation oncology patients and their attitudes to a new generation of health communication. Prostate Cancer Prostatic Dis 2005;8:189–93.

[8] Slack WV. A 67-year old man who e-mails his physician. JAMA 2004;292:2255–61.

[9] Website S. Retrieved from: <http://www.statista.com/statistics/264810/number-of-monthly-active-facebook-users-worldwide/>; 2016.

[10] Wiener L, Crum C, Grady C, Merchant M. To friend or not to friend: The use of social media in clinical oncology. J Oncol Pract 2012;8:103–6.

[11] Lewis MA, Dicker AP. Social media and oncology: the past, present, and future of electronic communication between physician and patient. Semin Oncol 2015;42(5):764–71.

[12] Konsbruck, R.L. Impacts of information technology on society in the new century 2001.
[13] Novella, S. Reliability of health information on the Web 2010.
[14] Society, A. C. Antiperspirants and breast cancer risk; Retrieved from: <https://www.cancer.org/cancer/cancer-causes/antiperspirants-and-breast-cancer-risk.html>; 2016.
[15] National Institute on Aging, U. S. D. o. H. a. H. S. Online Health Information: Can You Trust It? National Institutes of Health 2014. Page 323.
[16] Hawn C. Take two aspirin and tweet me in the morning: how Twitter, Facebook, and other social media are reshaping health care. Health Aff (Millwood) 2009;28(2):361–8.
[17] Mercuri JJ. The ethics of electronic health records. Clinical Correlations: the NYU Langone Online Journal of Medicine 2010.
[18] LeBlanc TW, Shulman LN, Yu PP, Hirsch BR, Abernethy AP. The ethics of health information technology in oncology: emerging issues from both local and global perspectives. ASCO Educ Book; 2013. p. 136–42.
[19] Harvey MJ, Harvey MG. Privacy and security issues for mobile health platforms. J Assoc Inf Sci Technol 2014;65:1305–18.
[20] R. King and Y. Danny, Windows XP: old platforms die hard, security risks live on, Wall St J 2014.
[21] Hiremane R. From Moore's Law to Intel innovation—prediction to reality. Technology@Intel Magazine; 2005.
[22] Lewis T. Human genome project marks 10th anniversary. Live Sci 2013.
[23] Office of the Press Secretary, T. W. H. FACT SHEET: at cancer moonshot summit, vice president Biden announces new actions to accelerate progress toward ending cancer as we know it 2016.
[24] Lowy DR, Collins FS. Aiming high—changing the trajectory for cancer. N Engl J Med 2016;374:1901–4.
[25] Maron DF. Can we truly "cure" cancer? Sci Am 2016.
[26] W. C. Institute. Oncogenes: Ras. Cancer Quest 2016.
[27] Mukherjee S. Joe Biden just announced a huge New National Cancer Database. Fortune 2016.
[28] S. Klein, M. Hostetter In focus: learning health care systems. Quality matters archive, August/September 2013 Issue.
[29] System UMH. (2016). Learning Health System (LHS) Vision.
[30] Website, V. Retrieved from: <https://www.vizientinc.com/>; 2016.
[31] Website HCSRN. Retrieved from: <http://www.hcsrn.org/en/>; 2016.
[32] Olsen L, Aisner D, McGinnis JM. The learning healthcare system: workshop summary. IOM Roundtable on Evidence-Based Medicine 2007.

Index

Note: Page numbers followed by "*f*," "*t*," and "*b*" refer to figures, tables, and boxes, respectively.

D

I

J

K

L

M